Modern Birkhäuser Classics

Many of the original research and survey monographs in pure and applied mathematics published by Birkhäuser in recent decades have been groundbreaking and have come to be regarded as foundational to the subject. Through the MBC Series, a select number of these modern classics, entirely uncorrected, are being re-released in paperback (and as eBooks) to ensure that these treasures remain accessible to new generations of students, scholars, and researchers.

Alberto Facchini

Module Theory

Endomorphism rings and direct sum
decompositions in some classes of modules

Reprint of the 1998 Edition

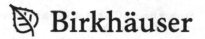 Birkhäuser

Alberto Facchini
Department of Mathematics
University of Padova
Padova
Italy

ISBN 978-3-0348-0302-1   e-ISBN 978-3-0348-0303-8
DOI 10.1007/978-3-0348-0303-8
Springer Basel Dordrecht Heidelberg London New York

Library of Congress Control Number: 2012930380

2010 Mathematics Subject Classification: 16D70, 16-02, 16S50, 16P20, 16P60, 16P70

© Springer Basel AG 1998
Reprint of the 1st edition 1998 by Birkhäuser Verlag, Switzerland
Originally published as volume 167 in the Progress in Mathematics series

Printed on acid-free paper

Springer Basel AG is part of Springer Science+Business Media
(www.birkhauser-science.com)

Contents

Preface

This expository monograph was written for three reasons. Firstly, we wanted to present the solution to a problem posed by Wolfgang Krull in 1932 [Krull 32]. He asked whether what we now call the "Krull-Schmidt Theorem" holds for artinian modules. The problem remained open for 63 years: its solution, a negative answer to Krull's question, was published only in 1995 (see [Facchini, Herbera, Levy and Vámos]). Secondly, we wanted to present the answer to a question posed by Warfield in 1975 [Warfield 75]. He proved that every finitely presented module over a serial ring is a direct sum of uniserial modules, and asked if such a decomposition was unique. In other words, Warfield asked whether the "Krull-Schmidt Theorem" holds for serial modules. The solution to this problem, a negative answer again, appeared in [Facchini 96]. Thirdly, the solution to Warfield's problem shows interesting behavior, a rare phenomenon in the history of Krull-Schmidt type theorems. Essentially, the Krull-Schmidt Theorem holds for some classes of modules and not for others. When it does hold, any two indecomposable decompositions are uniquely determined up to a permutation, and when it does not hold for a class of modules, this is proved via an example. For serial modules the Krull-Schmidt Theorem does not hold, but any two indecomposable decompositions are uniquely determined up to *two* permutations. We wanted to present such a phenomenon to a wider mathematical audience.

Apart from these three reasons, we present in this book various topics of module theory and ring theory, some of which are now considered classical (like Goldie dimension, semiperfect rings, Krull dimension, rings of quotients, and their applications) whereas others are more specialized (like dual Goldie dimension, semilocal endomorphism rings, serial rings and modules, exchange property, Σ-pure-injective modules).

We now consider the three reasons above in more detail.

1) Krull's problem. The classical Krull-Schmidt Theorem says that if M_R is a right module of finite composition length over a ring R and

$$M_R = A_1 \oplus \cdots \oplus A_n = B_1 \oplus \cdots \oplus B_m$$

are two decompositions of M_R as direct sums of indecomposable modules, then

$n = m$ and, after a suitable renumbering of the summands, $A_i \cong B_i$ for every $i = 1, \ldots, n$. In the paper [Krull 32, pp. 37–38] Krull recalls this theorem and asks whether the result remains true for artinian modules; that is, whether $A_1 \oplus \cdots \oplus A_n = B_1 \oplus \cdots \oplus B_m$ with each A_i and B_j an indecomposable artinian right module over a ring R, implies that $m = n$ and, after a renumbering of the summands, $A_i \cong B_i$ for each i.

During the years Krull's question was not forgotten (see for instance [Levy, p. 660]), and various partial results were proved. For instance, [Warfield 69a, Proposition 5] showed that the answer is "yes" when the ring R is either right noetherian or commutative (Proposition 2.63). He did this by showing that, over any ring, every artinian indecomposable module with Loewy length $\leq \omega$ has a local endomorphism ring. By the Krull-Schmidt-Remak-Azumaya Theorem, direct sums of indecomposable modules with local endomorphism rings have unique direct sum decompositions, even when the direct sum contains infinitely many terms, so that Warfield could conclude that the answer to Krull's question was positive if the base ring R was right noetherian or commutative.

Since direct sum decompositions of modules correspond to decompositions of their endomorphism ring in a natural way, Krull's problem is a particular case of the problem of determining what kinds of rings can occur as endomorphism rings of artinian modules. Rosa Camps and Warren Dicks [Camps and Dicks] showed that the endomorphism ring of any artinian module is semilocal, i.e., semisimple artinian modulo its Jacobson radical (Theorem 4.12). This allowed them to prove that artinian modules cancel from direct sums; that is, if $M \oplus A \cong M \oplus B$ with A and B arbitrary modules and M artinian, then $A \cong B$ (Corollary 4.6). All module-finite algebras over a semilocal noetherian commutative ring are semilocal, and applying a result of Camps and Menal it is possible to prove that all such module-finite algebras are isomorphic to endomorphism rings of artinian modules (Corollary 8.18). Therefore all decompositions of noetherian modules over the semilocal rings that occur in integral representation theory yield corresponding decompositions of artinian modules over suitable rings. Using this, it is possible to construct various examples. For instance, fix an integer $n \geq 2$. Then there is a ring R and an artinian right module M_R which is the direct sum of 2 indecomposable modules, and also the direct sum of 3 indecomposable modules, and also the direct sum of 4 indecomposable modules, and ..., and also the direct sum of n indecomposable modules (Example 8.21). There exists another ring R with four indecomposable, pairwise nonisomorphic, artinian modules M_1, M_2, M_3, M_4 such that $M_1 \oplus M_2 \cong M_3 \oplus M_4$ (Example 8.20).

These examples answer Krull's question: the Krull-Schmidt Theorem fails for artinian modules.

2) Warfield's problem. Recall that a module is said to be *uniserial* if for any submodules A and B of M either $A \subseteq B$ or $B \subseteq A$. A *serial* module is a module that is a direct sum of uniserial modules, and a ring R is *serial*

if the two modules R_R and $_RR$ are both serial modules. Important classes of
rings yield examples of serial rings. For instance, semisimple artinian rings,
commutative valuation rings, and rings of triangular matrices over a field are
serial rings.

In 1975 R. B. Warfield published a paper in which he described the structure of serial rings and proved that every finitely presented module over a serial
ring is a direct sum of uniserial modules [Warfield 75]. On page 189 of that paper, talking of the problems that remained open, he said that "... perhaps the
outstanding open problem is the uniqueness question for decompositions of a
finitely presented module into uniserial summands (proved in the commutative
case and in one noncommutative case by [Kaplansky 49])." In other words,
Warfield asked whether the Krull-Schmidt Theorem holds for direct sums of
uniserial modules.

Warfield's problem was solved completely in [Facchini 96] by giving a
counterexample: Krull-Schmidt fails for serial modules. For instance, fix an
integer $n \geq 2$. Then there exist $2n$ pairwise non-isomorphic finitely presented
uniserial modules $U_1, U_2, \ldots, U_n, V_1, V_2, \ldots, V_n$ over a suitable serial ring such
that $U_1 \oplus U_2 \oplus \cdots \oplus U_n \cong V_1 \oplus V_2 \oplus \cdots \oplus V_n$ (Example 9.21).

3) The weak Krull-Schmidt Theorem for serial modules. As we have just
said, the Krull-Schmidt Theorem does not hold for direct sums of uniserial
modules. Nevertheless, a weak form of the Krull-Schmidt Theorem still holds
for these modules. If A and B are modules, write $[A]_m = [B]_m$ if A and B
are in the same *monogeny class*, that is, if there is a monomorphism $A \to B$
and a monomorphism $B \to A$. Similarly, write $[A]_e = [B]_e$ if A and B are
in the same *epigeny class*, that is, if there is an epimorphism $A \to B$ and
an epimorphism $B \to A$. If $U_1, \ldots, U_n, V_1, \ldots, V_t$ are uniserial modules, then
$U_1 \oplus \cdots \oplus U_n \cong V_1 \oplus \cdots \oplus V_t$ if and only if $n = t$ and there are two permutations
σ, τ of $\{1, \ldots, n\}$ such that $[U_i]_m = [V_{\sigma(i)}]_m$ and $[U_i]_e = [V_{\tau(i)}]_e$ for every
$i = 1, \ldots, n$. This is a rare phenomenon: the isomorphism class of a serial
module is completely determined up to *two* permutations.

Such a weak form of the Krull-Schmidt Theorem holds not only for direct
sums of uniserial modules, but, more generally, for direct sums of *biuniform*
modules, i.e., modules that are uniform and couniform (Chapter 9).

This book also deals with a number of other topics as well. For instance,
we study the class of the rings that can be realized as endomorphism rings
of artinian modules, and serial rings belonging to this class are characterized.
We introduce Σ-pure-injective modules, because every artinian module is Σ-
pure-injective as a module over its endomorphism ring. In order to determine
whether a ring can be realized as the endomorphism ring of an artinian module,
we may look for sufficient information about the structure of its Σ-pure-injective
modules. We consider modules with the exchange property, semiperfect rings,
serial rings, their Krull dimension and their quotient rings.

The last chapter contains some open problems.

I am deeply indebted to my co-authors, the mathematicians with whom I have written papers containing some of the results presented in this book: Rosa Camps, Nguyen Viet Dung, Dolors Herbera, Larry Levy, Gennadi Puninski, Luigi Salce and Peter Vámos. I also thank Davide Fattori, Gino Favero and R. N. Gupta for several helpful comments on a preliminary version of this book. I wish to express my gratitude to the Department of Mathematics of the Universitat Autònoma de Barcelona, where I spent the academic year 1992/93, became acquainted with the results of Rosa Camps and Warren Dicks and begun my studies on the subjects with which this book deals; to the Centre de Recerca Matemàtica of the Institut d'Estudis Catalans, where I spent a sabbatical year in 1996/97 writing this book; and to my Faculty at the Università di Udine that gave me permission to take that sabbatical year. Finally, I would like to thank the Ministero dell'Università e della Ricerca Scientifica e Tecnologica (Italy), the Consiglio Nazionale delle Ricerche (Italy), and the Ministerio de Educación y Cultura (Spain) for financial support.

List of Symbols

R an associative ring with $1 \neq 0$.

M_R a right module over a ring R.

$_R M$ a left module over a ring R.

$J(R)$ the Jacobson radical of the ring R.

$N(R)$ the prime radical of the ring R.

$A \subseteq B$ A is a subset of B.

$A \subset B$ A is a proper subset of B.

$A \setminus B$ set-theoretic difference.

1_A the identity mapping $A \to A$, where A is a set.

$\mathbf{N}$ the set of non-negative integers.

ω the first infinite ordinal number.

$(0 :_M r) = \{y \in M \mid yr = 0\}$, where R is a ring, M_R is a right R-module and $r \in R$.

$(0 :_R x) = \{t \in R \mid xt = 0\}$, where R is a ring and x is an element of a right R-module M

$\mathrm{ann}_R(M) = \{t \in R \mid Mt = 0\}$, where R is a ring and M_R is a right R-module.

$\mathbf{M}_n(R)$ the ring of $n \times n$ matrices over the ring R.

$U(R)$ the group of invertible elements of the ring R.

$Z(R_R)$ the right singular ideal of the ring R.

$M^{(I)}$ the direct sum $\oplus_{i \in I} M_i$ with $M_i \cong M$ for all $i \in I$.

M^I the direct product $\prod_{i \in I} M_i$ with $M_i \cong M$ for all $i \in I$.

$|A|$ the cardinality of the set A.

$f|_A$ the restriction of a mapping $f \colon B \to C$ to a subset A of B.

$\mathrm{r.ann}_R(b) = \{a \in R \mid ba = 0\}$, where R is a ring and $b \in R$.

$\mathrm{l.ann}_R(b) = \{a \in R \mid ab = 0\}$, where R is a ring and $b \in R$.

Chapter 1

Basic Concepts

In this book we shall consider unital right modules over associative rings with identity $1 \neq 0$.

1.1 Semisimple rings and modules

In the two first sections of Chapter 1 we introduce some basic results that will be necessary in the book. All these results are elementary, well-known, and can be found in most text-books of ring theory or module theory, so that we shall not give a proof of all the statements. We refer the reader to the book of [Anderson and Fuller].

Recall that a module M is *simple* if 0 and M are the only submodules of M and $M \neq 0$. A ring R is *simple* if 0 and R are the only two-sided ideals of R. A module M over a ring R is said to be a *semisimple module* if it is the direct sum of a family of simple submodules.

Proposition 1.1 *Let M be a module over a ring R. The following conditions are equivalent:*

(a) *M is semisimple;*

(b) *M is the sum of a family of simple submodules;*

(c) *every submodule of M is a direct summand of M.* □

The class of all semisimple right R-modules is closed under submodules, homomorphic images and arbitrary direct sums.

For a module M the *(Jacobson) radical* of M is defined to be the intersection of all maximal submodules of M, and is denoted by $\mathrm{rad}(M)$. If M has no maximal submodules then we define $\mathrm{rad}(M) = M$. One has

$$\mathrm{rad}(M/\mathrm{rad}(M)) = 0$$

for any module M, and $\mathrm{rad}(M \oplus N) = \mathrm{rad}(M) \oplus \mathrm{rad}(N)$ for any M and N. In particular, if $e \in R$ is idempotent, then $\mathrm{rad}(eR) = e\,\mathrm{rad}(R)$.

For a ring R the intersection of all the maximal left ideals is equal to the intersection of all the maximal right ideals. Therefore this common intersection, which is the Jacobson radical of R both as a right and a left module over itself, is a two-sided ideal of R. It is called the *Jacobson radical* of R and denoted $J(R)$. The Jacobson radical $J(R)$ is the annihilator in R of the class of all simple right R-modules, that is, it is the set of all elements $r \in R$ such that $Mr = 0$ for every simple R-module M_R.

One has

$$
\begin{aligned}
J(R) \;&=\; \{\, x \in R \mid 1 - xr \text{ has a right inverse for all } r \in R \,\} \\
&=\; \{\, x \in R \mid 1 - rx \text{ has a left inverse for all } r \in R \,\} \\
&=\; \{\, x \in R \mid 1 - rxr' \text{ has a two-sided inverse for all } r, r' \in R \,\}.
\end{aligned}
$$

If $e \in R$ is idempotent, then $J(eRe) = eJ(R)e$. If I is an ideal of an arbitrary ring R and $I \subseteq J(R)$, then $J(R/I) = J(R)/I$.

Theorem 1.2 *Let R be a ring. The following conditions are equivalent:*

(a) *R_R is a semisimple module.*

(b) *R is the direct sum of a finite family of simple right ideals.*

(c) *Every right R-module is semisimple.*

(d) *Every right R-module is projective.*

(e) *Every right R-module is injective.*

(f) *The ring R is right artinian and $J(R) = 0$.*

(g) *There exist a finite number of division rings $D_1, \ldots, D_t$ and positive integers $n_1, \ldots, n_t$ such that $R \cong \prod_{i=1}^{t} \mathbf{M}_{n_i}(D_i)$. Here $\mathbf{M}_{n_i}(D_i)$ denotes the ring of $n_i \times n_i$ matrices over the division ring D_i.* □

Since condition (g) is left-right symmetric, it follows that "right" can be replaced by "left" everywhere in the conditions of Theorem 1.2, and a ring R that satisfies any of these equivalent conditions is said to be a *semisimple artinian ring*.

Lemma 1.3 *If M_R is a right module over a ring R, then $M J(R) \subseteq \mathrm{rad}(M_R)$.*

Proof. Let N be any maximal submodule of M_R. Since $J(R)$ annihilates all simple R-modules, $(M/N)J(R) = 0$. Therefore $M J(R) \subseteq N$. But N is an arbitrary maximal submodule, so that $M J(R) \subseteq \mathrm{rad}(M_R)$. □

Corollary 1.4 (Nakayama's Lemma) *Let M be a finitely generated module and let N be a submodule of M such that $N + MJ(R) = M$. Then $N = M$.*

Proof. Let M be a finitely generated module and let N be a proper submodule of M. Every non-zero finitely generated module has maximal submodules, so that $\mathrm{rad}(M/N) \subset M/N$. Lemma 1.3 implies that $(M/N)J(R) \subset M/N$, so that $MJ(R) + N \subset M$. $\square$

The class of module-finite algebras over a commutative ring will enable us to construct interesting examples. Let k be a commutative ring with identity. By a *k-algebra* we understand a ring R which is a k-module such that multiplication is bilinear. Equivalently, a k-algebra is a ring R with a homomorphism $\alpha \mapsto \alpha \cdot 1_R$ of k into the center of R. A *module-finite k-algebra* is any k-algebra R which is finitely generated as a k-module.

Proposition 1.5 *If R is a module-finite algebra over a commutative ring k, then $J(k)R \subseteq J(R)$.*

Proof. In order to show that $J(k)R \subseteq J(R)$ it is sufficient to prove that $MJ(k) = 0$ for every simple R-module M_R.

Now $MJ(k)$ is a submodule of M_R, because $MJ(k)R = MRJ(k) = MJ(k)$. Since M_R is simple, either $MJ(k) = M$ or $MJ(k) = 0$. But R_k is finitely generated and M_R is a homomorphic image of R_R, so that M_k is finitely generated. By Nakayama's Lemma, $MJ(k) = M$ implies $M = 0$, a contradiction. Therefore $M_R J(k) = 0$. $\square$

Let R be a ring. The ring R is called a *prime ring* if $aRb \neq 0$ for every non-zero $a, b \in R$. A proper ideal I of R is called *prime* if R/I is a prime ring. Equivalently, a proper ideal I of R is prime if and only if for each $a, b \in R$, $aRb \subseteq I$ implies $a \in I$ or $b \in I$, if and only if $AB \not\subseteq I$ whenever A and B are ideals of R not contained in I. The *prime radical* of R is the intersection of all prime ideals of R. We shall use $N(R)$ to denote the prime radical of R. The ring R is called a *semiprime ring* if $N(R) = 0$; and a proper ideal I of R is a *semiprime ideal* if R/I is a semiprime ring.

An element a of R is *nilpotent* if $a^n = 0$ for some $n \geq 1$, and an ideal I of R is *nilpotent* if $I^n = 0$ for some $n \geq 0$. The ideal I is *nil* if every element of I is nilpotent. Therefore every nilpotent ideal is nil. For a right artinian ring R the Jacobson radical $J(R)$ is a nilpotent ideal.

An element u of R is *strongly nilpotent* if, for each infinite sequence a_k, $k \geq 0$, such that $a_0 = a$ and $a_{k+1} \in a_k R a_k$ for every $k \geq 0$, there exists an integer n such that $a_n = 0$. Every strongly nilpotent element $a \in R$ is nilpotent (it is sufficient to consider the sequence defined by $a_k = a^{2^k}$ for every $k \geq 0$). The following result is due to [Levitzki].

Proposition 1.6 *An element a of a ring R is strongly nilpotent if and only if $a \in N(R)$.*

Proof. Let a be an element of R and suppose $a \notin N(R)$. Then there exists a prime ideal P of R such that $a \notin P$. Set $a_0 = a$. By induction on $k \geq 0$ construct a sequence a_k, $k \geq 0$, with the property that $a_k \in a_{k-1}Ra_{k-1}$ for every $k \geq 1$ and $a_k \notin P$ for every $k \geq 0$ as follows: if $a_k \notin P$ has already been defined, then $a_kRa_k \nsubseteq P$ because P is prime, so that there exists $a_{k+1} \in a_kRa_k \setminus P$. Since $a_k \notin P$ for every $k \geq 0$ we have that $a_k \neq 0$ for every $k \geq 0$, that is, a is not strongly nilpotent.

Conversely, suppose that $a \in R$ is not strongly nilpotent. Then there exists a sequence a_k of non-zero elements of R such that $a_0 = a$ and $a_{k+1} \in a_kRa_k$ for every $k \geq 0$. Set $S = \{ a_k \mid k \geq 0 \}$. Then $0 \notin S$, and by Zorn's Lemma, there exists an ideal P maximal in the set of all the ideals of R disjoint from S.

Let us show that P is prime. Let I, J be ideals of R not contained in P. Then $I + P$ and $J + P$ are ideals of R that contain P properly. It follows that $(I + P) \cap S \neq \emptyset$ and $(J + P) \cap S \neq \emptyset$. Hence $a_m \in I + P$ and $a_n \in J + P$ for suitable $m, n \geq 0$. If ℓ is the greatest index among m and n, then $a_\ell \in I + P$ and $a_\ell \in J + P$. Then $a_{\ell+1} \in a_\ell Ra_\ell \subseteq (I + P)(J + P) \subseteq IJ + P$. This shows that $(IJ + P) \cap S \neq \emptyset$, thus $IJ \nsubseteq P$ and P is prime. As $a \notin P$, we have that $a \notin N(R)$. □

Corollary 1.7 *The following conditions on a ring R are equivalent:*

(a) R *is semiprime.*

(b) R *has no non-zero nilpotent right ideal.*

(c) R *has no non-zero nilpotent ideal.*

Proof. (a) $\Rightarrow$ (b). Every nilpotent right ideal is contained in every prime ideal.

(b) $\Rightarrow$ (c). This is trivial.

(c) $\Rightarrow$ (a). Let R be a ring with no non-zero nilpotent ideal. We must prove that if $a \in R$ and $a \neq 0$, then a is not strongly nilpotent (Proposition 1.6). Let a be a non-zero element of R. Construct an infinite sequence of non-zero elements $a_k \in R$, $k \geq 0$, such that $a_0 = a$ and $a_{k+1} \in a_kRa_k$ for every $k \geq 0$ as follows. Induction on k. Set $a_0 = a$ and suppose that $a_k \neq 0$ has been defined. Then the ideal Ra_kR is non-zero, hence it is not nilpotent. In particular $a_kRa_k \neq 0$. Now let a_{k+1} be any non-zero element of a_kRa_k. □

We conclude this section with two results on the prime radical of a ring.

Proposition 1.8 *If R satisfies the ascending chain condition on two-sided ideals, the ideal $N(R)$ is nilpotent.*

Proof. If R satisfies the ascending chain condition on two-sided ideals, there is a maximal nilpotent ideal I. Then $I \subseteq N(R)$. Now the ring R/I is semiprime, otherwise it has a non-zero nilpotent ideal J/I, so that $J \supset I$ is a nilpotent ideal of R, contradiction. Thus R/I is semiprime, hence $I \supseteq N(R)$ and $N(R)$ is nilpotent. □

Proposition 1.9 *For an arbitrary ring R, $N(R) \subseteq J(R)$. For a right artinian ring R, $N(R) = J(R)$.*

Proof. If $x \in N(R)$ and $r \in R$, then $rx \in N(R)$, so that rx is strongly nilpotent, hence nilpotent. If $(rx)^n = 0$, then

$$(1 + rx + (rx)^2 + \cdots + (rx)^{n-1})(1 - rx) = 1,$$

so that $1 - rx$ has a left inverse in R. Therefore $x \in J(R)$.

If R is a right artinian ring, its Jacobson radical $J(R)$ is nilpotent. Hence $J(R)$ is contained in every prime ideal of R. Thus $J(R) \subseteq N(R)$. $\square$

1.2 Local and semilocal rings

A module is *local* if it has a greatest proper submodule. Equivalently, a module is local if and only if it is cyclic, non-zero, and has a unique maximal proper submodule. A module M is *indecomposable* if it is $\neq 0$ and has no direct summands $\neq 0$ and $\neq M$. Every local module is indecomposable. We begin this section with a well-known proposition.

Proposition 1.10 *The following conditions are equivalent for a ring R:*

(a) $R/J(R)$ *is a division ring;*

(b) R_R *is a local module (that is, R has a unique maximal proper right ideal);*

(c) *the sum of two non-invertible elements of R is non-invertible;*

(d) $J(R)$ *is a maximal right ideal;*

(e) $J(R)$ *is the set of all non-invertible elements of R.* $\square$

Since condition (a) is left-right symmetric, "right" can be replaced by "left" everywhere in the conditions of Proposition 1.10. Hence for any ring R the right module R_R is local if and only if the left module $_RR$ is local. In this case, that is, if R is a ring satisfying the equivalent conditions of Proposition 1.10, R is said to be a *local ring*.

Recall that a submodule N of a module M is *superfluous* in M if $N + K$ is a proper submodule of M for every proper submodule K of M. It can be proved that the radical $\mathrm{rad}(M_R)$ of an arbitrary module M_R is the sum of all superfluous submodules of M_R.

Proposition 1.11 *If e is an idempotent in a ring R, then eRe is a local ring if and only if the right R-module eR is local.*

Proof. Suppose that eRe is a local ring, so that in particular $e \neq 0$. Then the non-zero module eR has a maximal submodule N. In order to prove that eR is local, it is sufficient to show that N is superfluous in eR. Assume the contrary, so that there is a proper submodule K of eR with $K + N = eR$. Since $eR/N = K + N/N \cong K/K \cap N$, there is an isomorphism $\varphi \colon eR/N \to K/K \cap N$. Let $\pi \colon eR \to eR/N$ and $\pi' \colon K \to K/K \cap N$ be the canonical projections. The module eR is projective. Hence there is a homomorphism $\psi \colon eR \to K$ such that $\varphi\pi = \pi'\psi$. Since K is a proper submodule of eR, ψ can be viewed as a non-invertible element of the local ring $\operatorname{End}_R(eR) \cong eRe$, that is, ψ is left multiplication by an element of $J(eRe) = eJ(R)e$. Hence there exists $r \in J(R)$ such that $\psi(et) = eret$ for every $t \in R$. Now $K/K \cap N = \varphi\pi(eR) = \pi'\psi(eR) = ereR + K \cap N/K \cap N$, from which $ereR \nsubseteq K \cap N$. Since $ereR \subseteq K$, we have $ereR \nsubseteq N$. But $ereR \subseteq eJ(R) = \operatorname{rad}(eR) \subseteq N$, a contradiction.

Conversely, suppose that eR is a local R-module with maximal proper submodule N. In order to show that $eRe \cong \operatorname{End}(eR)$ is local, it suffices to prove that the sum of two non-invertible elements of $\operatorname{End}(eR)$ is non-invertible. For this, it is sufficient to show that if $f \in \operatorname{End}(eR)$, then f is non-invertible if and only if $f(eR) \subseteq N$. It is clear that if $f(eR) \subseteq N$, then f is non-invertible. Conversely, if $f(eR) \nsubseteq N$, then $f(eR) = eR$, so that $f \colon eR \to eR$ is surjective. Since eR is projective, $\ker f$ is a direct summand of eR. But every local module is indecomposable, and thus either $f = 0$ or f is injective. If $f = 0$, then $eR = 0$, which is not because eR is local. Thus f is injective, and so it is an invertible element of $\operatorname{End}(eR)$. $\qquad\square$

A ring R is *semilocal* if $R/J(R)$ is a semisimple artinian ring. Since $J(R/J(R)) = 0$ for every ring R, the next lemma follows immediately from Theorem 1.2.

Lemma 1.12 *The following conditions are equivalent for a ring R:*

(a) *R is semilocal;*

(b) *$R/J(R)$ is a right artinian ring;*

(c) *$R/J(R)$ is a left artinian ring.* $\qquad\square$

The study of semilocal rings will be one of the main themes of this monograph. We immediately give some easy first examples of semilocal rings.

(1) *Every right (or left) artinian ring is semilocal.*

(2) *Every local ring is semilocal.*

(3) *If R is a semilocal ring, the ring $\mathbf{M}_n(R)$ of $n \times n$ matrices over R is semilocal.*

This is easily seen from the following known facts:

(a) $J(\mathbf{M}_n(R)) = \mathbf{M}_n(J(R))$,

(b) $\mathbf{M}_n(R)/J(\mathbf{M}_n(R)) \cong \mathbf{M}_n(R/J(R))$,

(c) $\mathbf{M}_n(S)$ is a semisimple artinian ring for every semisimple artinian ring S.

(4) *The direct product $R_1 \times R_2$ of two semilocal rings R_1 and R_2 is semi-local.*

(5) *Every homomorphic image of a semilocal ring is semilocal.*

In fact, let I be an ideal of a semilocal ring R. Since every simple R/I-module is a simple R-module, if $\pi: R \to R/I$ is the canonical projection, then $\pi(J(R)) \subseteq J(R/I)$. Hence π induces a surjective homomorphism $R/J(R) \to (R/I)/J(R/I)$. But every homomorphic image of a semisimple artinian ring is a semisimple artinian ring, and thus R/I is semilocal.

(6) *A commutative ring is semilocal if and only if it has finitely many maximal ideals.*

This easily follows from the fact that a semisimple artinian commutative ring is a finite direct product of fields, and therefore it has finitely many maximal ideals.

(7) *Every module-finite algebra over a commutative semilocal ring is semi-local.*

If R is a module-finite algebra over a commutative semilocal ring k, then $J(k)R \subseteq J(R)$ (Proposition 1.5), so that $R/J(R)$ is an algebra over $k/J(k)$. Then $R/J(R)$ is a module-finite algebra over the artinian commutative ring $k/J(k)$, so that $R/J(R)$ is a (left and right) artinian ring.

An *artinian poset* (artinian partially ordered set) is a set A partially ordered by a relation $\leq$ satisfying the following equivalent conditions:

(a) Every decreasing sequence $a_0 \geq a_1 \geq a_2 \geq \ldots$ in A is stationary (i.e., there exists an index n such that $a_n = a_{n+1} = \ldots$).

(b) Every non-empty subset of A has a minimal element.

Proposition 1.13 *If R is a semilocal ring and e is a non-zero idempotent of R, then eRe is a semilocal ring.*

Proof. Let $\mathcal{L}(R/J(R))$ denote the lattice of all right ideals of the ring $R/J(R)$, and $\mathcal{L}(eRe/J(eRe))$ denote the lattice of all right ideals of the ring $eRe/J(eRe)$. In order to show that R semilocal implies eRe semilocal, it is sufficient to show that if $\mathcal{L}(R/J(R))$ is an artinian poset then $\mathcal{L}(eRe/J(eRe))$ is an artinian poset, and for this it is enough to prove that there is an order preserving injective mapping $\mathcal{L}(eRe/J(eRe)) \to \mathcal{L}(R/J(R))$. Hence the proposition will be proved if we show that $(IR + J(R)) \cap eRe = I$ for every right ideal I of eRe containing $J(eRe)$. Let $\sum_k i_k r_k + j = ere$ be an element of $(IR + J(R)) \cap eRe$, with $i_k \in I$, $r, r_k \in R$ and $j \in J(R)$. Then $ere = e\left(\sum_k i_k r_k + j\right)e = \sum_k ei_k er_k e + eje \in IeRe + eJ(R)e \subseteq I$. $\square$

We conclude this section with an elementary lemma about finitely generated modules over a semilocal ring. Recall that a module M is *of finite (composition) length* if it has a finite chain of submodules

$$0 = M_0 \subset M_1 \subset M_2 \subset \cdots \subset M_n = M$$

which has no refinement. In this case the integer n is said to be the *length* of the module M. A module has finite length if and only if it is noetherian and artinian.

Lemma 1.14 *Let R be a semilocal ring and let M be a finitely generated right R-module. Then*

(a) *$M/MJ(R)$ is a semisimple R-module of finite length.*

(b) *M is local if and only if $M/MJ(R)$ is simple.*

(c) *M is a direct sum of indecomposable modules.*

Proof. (a) This is trivial.

(b) If M is local, $M/MJ(R)$ is a local module over the semisimple artinian ring $R/J(R)$. Hence $M/MJ(R)$ is simple. Conversely, if $M/MJ(R)$ is simple, $MJ(R)$ is a maximal submodule of M. By Lemma 1.3 $MJ(R) \subseteq \mathrm{rad}(M)$. Since M is finitely generated, $\mathrm{rad}(M)$ is a proper submodule of M, so that $MJ(R) = \mathrm{rad}(M)$ is the unique maximal submodule of M. Finally, since M is non-zero and finitely generated, every proper submodule of M is contained in a maximal submodule, and therefore $MJ(R) = \mathrm{rad}(M)$ is the greatest proper submodule of M.

(c) Suppose not; then the set of direct summands of M for which (c) is false is not empty. Hence this set has an element N for which the composition length of the semisimple R-module $N/NJ(R)$ is minimal. Since N is not a direct sum of indecomposable modules, N itself is not indecomposable, so that $N = N' \oplus N''$, N, N' non-zero finitely generated R-modules. Then

$$N/NJ(R) \cong N'/N'J(R) \oplus N''/N''J(R),$$

and the semisimple modules $N'/N'J(R)$ and $N''/N''J(R)$ have non-zero length by Nakayama's Lemma 1.4. Therefore N' and N'' are direct sums of indecomposable modules, so that N is a direct sum of indecomposable modules; contradiction. $\square$

1.3   Serial rings and modules

Recall that an R-module M is said to be *uniserial* if for any submodules A and B of M we have $A \subseteq B$ or $B \subseteq A$. Hence a module M is uniserial if and only if the lattice $\mathcal{L}(M)$ of its submodules is linearly ordered under set inclusion. Every submodule and every homomorphic image of a uniserial module is uniserial.

A uniserial module M is *shrinkable* if M is isomorphic to K/H for suitable submodules $0 \subset H \subset K \subset M$. Otherwise the uniserial module M is said to be *unshrinkable*.

Example 1.15 *Artinian uniserial modules and noetherian uniserial modules are unshrinkable.*

In fact, if M is an artinian uniserial module, its lattice of submodules $\mathcal{L}(M)$ is well-ordered under inclusion, i.e., it is order isomorphic to an ordinal number α. If M is shrinkable, that is, there exist submodules $0 \subset H \subset K \subset M$ such that M is isomorphic to K/H, then there exist non-zero ordinal numbers β and γ such that $\beta + \alpha + \gamma = \alpha$. This is not possible. Therefore M is unshrinkable.

Dually for the noetherian case. $\square$

Example 1.16 *There exist shrinkable uniserial modules.*

The following example of a shrinkable uniserial module is taken from [Facchini and Salce, p. 502]. Let $\mathbf{Z}$ be the ring of integers, p and q be distinct prime numbers, $\mathbf{Z}_p$ and $\mathbf{Z}_q$ be the localizations of $\mathbf{Z}$ at the principal prime ideals generated by p and q respectively, so that $\mathbf{Z}_p$ and $\mathbf{Z}_q$ are discrete valuation rings, and let $\mathbf{Q}$ be the field of rational numbers. Consider the ring

$$R = \begin{pmatrix} \mathbf{Z}_p & 0 \\ \mathbf{Q} & \mathbf{Z}_q \end{pmatrix}$$

and its right ideal

$$H = \begin{pmatrix} \mathbf{Z}_p & 0 \\ \mathbf{Z}_p & 0 \end{pmatrix}.$$

The operations in the ring are the ordinary matrix addition and multiplication. It is easily seen that the right ideals of R containing H are the ideals

$$\begin{pmatrix} \mathbf{Z}_p & 0 \\ J & 0 \end{pmatrix} \quad \text{and} \quad \begin{pmatrix} \mathbf{Z}_p & 0 \\ \mathbf{Q} & I \end{pmatrix},$$

where J is any $\mathbf{Z}_p$-submodule of $\mathbf{Q}$ containing $\mathbf{Z}_p$ and I is any ideal of $\mathbf{Z}_q$. Therefore R/H is a uniserial right R-module.

Now we shall prove that R/H is a shrinkable R-module. Consider the element

$$t = \begin{pmatrix} 1 & 0 \\ 0 & q/p \end{pmatrix} \in R.$$

and the endomorphism $\varphi\colon R_R \to R_R$ given by left multiplication by t. It is easily seen that

$$tH = \left(\begin{array}{cc} \mathbf{Z}_p & 0 \\ p^{-1}\mathbf{Z}_p & 0 \end{array} \right) \qquad \text{and} \qquad tR = \left(\begin{array}{cc} \mathbf{Z}_p & 0 \\ \mathbf{Q} & q\mathbf{Z}_q \end{array} \right),$$

so that $H \subset tH \subset tR \subset R$. Since $\varphi^{-1}(tH) = H$, φ induces an isomorphism of right R-modules $R/H \to tR/tH$. Therefore R/H is a shrinkable uniserial R-module. $\qquad\square$

In Proposition 1.18 we shall see that every uniserial right module over a ring that is either commutative or right noetherian is unshrinkable. To prove this we need the following lemma.

Lemma 1.17 *Every shrinkable uniserial right module contains a shrinkable cyclic submodule.*

Proof. Let M be a shrinkable uniserial right module over a ring R. Then there exist submodules $0 \subset H \subset K \subset M$ such that there is an isomorphism $\varphi\colon M \to K/H$. Fix an element $x \in M$, $x \notin K$ and suppose $\varphi(x) = y + H$. Then φ induces an isomorphism $xR \to yR/H$ and $H \subset yR \subseteq K \subset xR$. Therefore xR is a shrinkable R-module. $\qquad\square$

Proposition 1.18 *Let R be a ring that is either commutative or right noetherian. Then every uniserial right R-module is unshrinkable.*

Proof. Let M be a shrinkable uniserial right R-module. By Lemma 1.17 the module M contains a cyclic shrinkable submodule N.

If the ring R is right noetherian, then N is a uniserial noetherian shrinkable module. This contradicts what we have seen in Example 1.15.

If R is commutative, then $N \cong R/I$ for some ideal I of R, and $R/I \cong xR/J$ for some ideals $I \subset J \subset xR \subset R$. Let $r \in J$ be an element that does not belong to I. Then r annihilates xR/J, but it does not annihilate R/I. This contradicts the fact that xR/J and R/I are isomorphic. $\qquad\square$

A ring R is said to be a *right chain ring* if it is a uniserial module as a right module over itself. A *left chain ring* is defined similarly. A ring R is a *chain ring* if it is both a right and a left chain ring. A *valuation ring* is a commutative chain ring, not necessarily a domain.

Lemma 1.19 *Let R be a ring and $e \in R$ be a non-zero idempotent element. If the module eR is a uniserial module, then the ring eRe is a right chain ring.*

Proof. Let $eaeRe, ebeRe$ be two principal right ideals of the ring eRe, $a, b \in R$. It is sufficient to prove that either $eaeRe \subseteq ebeRe$ or $ebeRe \subseteq eaeRe$. If eR is a uniserial module, either $eaeR \subseteq ebeR$ or $ebeR \subseteq eaeR$. Suppose $eaeR \subseteq ebeR$. Then $eaeRe \subseteq ebeRe$. $\qquad\square$

A *serial* module is a module that is a direct sum of uniserial modules. A ring R is said to be *right (left) serial* if it is a serial module as a right (left) module over itself. If both conditions hold R is a *serial ring*. For example, every simple module is uniserial, so that every semisimple module is serial. Therefore every semisimple artinian ring is a serial ring. Every (right) chain ring is a (right) serial ring. In particular, every commutative valuation ring is serial. The rest of this section is devoted to giving further examples of (right) serial rings.

Recall that an idempotent e of a ring R is *primitive* if it is non-zero and cannot be written as a sum of two non-zero orthogonal idempotents. Equivalently, e is primitive if and only if the direct summand eR of R_R is indecomposable, if and only if the direct summand Re of $_RR$ is indecomposable. A *complete set of (pairwise) orthogonal idempotents* in R is a finite set $\{e_1, \ldots, e_n\}$ of idempotents of R such that $e_i e_j = 0$ for $i \neq j$ and $e_1 + \cdots + e_n = 1$ (equivalently, such that $R_R = e_1 R \oplus \cdots \oplus e_n R$).

Let R be a right serial ring. Since R_R is serial, there exists a complete set $\{e_1, \ldots, e_n\}$ of pairwise orthogonal idempotents in R for which every $e_i R$ is uniserial as a right R-module. Then $e_1, \ldots, e_n$ are primitive idempotents in R.

Lemma 1.20 *Every homomorphic image of a right serial ring is a right serial ring.*

Proof. If R is a right serial ring, there exists a complete set $\{e_1, \ldots, e_n\}$ of pairwise orthogonal idempotents in R such that $R_R = e_1 R \oplus \cdots \oplus e_n R$ and each $e_i R$ is serial as a right R-module. If I is an ideal of R, then

$$R/I \cong (e_1 R/e_1 I) \oplus \cdots \oplus (e_n R/e_n I).$$

Each $e_i R/e_i I$ is a homomorphic image of the uniserial right module $e_i R$, and so is itself uniserial. □

In this proof we saw that if I is an ideal of a right serial ring R, then $\{ e_i + I \mid i = 1, \ldots, n,\ e_i \notin I \}$ is a complete set of pairwise orthogonal primitive idempotents in R/I.

Example 1.21 *If k is a field and $\mathbf{UT}_n(k)$ is the ring of $n \times n$ upper triangular matrices over k, then $\mathbf{UT}_n(k)$ is a serial ring.*

Set $R = \mathbf{UT}_n(k)$. Let $E_1, \ldots, E_n$ be the primitive diagonal idempotents of $R = \mathbf{UT}_n(k)$. Then $R_R = E_1 R \oplus \cdots \oplus E_n R$ and $_RR = RE_1 \oplus \cdots \oplus RE_n$. It is easily seen that RE_i is a uniserial left R-module of finite length i and $E_i R$ is a uniserial right R-module of finite length $n + 1 - i$. □

We conclude this section showing that right serial rings are semilocal.

Proposition 1.22 *Every right serial ring is a semilocal ring.*

Proof. Let R be a right serial ring and $\{e_1, \ldots, e_n\}$ a complete set of pairwise orthogonal idempotents in R such that each $e_i R$ is serial. Then

$$\begin{aligned} R/J(R) &\cong (e_1 R/e_1 J(R)) \oplus \cdots \oplus (e_n R/e_n J(R)) \\ &= (e_1 R/\mathrm{rad}(e_1 R)) \oplus \cdots \oplus (e_n R/\mathrm{rad}(e_n R)), \end{aligned}$$

and every $e_i R/\mathrm{rad}(e_i R)$ is a simple right R-module. Therefore $R/J(R)$ is a semisimple right R-module of finite length. $\square$

1.4 Pure exact sequences

The rest of this chapter is devoted to illustrating the concepts of pure submodule of a module, pure exact sequence, pure-projective module and pure-injective module.

Let R be a ring. A short exact sequence $0 \to A_R \to B_R \to C_R \to 0$ of right R-modules is *pure* if the induced sequence of abelian groups

$$0 \to \mathrm{Hom}(E_R, A_R) \to \mathrm{Hom}(E_R, B_R) \to \mathrm{Hom}(E_R, C_R) \to 0$$

is exact for every finitely presented right R-module E_R. A submodule A of a right R-module B is a *pure submodule* of B if the canonical exact sequence $0 \to A \to B \to B/A \to 0$ is pure. A *pure* monomorphism is a monomorphism $A \to B$ whose image is a pure submodule of B.

A right R-module P_R is called *pure-projective* if the sequence

$$0 \to \mathrm{Hom}(P_R, A_R) \to \mathrm{Hom}(P_R, B_R) \to \mathrm{Hom}(P_R, C_R) \to 0$$

is exact for every pure exact sequence $0 \to A_R \to B_R \to C_R \to 0$ of right R-modules. For instance, finitely presented modules are pure-projective. An R-module M_R is *pure-injective* if the sequence

$$0 \to \mathrm{Hom}(C_R, M_R) \to \mathrm{Hom}(B_R, M_R) \to \mathrm{Hom}(A_R, M_R) \to 0$$

is exact for every pure exact sequence $0 \to A_R \to B_R \to C_R \to 0$.

Proposition 1.23 *For every module M_R there exist pure exact sequences*

$$0 \to A_R \xrightarrow{f} B_R \xrightarrow{g} M_R \to 0$$

where B_R is a direct sum of finitely presented modules.

Proof. Let $\mathcal{S}_R$ be a set of finitely presented right R-modules such that every finitely presented right R-module is isomorphic to an element of $\mathcal{S}_R$. For every $N_R \in \mathcal{S}_R$ let $N^{(\mathrm{Hom}(N_R, M_R))}$ be the direct sum of a set of copies of N_R indexed in $\mathrm{Hom}(N_R, M_R)$. Define $B_R = \oplus_{N_R \in \mathcal{S}_R} N^{(\mathrm{Hom}(N_R, M_R))}$. Let $g: B_R \to M_R$

be the homomorphism whose restriction to the copy of N_R indexed by $\varphi \in$ Hom(N_R, M_R) is φ for every $N_R \in \mathcal{S}_R$ and every $\varphi \in$ Hom(N_R, M_R). Since R_R is finitely presented, there is a module in $\mathcal{S}_R$ isomorphic to R_R. Hence g is surjective. Let A_R be the kernel of g and $f \colon A_R \to B_R$ be the inclusion, so that the sequence $0 \to A_R \xrightarrow{f} B_R \xrightarrow{g} M_R \to 0$ is exact.

In order to prove that this exact sequence is pure we must show that for every finitely presented module E_R and every homomorphism $\psi \colon E_R \to M_R$ there is a homomorphism $\psi' \colon E_R \to B_R$ such that $\psi = g\psi'$. Now there exists $N_R \in \mathcal{S}_R$ isomorphic to E_R. Let $\alpha \colon N_R \to E_R$ be such an isomorphism and let $\varepsilon \colon N_R \to B_R$ be the embedding of N_R into the direct summand of B_R indexed by $\psi\alpha \in$ Hom(N_R, M_R). Then $\psi' = \varepsilon\alpha^{-1}$ has the required property. $\qquad\square$

From Proposition 1.23 we immediately obtain the characterizations of pure-projective modules of the next proposition.

Proposition 1.24 *The following conditions are equivalent for a right R-module P_R.*

(a) *P_R is pure-projective.*

(b) *Every pure exact sequence of right R-modules $0 \to A_R \to B_R \to P_R \to 0$ is split.*

(c) *P_R is isomorphic to a direct summand of a direct sum $\oplus_{i \in I} N_i$ of finitely presented right R-modules N_i.*

Proof. (a) $\Rightarrow$ (b). This follows from the fact that if

$$0 \to A_R \xrightarrow{f} B_R \xrightarrow{g} P_R \to 0$$

is a pure exact sequence and P_R is pure-projective, then the induced sequence

$$0 \to \text{Hom}(P_R, A_R) \to \text{Hom}(P_R, B_R) \to \text{Hom}(P_R, P_R) \to 0$$

is exact. Hence there exists $\varphi \in$ Hom(P_R, B_R) such that $g\varphi$ is the identity of P_R.

(b) $\Rightarrow$ (c). This follows from Proposition 1.23.

(c) $\Rightarrow$ (a). This is trivial. $\qquad\square$

Note how pure-projective modules extend the notion of projective module, because a module is projective if and only if the sequence

$$0 \to \text{Hom}_R(P, A) \to \text{Hom}_R(P, B) \to \text{Hom}_R(P, C) \to 0$$

is exact for every exact sequence $0 \to A_R \to B_R \to C_R \to 0$, if and only if every exact sequence of right R-modules $0 \to A_R \to B_R \to P_R \to 0$ is split, if and only if P is isomorphic to a direct summand of a direct sum $\oplus_{i \in I} F_i$ of right R-modules $F_i \cong R_R$, that is, isomorphic to a direct summand of a free R-module.

For the proof of Lemma 1.26 we need the following result whose proof can be easily obtained by chasing the diagram.

Lemma 1.25 *Let*

$$\begin{array}{ccccc}
G' & \xrightarrow{\alpha} & G & \xrightarrow{\alpha'} & G'' \\
\gamma' \downarrow & & \gamma \downarrow & & \gamma'' \downarrow \\
H' & \xrightarrow{\beta} & H & \xrightarrow{\beta'} & H''
\end{array}$$

be a commutative diagram of abelian groups with exact rows.

(a) *Suppose γ is a monomorphism and α' is an epimorphism. Then γ'' is a monomorphism if and only if $\beta(H') \cap \gamma(G) \subseteq \gamma\alpha(G')$.*

(b) *Suppose γ is an epimorphism and β is a monomorphism. Then γ' is an epimorphism if and only if $\alpha'(G) \cap \ker \gamma'' \subseteq \alpha'(\ker \gamma)$.*   □

Now let $M = (r_{ij})$ be a $n \times m$ matrix with entries in a ring R. If we look at elements of R_R^m and R_R^n as $m \times 1$ and $n \times 1$ matrices respectively, left multiplication by M is a homomorphism $\alpha \colon R_R^m \to R_R^n$. Let E_R denote the cokernel of α, so that E_R is a finitely presented right R-module and every finitely presented right R-module is isomorphic to a module of this type. Similarly, looking at elements of $_RR^n$ and $_RR^m$ as $1 \times n$, $1 \times m$ matrices respectively, right multiplication by M is a homomorphism $\alpha^* \colon {}_RR^n \to {}_RR^m$. Let $_RF$ denote the cokernel of α^*. Then $_RF$ is a finitely presented left R-module and every finitely presented right R-module is isomorphic to a module constructed in this way.

Lemma 1.26 *In the notation above the following conditions are equivalent for an exact sequence $0 \to A_R \xrightarrow{f} B_R \xrightarrow{g} C_R \to 0$ of right R-modules.*

(a) *The mapping $\mathrm{Hom}(E_R, g) \colon \mathrm{Hom}(E_R, B_R) \to \mathrm{Hom}(E_R, C_R)$ is surjective.*

(b) *The mapping $f \otimes_R F \colon A \otimes_R F \to B \otimes_R F$ is injective.*

(c) *For every $a'_1, \ldots, a'_m \in f(A)$, if the system of m linear equations*

$$(x_1 \ \ldots \ x_n)M = (a'_1 \ \ldots \ a'_m)$$

has a solution in B^n, it also has a solution in $(f(A))^n$.

Proof. Without loss of generality we may suppose $A_R \subseteq B_R$, $f \colon A_R \to B_R$ the embedding and $g \colon B_R \to C_R$ the canonical projection onto $C_R = B_R/A_R$. This will simplify the notation.

(a) $\Leftrightarrow$ (c). From the exact sequence $R_R^m \xrightarrow{\alpha} R_R^n \to E_R \to 0$ we obtain a commutative diagram

$$
\begin{array}{ccccccc}
0 & \to & \mathrm{Hom}(E_R, B_R) & \to & \mathrm{Hom}(R_R^n, B_R) & \to & \mathrm{Hom}(R_R^m, B_R) \\
& & \downarrow & & \downarrow & & \downarrow \\
0 & \to & \mathrm{Hom}(E_R, C_R) & \to & \mathrm{Hom}(R_R^n, C_R) & \to & \mathrm{Hom}(R_R^m, C_R)
\end{array}
$$

in which the rows are exact and the last two vertical arrows are epimorphisms. We shall apply Lemma 1.25(b). The first vertical arrow in the diagram is the mapping $\mathrm{Hom}(E_R, g)\colon \mathrm{Hom}(E_R, B_R) \to \mathrm{Hom}(E_R, C_R)$. Up to natural identification the right-hand square in the diagram can be rewritten as

$$
\begin{array}{ccc}
B^n & \to & B^m \\
\downarrow & & \downarrow \\
(B/A)^n & \to & (B/A)^m
\end{array},
$$

where the vertical arrows are the canonical projections and the horizontal arrows are right multiplication by the matrix M. Hence in this diagram the image of the upper horizontal arrow is $B^n M$, the kernel of the vertical arrow on the right is A^m, the kernel of the vertical arrow on the left is A^n, and its image via the upper horizontal arrow is $A^n M$. Hence, the condition $\alpha'(G) \cap \ker \gamma'' \subseteq \alpha'(\ker \gamma)$ in Lemma 1.25(b) becomes $B^n M \cap A^m \subseteq A^n M$. Thus $\mathrm{Hom}(E_R, g)\colon \mathrm{Hom}(E_R, B_R) \to \mathrm{Hom}(E_R, C_R)$ is surjective if and only if $B^n M \cap A^m \subseteq A^n M$, and this is equivalent to (c).

The proof of (b) $\Leftrightarrow$ (c) is analogous to the proof of (a) $\Leftrightarrow$ (c), starting from the exact sequence ${}_R R^n \xrightarrow{\alpha^*} {}_R R^m \to {}_R F \to 0$, applying the functor $\otimes_R$ and using Lemma 1.25(a). $\qquad\square$

We are ready to prove some important characterizations of pure exact sequences.

Theorem 1.27 *Let R be a ring and let $0 \to A_R \to B_R \to C_R \to 0$ be an exact sequence of right R-modules. The following conditions are equivalent:*

(a) *The sequence $0 \to A_R \to B_R \to C_R \to 0$ is pure.*

(b) *For every finitely presented left R-module ${}_R F$ the induced sequence of abelian groups*

$$
0 \to A \otimes_R F \to B \otimes_R F \to C \otimes_R F \to 0
$$

is exact.

(c) *For every left R-module ${}_R F$ the induced sequence of abelian groups*

$$
0 \to A \otimes_R F \to B \otimes_R F \to C \otimes_R F \to 0
$$

is exact.

(d) *Every system of m linear equations*

$$\sum_{i=1}^{n} x_i r_{ij} = a'_j, \quad j = 1, 2, \ldots, m,$$

with $r_{ij} \in R$ and $a'_j \in f(A)$ $(i = 1, 2, \ldots, n$, $j = 1, 2, \ldots, m)$, which has a solution in B^n, also has a solution in $(f(A))^n$.

Proof. The equivalence (a) $\Leftrightarrow$ (b) $\Leftrightarrow$ (d) is just Lemma 1.26 stated for an arbitrary matrix M. The equivalence (b) $\Leftrightarrow$ (c) follows from the fact that every module is a direct limit of finitely presented modules over a suitable directed poset and that tensor products commute with direct limits. $\square$

Recall that a module N *cogenerates* a module B if B is isomorphic to a submodule of a direct product N^I of copies of N. Equivalently, N cogenerates B if for every non-zero $x \in B$ there is a homomorphism $\varphi_x : B \to N$ such that $\varphi_x(x) \neq 0$. If R is a ring, a *cogenerator in* Mod-R, the category of all right R-modules, is a module C_R that cogenerates every right R-module. If C_R is a cogenerator in Mod-R, S_R is a simple R-module and $E(S_R)$ is the injective envelope of S_R, then there is a homomorphism $\varphi : E(S_R) \to C_R$ such that $\varphi(S_R) \neq 0$. Hence $\ker(\varphi) \cap S_R = 0$, i.e, φ is a monomorphism. Thus every cogenerator in Mod-R contains a submodule isomorphic to $E(S_R)$ for every simple right R-module S_R. In particular, if U_R is an injective cogenerator, that is, a cogenerator U_R in Mod-R that is also injective, then U_R must contain a direct summand isomorphic to $E(\oplus_{\lambda \in \Lambda} S_\lambda)$, where $\{\, S_\lambda \mid \lambda \in \Lambda \,\}$ is a set of representatives of the simple right R-modules up to isomorphism. Conversely, it is not difficult to see that $E(\oplus_{\lambda \in \Lambda} S_\lambda)$ is a cogenerator in Mod-R. For this reason $E(\oplus_{\lambda \in \Lambda} S_\lambda)$ is said to be the *minimal injective cogenerator* in Mod-R. Cogenerators and minimal injective cogenerators in R-Mod, the category of all left R-modules, are defined using left modules instead of right modules.

Let k be a commutative ring, let R be a k-algebra and let U be an injective cogenerator in k-Mod. If A_R is a right R-module, then $A^* = \operatorname{Hom}_k(A_R, U)$ is a left R-module, and if $_R A$ is a left R-module, then $A^* = \operatorname{Hom}_k(_R A, U)$ is a right R-module.

Corollary 1.28 *Let k be a commutative ring, let R be a k-algebra and let $_k U$ be an injective cogenerator in k-Mod. An exact sequence*

$$0 \to A_R \to B_R \to C_R \to 0$$

of right R-modules is pure if and only if the induced exact sequence of left R-modules $0 \to C^* \to B^* \to A^* \to 0$ is split.

Proof. The exact sequence $0 \to A_R \to B_R \to C_R \to 0$ of right R-modules is pure if and only if the induced sequence of k-modules

$$0 \to A \otimes_R F \to B \otimes_R F \to C \otimes_R F \to 0$$

is exact for every left R-module $_RF$. Since $_kU$ is an injective cogenerator in k-Mod, this happens if and only if the sequence

$$0 \to \mathrm{Hom}_k(_kC \otimes_R F, {}_kU) \to \mathrm{Hom}_k(_kB \otimes_R F, {}_kU) \to \mathrm{Hom}_k(_kA \otimes_R F, {}_kU) \to 0$$

is exact, that is, if and only if the sequence

$$0 \to \mathrm{Hom}_R(_RF, \mathrm{Hom}_k(_kC_R, {}_kU)) \to \mathrm{Hom}_R(_RF, \mathrm{Hom}_k(_kB_R, {}_kU))$$
$$\to \mathrm{Hom}_R(_RF, \mathrm{Hom}_k(_kA_R, {}_kU)) \to 0$$

is exact for every left R-module $_RF$, because there is a natural isomorphism $\mathrm{Hom}_k(- \otimes_R F, {}_kU) \cong \mathrm{Hom}_R(_RF, \mathrm{Hom}_k(-, {}_kU))$. And this happens if and only if the exact sequence

$$0 \to \mathrm{Hom}_k(_kC_R, {}_kU) \to \mathrm{Hom}_k(_kB_R, {}_kU) \to \mathrm{Hom}_k(_kA_R, {}_kU) \to 0$$

is split. $\qquad\square$

Since every ring is an algebra over the ring $\mathbf{Z}$ of integers and $\mathbf{Q}/\mathbf{Z}$ is an injective cogenerator in $\mathbf{Z}$-Mod, we immediately obtain the following corollary.

Corollary 1.29 *Let R be a ring. An exact sequence $0 \to A_R \to B_R \to C_R \to 0$ of right R-modules is pure if and only if the exact sequence of left R-modules $0 \to \mathrm{Hom}_{\mathbf{Z}}(C, \mathbf{Q}/\mathbf{Z}) \to \mathrm{Hom}_{\mathbf{Z}}(B, \mathbf{Q}/\mathbf{Z}) \to \mathrm{Hom}_{\mathbf{Z}}(A, \mathbf{Q}/\mathbf{Z}) \to 0$ is split.* $\qquad\square$

If R is an algebra over a commutative ring k, U is an injective cogenerator in k-Mod and $*$ denotes the functor $\mathrm{Hom}_k(-, U)$, then for every right (or left) R-module A there is a canonical monomorphism $\varepsilon_A \colon A \to A^{**}$ defined by $(\varepsilon_A(a))(\alpha) = \alpha(a)$ for every $a \in A$, $\alpha \in A^*$. This canonical homomorphism ε_A is a morphism of right (respectively left) R-modules.

Corollary 1.30 *Let k be a commutative ring, let R be a k-algebra, let $_kU$ be an injective cogenerator in k-Mod and let A_R be an R-module. Then the canonical monomorphism $\varepsilon_A \colon A \to A^{**}$ is a pure monomorphism of right R-modules.*

Proof. In order to prove that the canonical monomorphism $\varepsilon_A \colon A \to A^{**}$ is a pure monomorphism it suffices to show that the kernel of $(\varepsilon_A)^* \colon A^{***} \to A^*$ is a direct summand of A^{***} (Corollary 1.28). This follows from the fact that the composite mapping $(\varepsilon_A)^* \circ \varepsilon_{A^*}$ of $\varepsilon_{A^*} \colon A^* \to A^{***}$ and $(\varepsilon_A)^* \colon A^{***} \to A^*$ is the identity of A^*, as is easily seen. $\qquad\square$

As another application of characterization (c) of Theorem 1.27 we leave the proof of the following proposition to the reader as an easy exercise.

Proposition 1.31   (a) *If $\{\, A_j \mid j \in J \,\}$ is a chain of submodules of B and each A_j is pure in B, then $\bigcup_{j \in J} A_j$ is a pure submodule of B.*

(b) *If $A \subseteq B \subseteq C$ are modules, A is pure in C and B/A is pure in C/A, then B is pure in C.*

(c) *If $A \subseteq B \subseteq C$ are modules and A is pure in C, then A is pure in B.* $\qquad\square$

1.5 Finitely definable subgroups and pure-injective modules

In order to study pure-injective modules we shall now define the concept of finitely definable subgroup.

Let $M = (r_{ij})$ be a $n \times m$ matrix with entries in a ring R, and let A_R be a right R-module. Consider the n-tuples $X = (x_1, \ldots, x_n) \in A^n$ such that $XM = 0$ (these n-tuples are the solutions of the corresponding system of m homogeneous linear equations). The set $S_M = \{ X \in A^n \mid XM = 0 \}$ is an additive subgroup of A^n (it is not a submodule, in general). Hence if $\pi_1 \colon A^n \to A$ denotes the canonical projection onto the first summand, the image $\pi_1(S_M)$ of S_M is an additive subgroup of A. The additive subgroups $\pi_1(S_M)$ of A arising in this fashion are called *finitely definable subgroups* of A_R (or *finite matrix subgroups* of A_R). An additive subgroup G of A_R is a *definable subgroup* of A_R if it is an intersection of finitely definable subgroups.

If $_T A_R$ is a bimodule, every finitely definable subgroup of A_R is a submodule of $_T A$. In particular, the finitely definable subgroups of a right R-module A_R are submodules of the left $\mathrm{End}(A_R)$-module A. If the ring R is commutative, every finitely definable subgroup of A_R is a submodule of A_R.

The first trivial examples of finitely definable subgroups of an R-module A_R are the subgroups 0 and A (take as M the 1×1 matrices $M = (1)$ and $M = (0)$ respectively).

Proposition 1.32 *Let A_R be a right module over a ring R and let r be an element of R.*

(a) *The groups Gr and $(G :_A r) = \{ a \in A \mid ar \in G \}$ are finitely definable subgroups of A_R for every finitely definable subgroup G of A_R.*

(b) *There is a one-to-one correspondence between the set of all finitely definable subgroups G of A_R that contain $(0 :_A r)$ and the set of all finitely definable subgroups G' of A_R that are contained in Ar defined by*

$$G \mapsto G' = Gr.$$

Proof. Given a finitely definable subgroup G of A_R, there is a $n \times m$ matrix M with entries in R such that $G = \pi_1(S)$, where

$$S = \{X \in A^n \mid XM = 0\}$$

and $\pi_1 \colon A^n \to A$ is the canonical projection onto the first summand. It is easily seen that if $\pi_1' \colon A^{n+1} \to A$ is the canonical projection onto the first summand, then $Gr = \pi'_1(S')$, where

$$S' = \left\{ Y \in A^{n+1} \,\middle|\, Y \begin{pmatrix} 1 & 0 & 0 & 0 & \dots & 0 \\ -r & & & & & \\ 0 & & & & & \\ \vdots & & & A & & \\ 0 & & & & & \end{pmatrix} = 0 \right\},$$

and $(G :_A r) = \pi'_1(S'')$, where

$$S'' = \left\{ Y \in A^{n+1} \,\middle|\, Y \begin{pmatrix} r & 0 & 0 & 0 & \dots & 0 \\ -1 & & & & & \\ 0 & & & & & \\ \vdots & & & A & & \\ 0 & & & & & \end{pmatrix} = 0 \right\}.$$

This proves (a), and (b) follows easily from (a). □

In particular, if A_R is an R-module and r is an element of R, then Ar and $(0 :_A r)$ are finitely definable subgroups of A_R.

Lemma 1.33 *The sum and the intersection of two finitely definable subgroups of A_R are finitely definable subgroups of A_R.* □

The proof of Lemma 1.33 is left to the reader. (Hint: Suppose $G = \pi_1(S)$ with $S = \{X \in A^n \mid XM = 0\}$, and $G' = \pi'_1(S')$ with $S' = \{Y \in A^m \mid YM' = 0\}$. Then $y \in G + G'$ if and only if there exist $x_1 \in G$ and $x'_1 \in G'$ such that $y = x_1 + x'_1$, that is, if and only if there exist $x_1, \dots, x_n, x'_1, \dots, x'_m \in A$ such that $y = x_1 + x'_1$, $(x_1 \dots x_n)M = 0$, $(x'_1 \dots x'_m)M' = 0$. And $z \in G \cap G'$ if and only if $z \in G$ and $z \in G'$, i.e., if and only if there exist $x_2, \dots, x_n, x'_2, \dots, x'_m \in A$ such that $(y\ x_2\ \dots\ x_n)M = 0$ and $(y\ x'_2\ \dots\ x'_m)M' = 0$.)

From Lemma 1.33 we obtain that if A_R is an R-module and F is a finite subset of R, the additive subgroup $(0 :_A F)$ is a finitely definable subgroup of A_R.

Lemma 1.34 *If A_R is a pure submodule of B_R, then for every finitely definable subgroup G of A_R there exists a finitely definable subgroup H of B_R such that $G = H \cap A$.*

Proof. If G is a finitely definable subgroup of A_R, then there exists a $n \times m$ matrix M with entries in R such that $G = \pi_1(S)$, where $S = \{X \in A^n \mid XM = 0\}$ and $\pi_1 : A^n \to A$ is the canonical projection onto the first summand. Let $S' = \{Y \in B^n \mid YM = 0\}$, let $\pi'_1 : B^n \to B$ be the canonical projection onto the first summand, and let $H = \pi'_1(S')$, so that H is a finitely definable subgroup of B_R. Obviously $G \subseteq H \cap A$. The other inclusion follows from the characterization of pure submodules given in Theorem 1.27(d). □

Let R be a ring. A *compact topological R-module* (N_R, τ) is a right R-module N_R with a compact Hausdorff topology τ on the set N with the property that addition $+: N \times N \to N$ and multiplication $\cdot: N \times R \to N$ are continuous. Here $N \times N$ is endowed with the product topology, and $N \times R$ is endowed with the product topology of the topology τ on N and the discrete topology on R.

For instance, let $\mathbf{R}$ be the additive group of real numbers with the usual topology and let $\mathbf{T} = \mathbf{R}/\mathbf{Z}$ be endowed with the quotient topology. Then $\mathbf{T}$ is a compact topological $\mathbf{Z}$-module. More generally, let $_RA$ be an arbitrary left R-module, so that $\mathrm{Hom}_{\mathbf{Z}}(_RA, \mathbf{T})$ is a right R-module. The set $\mathrm{Hom}_{\mathbf{Z}}(_RA, \mathbf{T})$ is a subset of the set $\mathbf{T}^A$ of all mappings $A \to \mathbf{T}$. Since $\mathbf{T}$ is compact, the product topology on $\mathbf{T}^A$ is compact by the Tychonoff Theorem, and $\mathrm{Hom}_{\mathbf{Z}}(_RA, \mathbf{T})$ turns out to be a closed subset of the compact topological space $\mathbf{T}^A$. Hence $\mathbf{T}^A$ induces a compact Hausdorff topology τ on the set $\mathrm{Hom}_{\mathbf{Z}}(_RA, \mathbf{T})$, and $(\mathrm{Hom}_{\mathbf{Z}}(_RA, \mathbf{T}), \tau)$ is a compact topological R-module.

The reason why we are introducing compact topological R-modules is that we shall see in Theorem 1.35 that if (N_R, τ) is a compact topological R-module, then N_R is a pure-injective R-module. Conversely, given a pure-injective R-module M_R, there exists a module N_R such that M_R is isomorphic to a direct summand of N_R, and N_R can be endowed with a topology τ in such a way that (N_R, τ) becomes a compact topological R-module.

In Theorem 1.35 we shall give other characterizations of pure-injective modules. In characterization (d) we shall consider a system

$$\sum_{i \in I} x_i r_{ij} = m_j, \quad j \in J,$$

of linear equations with the coefficients r_{ij} in a ring R and the m_j in a right module M_R. This means that we have a set of variables $\{ x_i \mid i \in I \}$, and for every $j \in J$ there are at most finitely many $i \in I$ with $r_{ij} \neq 0$, that is, any equation of the system has only a finite number of non-zero coefficients. Also recall that a family $\mathcal{F}$ of subsets of a set M is said to have the *finite intersection property* if the intersection of every finite subfamily of the family $\mathcal{F}$ is non-empty.

Theorem 1.35 *The following conditions are equivalent for a right module M_R over an arbitrary ring R.*

(a) *M_R is pure-injective.*

(b) *Every pure exact sequence $0 \to M_R \to B_R \to C_R \to 0$ of right R-modules is split.*

(c) *There exists a compact topological R-module (N_R, τ) such that M_R is isomorphic to a direct summand of N_R.*

(d) *A system* $\sum_{i \in I} x_i r_{ij} = m_j$, $j \in J$, *of linear equations in* M_R *is soluble in* M_R *whenever it is finitely soluble in* M_R *(here* $r_{ij} \in R$ *and* $m_j \in M$ *for every* i, j, *and for every* $j \in J$ *there are at most finitely many* $i \in I$ *with* $r_{ij} \neq 0$).

(e) *If* $m_i \in M$ *and* G_i *is a finitely definable subgroup of* M_R *for every* $i \in I$, *and the family* $\{ m_i + G_i \mid i \in I \}$ *has the finite intersection property, then* $\bigcap_{i \in I} m_i + G_i \neq \emptyset$.

Proof. The proof that (a) $\Rightarrow$ (b) is the dual of the proof of (a) $\Rightarrow$ (b) in Proposition 1.24.

(b) $\Rightarrow$ (c). The ring R is a **Z**-algebra, and the abelian group $\mathbf{T} = \mathbf{R}/\mathbf{Z}$ is an injective cogenerator in **Z**-Mod. By Corollary 1.30 the canonical monomorphism $\varepsilon_M \colon M \to M^{**}$, where $*$ denotes the functor $\mathrm{Hom}_{\mathbf{Z}}(-, \mathbf{T})$, is a pure monomorphism of right R-modules. If (b) holds, then M_R is isomorphic to a direct summand of M^{**}, and on page 20 we have already remarked that there is a topology τ on $\mathrm{Hom}_{\mathbf{Z}}(M^*, \mathbf{T}) = M^{**}$ with respect to which $(\mathrm{Hom}_{\mathbf{Z}}(M^*, \mathbf{T}), \tau)$ is a compact topological R-module.

(c) $\Rightarrow$ (d). It is clear that if a module N_R satisfies property (d), then every direct summand of N_R satisfies (d). Hence it is sufficient to prove that if (N_R, τ) is a compact topological R-module, then N_R satisfies (d). Let

$$\sum_{i \in I} x_i r_{ij} = m_j, \quad j \in J,$$

be a finitely soluble system of linear equations, where $r_{ij} \in R$ and $m_j \in N_R$ for every i, j. Note that the solutions of this system are elements of the direct product N^I. Endow N^I with the product topology of the topologies τ. For every equation $\sum_{i \in I} x_i r_{ij} = m_j$, the mapping $\varphi_j \colon N^I \to N$, defined by

$$\varphi_j((a_i)_{i \in I}) = \sum_{i \in I} a_i r_{ij}$$

for every $(a_i)_{i \in I} \in N^I$, is a continuous mapping. Since τ is Hausdorff, the set $S_j = \varphi_j^{-1}(m_j)$ of solutions of the equation is a closed subset of N^I. Since the system is finitely soluble, the family of closed sets $\{ S_j \mid j \in J \}$ has the finite intersection property. As N^I is compact, the intersection of the family is non-empty. Hence the equations of the system have a common solution.

(d) $\Rightarrow$ (a). Let M_R be a module for which (d) holds, let A_R be a pure submodule of an R-module B_R, and let $\varphi \colon A_R \to M_R$ be a homomorphism. We must show that φ extends to a homomorphism $\psi \colon B_R \to M_R$. Let $\{ x_b \mid b \in B_R \}$ be a set of variables indexed in B_R. Consider the system $\mathcal{S}$ consisting of one equation $x_b + x_{b'} - x_{b+b'} = 0$ for each pair of elements $b, b' \in B_R$, one equation $x_b r - x_{br} = 0$ for each $b \in B_R$ and each $r \in R$, and one equation $x_a = \varphi(a)$ for each $a \in A_R$. In order to show that this system $\mathcal{S}$ is finitely soluble in M_R, fix a finite subsystem S' of S. Let S'' be the linear system in A_R having the same number of equations as S' and constructed in the following way: for each equation $x_b + x_{b'} - x_{b+b'} = 0$ of S', let $x_b + x_{b'} - x_{b+b'} = 0$ be an equation of S''; for each equation $x_b r - x_{br} = 0$ of S', let $x_b r - x_{br} = 0$ be an equation of S''; for each equation $x_a = \varphi(a)$ of S', let $x_a = a$ be an equation of S''. Then S'' is linear system in A_R that in B_R has the trivial solution $x_b = b$ for every $b \in B_R$. Since A_R is pure in B_R, S'' has a solution $x_b = a_b$ with $a_b \in A_R$ for every b (Lemma 1.26). Then $x_b = \varphi(a_b)$ for every b is a solution of S' in M_R. This proves that S is finitely soluble in M_R. By (d) the system S has a solution $x_b = \psi(b)$ with $\psi(b) \in M_R$ for every $b \in B_R$. Then $\psi: B_R \to M_R$ is a homomorphism and $\psi(a) = \varphi(a)$ for every $a \in A$. Hence M_R is pure-injective.

(d) $\Rightarrow$ (e). Let M_R be a module for which (d) holds. Let

$$\{\, m_i + G_i \mid i \in I \,\}$$

be a family with the finite intersection property, where $m_i \in M$ and G_i is a finitely definable subgroup of M_R for every $i \in I$. Every subgroup G_i is obtained from a matrix with entries in R. Denote the entries in the first row of this matrix by $r_{\ell,i}$ and the entries in the other rows of the matrix by $s_{j,\ell,i}$. Hence an element x of M_R belongs to G_i if and only if the system

$$x r_{\ell,i} + \sum_{j \in F_i} x_j s_{j,\ell,i} = 0, \quad \ell \in E_i$$

in the variables x_j, $j \in F_i$, is soluble in M_R. Thus an element y of M_R belongs to $m_i + G_i$ if and only if $y - m_i \in G_i$, if and only if the system

$$(y - m_i) r_{\ell,i} + \sum_{j \in F_i} x_j s_{j,\ell,i} = 0, \quad \ell \in E_i$$

is soluble in M_R, that is, if and only if the system

$$y r_{\ell,i} + \sum_{j \in F_i} x_j s_{j,\ell,i} = m_i r_{\ell,i}, \quad \ell \in E_i$$

is soluble. Let K be an arbitrary subset of I. Then $\bigcap_{i \in K} m_i + G_i \neq \emptyset$ if and only if there exists $y \in M_R$ that belongs to $m_i + G_i$ for every $i \in K$, i.e., such that the system $y r_{\ell,i} + \sum_{j \in F_i} x_j s_{j,\ell,i} = m_i r_{\ell,i}$, $i \in K, \ell \in E_i$ is soluble, that is, if and only if the system

$$S(K): \qquad z r_{\ell,i} + \sum_{j \in F_i} x_j s_{j,\ell,i} = m_i r_{\ell,i}, \quad i \in K, \ell \in E_i$$

in the variables z and x_j, $j \in \bigcup_{i \in K} F_i$, is soluble in M_R. It is now clear that if the family $\{\, m_i + G_i \mid i \in I \,\}$ has the finite intersection property, then the system $S(K)$ is soluble for every finite subset K of I, so that the system $S(I)$

is finitely soluble. In this case, by (d), the system $\mathcal{S}(I)$ is soluble in M_R, i.e., $\bigcap_{i \in I} m_i + G_i \neq \emptyset$.

(e) $\Rightarrow$ (d). We claim that if M_R is a module for which (e) holds,

$$\sum_{i \in I} x_i r_{ij} = m_j, \quad j \in J, \tag{1.1}$$

is a finitely soluble system in the variables x_i, $i \in I$, with $J \neq \emptyset$, and $k \in I$ is a fixed index, then there exists $n_k \in M_R$ such that the system

$$\sum_{i \in I \setminus \{k\}} x_i r_{ij} = m_j - n_k r_{kj}, \quad j \in J,$$

is still finitely soluble.

In order to prove the claim, set $I' = I \setminus \{k\}$. Let $\mathcal{F}$ be the set of all finite non-empty subsets of J. For every $L \in \mathcal{F}$ choose a solution $(n_i^L)_{i \in I}$ of the corresponding finite system

$$\sum_{i \in I} x_i r_{ij} = m_j, \quad j \in L. \tag{1.2}$$

Let S_L be the set of all $(p_i)_{i \in I} \in M^I$ that are solutions of the homogeneous system $\sum_{i \in I} x_i r_{ij} = 0$, $j \in L$, and set $G_L = \pi_k(S_L)$, so that G_L is a finitely definable subgroup of M_R. Note that if $(q_i)_{i \in I}$ is a solution of system (1.2), then $(q_i)_i - (n_i^L)_i \in S_L$, so that $q_k - n_k^L \in G_L$, that is, $q_k \in n_k^L + G_L$. In particular, since system (1.1) is finitely soluble, the family $\{ n_k^L + G_L \mid L \in \mathcal{F} \}$ has the finite intersection property. From (e) it follows that there exists an element n_k that belongs to $n_k^L + G_L$ for every $L \in \mathcal{F}$. In order to prove that this element n_k has the property required in the claim, we must show that for every $L \in \mathcal{F}$ the system

$$\sum_{i \in I'} x_i r_{ij} = m_j - n_k r_{kj}, \quad j \in L, \tag{1.3}$$

is soluble. Now $n_k \in n_k^L + G_L$, that is, $n_k - n_k^L \in G_L$, so that there exist elements p_i, $i \in I'$, of M_R with $(n_k - n_k^L) r_{kj} + \sum_{i \in I'} p_i r_{ij} = 0$ for every $j \in L$. Since $(n_i^L)_i$ is a solution of system (1.2), we know that $n_k^L r_{kj} + \sum_{i \in I'} n_i^L r_{ij} = m_j$ for every $j \in L$. If we add these two equalities we obtain that

$$n_k r_{kj} + \sum_{i \in I'} (p_i + n_i^L) r_{ij} = m_j$$

for every $j \in L$. Hence $(p_i + n_i^L)_{i \in I'}$ is a solution of system (1.3). This proves the claim.

Now let M_R be a module for which (e) holds and consider a finitely soluble system (1.1). Let $\mathcal{L}$ be the set of all pairs (C, g), where C is a subset of I, $g \in M^C$ and the system of equations

$$\mathcal{S}(C, g): \quad \sum_{i \in I \setminus C} x_i r_{ij} = m_j - \sum_{i \in C} g(i) r_{ij}, \quad j \in J,$$

is finitely soluble. By the claim the set $\mathcal{L}$ is non-empty. Partially order $\mathcal{L}$ by setting $(C, g) \leq (C', g')$ if $C \subseteq C'$ and $g(i) = g'(i)$ for every $i \in C$. Let $(C_a, g_a)_{a \in A}$ be a chain in $\mathcal{L}$. Set $C^* = \bigcup_{a \in A} C_a$ and let $g^* \in M^{C^*}$ be defined by $g^*(i) = g_a(i)$ for every $i \in C_a$, $a \in A$. In order to prove that $\mathcal{S}(C^*, g^*)$ is finitely soluble, fix a finite subset L of J. Let $I(L)$ be the finite set of all $i \in I$ for which there exists $j \in L$ with $r_{ij} \neq 0$. Since $I(L) \cap C^*$ is a finite subset of C^* and $(C_a)_{a \in A}$ is a chain of subsets of C^*, there exists $b \in A$ such that $I(L) \cap C^* \subseteq C_b$. Note that $I(L) \cap C^* = I(L) \cap C_b$ and $I(L) \setminus C^* = I(L) \setminus C_b$. Since $\mathcal{S}(C_b, g_b)$ is finitely soluble, we have that the system $\sum_{i \in I \setminus C_b} x_i r_{ij} = m_j - \sum_{i \in C_b} g_b(i) r_{ij}$, $j \in L$, is soluble, that is, the system $\sum_{i \in I \setminus C^*} x_i r_{ij} = m_j - \sum_{i \in C^*} g^*(i) r_{ij}$, $j \in L$, is soluble. Hence $(C^*, g^*) \in \mathcal{L}$ and the partially ordered set $\mathcal{L}$ is inductive. By Zorn's Lemma $\mathcal{L}$ has a maximal element $(\overline{C}, \overline{g})$. If $\overline{C}$ is a proper subset of I, apply the claim to the finitely soluble system $\mathcal{S}(\overline{C}, \overline{g})$ and to an element $k \in I \setminus \overline{C}$. By the claim there exists an element $(\overline{C} \cup \{k\}, g^{**})$ in $\mathcal{L}$ strictly greater than $(\overline{C}, \overline{g})$. This contradiction shows that $\overline{C} = I$. Hence $\overline{g}$ is a solution of system (1.1). $\qquad\square$

A module satisfying condition (d) in Theorem 1.35 is said to be an *algebraically compact* module. Hence a module is algebraically compact if and only if it is pure-injective. As an immediate application of condition (d) of Theorem 1.35 we obtain the following corollary.

Corollary 1.36 *If $\varphi \colon R \to S$ is a ring homomorphism, every pure-injective right S-module is pure-injective as a right R-module.* $\qquad\square$

We conclude the section with a technical result that will be used in the sequel.

Proposition 1.37 *Let e be an idempotent element in a ring R. Let M_R, N_R be two R-modules, and assume $m \in Me$, $n \in Ne$. Suppose N_R pure-injective. The following conditions are equivalent:*

(a) *for every finitely presented module F_R, every element $a \in Fe$ and every homomorphism $f \colon F_R \to M_R$ such that $f(a) = m$ there exists a homomorphism $f' \colon F_R \to N_R$ such that $f'(a) = n$;*

(b) *there exists a homomorphism $g \colon M_R \to N_R$ such that $g(m) = n$.*

Proof. For the implication (b) $\Rightarrow$ (a) it is sufficient to take $f' = gf$. For the converse, suppose (a) holds. Consider a variable x_p for every element $p \in M_R$. Let $\mathcal{S}$ be the linear system consisting of one linear equation $x_p + x_q - x_{p+q} = 0$ for each pair $p, q \in M_R$, one linear equation $x_p r - x_{pr} = 0$ for each $p \in M_R$, $r \in R$, and the equation $x_m = n$. In order to prove that this system $\mathcal{S}$ is finitely soluble in N_R, let $\mathcal{S}'$ be a finite subsystem of $\mathcal{S}$. If the equation $x_m = n$ does not appear in $\mathcal{S}'$, then $x_p = 0$ for every p is a solution of the finite subsystem $\mathcal{S}'$. Suppose that the equation $x_m = n$ appears in $\mathcal{S}'$. Only finitely many variables appear in $\mathcal{S}'$. Hence there exists a finite subset $I \subseteq M_R$ such that all the variables that appear in $\mathcal{S}'$ belong to the set $\{\, x_p \mid p \in I \,\}$. Note that $m \in I$. Let F_R be the finitely presented R-module having the set $\{\, x_p \mid p \in I \,\}$ as a set of generators and the equations in $\mathcal{S}'$ different from the equation $x_m = n$ as a set of relations. Let $f\colon F_R \to M_R$ be the homomorphism defined by $f(x_p) - p$ for every $p \in I$. Then $f(x_m) - m$ and $x_m = x_{me} = x_m e \in Fe$, so that by (a) there exists a homomorphism $f'\colon F_R \to N_R$ such that $f'(x_m) = n$. It is now easily seen that $x_p = f'(x_p)$ for every p is a solution of the finite subsystem $\mathcal{S}'$ in N_R. Since N_R is pure-injective, the system $\mathcal{S}$ is soluble (Theorem 1.35(d)). Hence for every $p \in M_R$ there is an element $g(p) \in N_R$ such that $g(p) + g(q) = g(p + q)$ for every $p, q \in M_R$, $g(p)r = g(pr)$ for every $p \in M_R$, $r \in R$, and $g(m) = n$. Thus $g\colon M_R \to N_R$ is a homomorphism and $g(m) = n$. $\qquad\square$

1.6   The category ($_R$**FP, Ab**)

In this book we shall deal only with elementary category theory. We introduce it in order to present pure-injective modules as injective objects of a suitable Grothendieck category. Recall that a *Grothendieck category* is an abelian category with a generator which has arbitrary direct sums and in which direct limits are exact. Every object in a Grothendieck category is a subobject of an injective object.

Let R be a ring, $_R$FP the full subcategory of R-Mod whose objects are the finitely presented left R-modules and Ab the category of abelian groups. Recall that if $F, G\colon {}_R$FP$\to$Ab are additive functors, a *natural transformation* (or *functorial morphism*) u from F to G consists of a homomorphism of abelian groups $u_X\colon F(X) \to G(X)$ for every object X of $_R$FP such that the diagram

$$\begin{array}{ccc} F(X) & \xrightarrow{\;u_X\;} & G(X) \\ F(\alpha) \downarrow & & \downarrow G(\alpha) \\ F(Y) & \xrightarrow{\;u_Y\;} & G(Y) \end{array}$$

is commutative for every morphism $\alpha\colon X \to Y$ in $_R$FP. Let ($_R$FP, Ab) be the category of all additive functors from $_R$FP to Ab, that is, the category whose objects are all the additive functors from $_R$FP to Ab and whose morphisms from an additive functor $F\colon {}_R$FP $\to$Ab to an additive functor $G\colon {}_R$FP$\to$Ab are the natural transformations from F to G. If $u\colon F \to G$ is a morphism in ($_R$FP, Ab), then the kernel $\ker u$ of u turns out to be defined as $(\ker u)(X) = \ker u_X$ for every finitely presented left R-module X, the image $u(F)$ of u turns out to be defined as $(u(F))(X) = u_X(X)$, and the cokernel $\operatorname{coker} u$ of u turns out to be defined as $(\operatorname{coker} u)(X) = G(X)/u_X(X)$.

A morphism $u\colon F \to G$ is a monomorphism in ($_R$FP, Ab) if and only if $u_X\colon F(X) \to G(X)$ is a monomorphism of abelian groups for every object X in

$_R$FP, and is an epimorphism if and only if $u_X\colon F(X) \to G(X)$ is an epimorphism of abelian groups for every object X. More generally, if $F\overset{u}{\longrightarrow}G\overset{v}{\longrightarrow}H$ is a sequence of additive functors and natural transformations, the sequence is exact in the category $(_R\mathrm{FP}, \mathrm{Ab})$ if and only if the sequence of abelian groups $F(X)\overset{u_X}{\longrightarrow}G(X)\overset{v_X}{\longrightarrow}H(X)$ is exact for every finitely presented left R-module X. The category $(_R\mathrm{FP}, \mathrm{Ab})$ is a Grothendieck category.

If M_R is an object of Mod-R, then the functor $M \otimes_R -\colon {_R}\mathrm{FP}\to\mathrm{Ab}$ is an object of $(_R\mathrm{FP}, \mathrm{Ab})$. Therefore there is a functor

$$\boldsymbol{\Psi}\colon \mathrm{Mod}\text{-}R \to (_R\mathrm{FP}, \mathrm{Ab})$$

defined by $\boldsymbol{\Psi}(M_R) = M \otimes_R -$ for every right R-module M_R. This functor $\boldsymbol{\Psi}\colon \mathrm{Mod}\text{-}R \to (_R\mathrm{FP}, \mathrm{Ab})$ is full and faithful, that is, for each pair M_R, M'_R of objects of Mod-R, the mapping

$$\mathrm{Hom}(M_R, M'_R) \to \mathrm{Hom}_{(_R\mathrm{FP,\ Ab})}(M \otimes_R -, M' \otimes_R -)$$

is both surjective and injective. Hence $\boldsymbol{\Psi}\colon \mathrm{Mod}\text{-}R \to (_R\mathrm{FP}, \mathrm{Ab})$ induces an equivalence between the category Mod-R and a full subcategory of $(_R\mathrm{FP}, \mathrm{Ab})$. The characterization of pure exact sequences given in Theorem 1.27(b) yields the following lemma.

Lemma 1.38 *A sequence* $0 \to A_R \to B_R \to C_R \to 0$ *of right R-modules is exact and pure in* Mod-R *if and only if the induced sequence*

$$0 \to A \otimes_R - \to B \otimes_R - \to C \otimes_R - \to 0$$

is exact in $(_R\mathrm{FP}, \mathrm{Ab})$.   □

We are ready for the main result of this section.

Proposition 1.39 *An object F of* $(_R\mathrm{FP}, \mathrm{Ab})$ *is injective if and only if it is isomorphic to* $\boldsymbol{\Psi}(M_R) = M \otimes_R -$ *for some pure-injective right R-module M_R.*

Hence the full subcategory of Mod-R *whose objects are the pure-injective right R-modules is equivalent to the full subcategory of* $(_R\mathrm{FP}, \mathrm{Ab})$ *whose objects are the injective objects of* $(_R\mathrm{FP}, \mathrm{Ab})$.

Proof. Let F be an injective object of $(_R\mathrm{FP}, \mathrm{Ab})$. We claim that if X, Y, Z are finitely presented left R-modules and $X \to Y \to Z \to 0$ is an exact sequence in R-Mod, then the induced sequence of abelian groups

$$F(X) \to F(Y) \to F(Z) \to 0$$

is exact. In order to prove the claim, let $_RX, {_R}Y, {_R}Z$ be finitely presented left R-modules and let $_RX \to {_R}Y \to {_R}Z \to 0$ be an exact sequence. Then for every object $_RN$ of $_R\mathrm{FP}$ the sequence of abelian groups

$$0 \to \mathrm{Hom}(_RZ, {_R}N) \to \mathrm{Hom}(_RY, {_R}N) \to \mathrm{Hom}(_RX, {_R}N)$$

is exact, that is,

$$0 \to \text{Hom}(_RZ, -) \to \text{Hom}(_RY, -) \to \text{Hom}(_RX, -)$$

is an exact sequence in the category $(_R\text{FP},\ Ab)$. Since F is an injective object in the category $(_R\text{FP},\ Ab)$, the functor $\text{Hom}_{(_R\text{FP},\ Ab)}(-, F)$ is exact. Hence the sequence of abelian groups

$$\text{Hom}_{(_R\text{FP},\ Ab)}(\text{Hom}(_RX, -), F) \to \text{Hom}_{(_R\text{FP},\ Ab)}(\text{Hom}(_RY, -), F)$$
$$\to \text{Hom}_{(_R\text{FP},\ Ab)}(\text{Hom}(_RZ, -), F) \to 0$$

is exact. By the Yoneda Lemma $\text{Hom}_{(_R\text{FP},\ Ab)}(\text{Hom}(_RW, -), F) \cong F(_RW)$ for every object $_RW$, so that the sequence of abelian groups

$$F(X) \to F(Y) \to F(Z) \to 0$$

is exact, which proves the claim.

Now let M be the abelian group $F(_RR)$. Since F induces a homomorphism $R \cong \text{Hom}(_RR, {}_RR) \to \text{Hom}(F(_RR), F(_RR)) = \text{End}_{\mathbf{Z}}(M)$, the abelian group M has a right R-module structure. As the functor F is additive, there is a canonical isomorphism $F(_RR^n) \cong M \otimes_R R^n$ for every integer $n \geq 0$. If $_RX$ is a finitely presented left R-module, there is an exact sequence $_RR^n \to {}_RR^m \to {}_RX \to 0$, so that by the claim, the first row in the commutative diagram

$$
\begin{array}{ccccccc}
F(_RR^n) & \to & F(_RR^m) & \to & F(_RX) & \to & 0 \\
\downarrow & & \downarrow & & & & \\
M \otimes_R R^n & \to & M \otimes_R R^m & \to & M \otimes_R X & \to & 0
\end{array}
$$

is exact. Since the second row also is exact and the vertical arrows are the canonical isomorphisms, there is an isomorphism $F(_RX) \cong M \otimes_R X$ for every finitely presented left R-module $_RX$, and we leave the verification that this isomorphism is natural, i.e., that $F \cong M \otimes_R -$ in $(_R\text{FP},\ Ab)$, to the reader.

We shall now show that the module M_R is pure-injective. Let A_R be a pure submodule of a module B_R, let $\varepsilon \colon A_R \to B_R$ be the embedding, and let $f \colon A_R \to M_R$ be a homomorphism. Then $\varepsilon \otimes - \colon A \otimes_R - \to B \otimes_R -$ is a monomorphism in $(_R\text{FP},\ Ab)$ by Lemma 1.38, hence the morphism $f \otimes - \colon A \otimes_R - \to M \otimes_R -$ into the injective object $M \otimes_R - \cong F$ extends to a morphism $u \colon B \otimes_R - \to M \otimes_R -$, that is, $u \circ (\varepsilon \otimes -) = f \otimes -$. Since the functor $\boldsymbol{\Psi} \colon \text{Mod-}R \to (_R\text{FP},\ Ab)$ is full and faithful, there exists a unique $g \colon B_R \to M_R$ such that $u = g \otimes -$. In particular, $(g \otimes 1_R) \circ (\varepsilon \otimes 1_R) = f \otimes 1_R$, that is, $g \circ \varepsilon = f$. This proves that M_R is pure-injective.

Conversely, let M_R be a pure-injective module. Then the functor $M \otimes_R -$ is an object in the Grothendieck category $(_R\text{FP},\ Ab)$, hence $M \otimes_R -$ is a subobject of an injective object G of $(_R\text{FP},\ Ab)$. By the first part of this proof, there exists a pure-injective module E_R such that $G \cong E \otimes_R -$. Hence there is a monomorphism $M \otimes_R - \to E \otimes_R -$ in $(_R\mathrm{FP}, \mathrm{Ab})$, which is of the type $f \otimes -$ for a unique $f: M_R \to E_R$ because the functor $\Psi: \mathrm{Mod}\text{-}R \to (_R\mathrm{FP}, \mathrm{Ab})$ is full and faithful. By Lemma 1.38 f is a pure monomorphism. But M_R is pure-injective, so that M_R is isomorphic to a direct summand of E_R. Hence the functor $M \otimes_R -$ is isomorphic to a direct summand of the injective object $G \cong E \otimes_R -$ of $(_R\mathrm{FP}, \mathrm{Ab})$. In particular, $M \otimes_R -$ is an injective object in $(_R\mathrm{FP}, \mathrm{Ab})$. This proves the first part of the statement.

It is now obvious that the full and faithful functor $\Psi: \mathrm{Mod}\text{-}R \to (_R\mathrm{FP}, \mathrm{Ab})$ induces an equivalence between the full subcategory of $\mathrm{Mod}\text{-}R$ whose objects are the pure-injective right R-modules and the full subcategory of $(_R\mathrm{FP}, \mathrm{Ab})$ whose objects are the injective objects of $(_R\mathrm{FP}, \mathrm{Ab})$. $\qquad\square$

1.7 Σ-pure-injective modules

A right R-module A is said to be Σ-*pure-injective* (or Σ-*algebraically compact*) if for every index set I the direct sum $A^{(I)}$ is a pure-injective R-module.

Theorem 1.40 *The following conditions are equivalent for a right module A_R over an arbitrary ring R.*

(a) A_R *is Σ-pure-injective.*

(b) *For every index set I the submodule $A^{(I)}$ of the R-module A^I is a direct summand of A^I.*

(c) *The submodule $A^{(\mathbf{N})}$ of the R-module $A^{\mathbf{N}}$ is a direct summand of $A^{\mathbf{N}}$.*

(d) A_R *satisfies the descending chain condition (d.c.c.) on finitely definable subgroups.*

Proof. For (a) $\Rightarrow$ (b) it is sufficient to note that $A^{(I)}$ is a pure submodule of A^I. The implication (b) $\Rightarrow$ (c) is trivial.

(c) $\Rightarrow$ (d). Suppose $A^{(\mathbf{N})}$ is a direct summand of $A^{\mathbf{N}}$ and there exists a strictly descending chain of finitely definable subgroups

$$G_0 \supset G_1 \supset G_2 \supset \dots$$

of A_R. For every $n \geq 0$ let g_n be an element of $G_n \setminus G_{n+1}$. Set

$$g = (g_0, g_1, g_2, \dots) \in A^{\mathbf{N}}.$$

Let S be the endomorphism ring of the R-module $A^{\mathbf{N}}$. Since $A^{(\mathbf{N})}$ is a direct summand of $A^{\mathbf{N}}$, there exists an endomorphism $\varphi \in S$ that is the identity on $A^{(\mathbf{N})}$ and whose image is $A^{(\mathbf{N})}$. Then $\varphi(g) \in A^{(\mathbf{N})}$, so that $\varphi(g) =$

$(a_0, a_1, a_2, \ldots, a_{n-1}, 0, 0, \ldots)$ for a suitable $n \geq 1$. In particular,

$$
\begin{aligned}
(a_0, a_1, a_2, \ldots, a_{n-1}, 0, 0, \ldots) &= \varphi(g) \\
&= \varphi(g_0, g_1, g_2, \ldots, g_n, 0, 0, \ldots) + \varphi(0, 0, \ldots, 0, g_{n+1}, g_{n+2}, \ldots) \\
&= (g_0, g_1, g_2, \ldots, g_n, 0, 0, \ldots) + \varphi(0, 0, \ldots, 0, g_{n+1}, g_{n+2}, \ldots). \quad (1.4)
\end{aligned}
$$

Now if M is the matrix that defines the finitely definable subgroup G_{n+1} of A_R, M defines the finitely definable subgroup $G_{n+1}^{\mathbf{N}}$ of $M^{\mathbf{N}}$, and obviously $(0, 0, \ldots, 0, g_{n+1}, g_{n+2}, \ldots) \in G_{n+1}^{\mathbf{N}}$. Since $A^{\mathbf{N}}$ is an S-R-bimodule, finitely definable subgroups of $A^{\mathbf{N}}$ are S-submodules. In particular, $\varphi(G_{n+1}^{\mathbf{N}}) \subseteq G_{n+1}^{\mathbf{N}}$, so that $\varphi(0, 0, \ldots, 0, g_{n+1}, g_{n+2}, \ldots) \in G_{n+1}^{\mathbf{N}}$. Comparing the n-th coordinate in relation (1.4) we obtain that $g_n \in G_{n+1}$, contradiction.

(d) $\Rightarrow$ (a). Suppose that A_R satisfies the d.c.c. on finitely definable subgroups. If a matrix M defines a finitely definable subgroup G in A_R, then M defines the finitely definable subgroup $G^{(I)}$ in $A^{(I)}$. In other words, the finitely definable subgroups of $A^{(I)}$ are exactly those of the form $G^{(I)}$ for some finitely definable subgroup G of A_R. Hence if A_R satisfies the d.c.c. on finitely definable subgroups, then $A^{(I)}$ also satisfies the d.c.c. on finitely definable subgroups. Thus in order to conclude it suffices to show that if a module A_R satisfies the d.c.c. on finitely definable subgroups, then A_R is pure-injective.

We shall apply Theorem 1.35. Let $\{\, a_i + G_i \mid i \in I \,\}$ be a family with the finite intersection property, where $a_i \in A_R$ and G_i is a finitely definable subgroup of A_R for every $i \in I$. Any intersection of finitely many subgroups G_i is a finitely definable subgroup of A_R. By the d.c.c. there is an intersection of finitely many G_i that is contained in all the other G_i, i.e., there is a finite subset F of I such that $\bigcap_{j \in F} G_j \subseteq G_i$ for every $i \in I$. Now for every index i there exists $b_i \in (a_i + G_i) \cap \left(\bigcap_{j \in F} (a_j + G_j) \right)$. Then $b_i + G_i = a_i + G_i$ and $b_i + G_j = a_j + G_j$ for every $j \in F$. Then

$$
\bigcap_{j \in F} a_j + G_j = \bigcap_{j \in F} b_i + G_j = b_i + \bigcap_{j \in F} G_j \subseteq b_i + G_i = a_i + G_i.
$$

Thus $\bigcap_{i \in I} a_i + G_i \neq \emptyset$. $\qquad\square$

We have already remarked that if $_S A_R$ is a bimodule, every finitely definable subgroup of A_R is a submodule of $_S A$. Hence if $_S A_R$ is a bimodule and $_S A$ is artinian, then A_R is Σ-pure-injective. In particular, every artinian module over a commutative ring is Σ-pure-injective.

Example 1.41 *If R is an arbitrary ring and A_R is a right R-module that is artinian as a left $\mathrm{End}(A_R)$-module, then A_R is Σ-pure-injective.*

If R is an arbitrary ring and A_R is an artinian right R-module, then A is Σ-pure-injective as a left $\mathrm{End}(A_R)$-module. $\qquad\square$

We conclude the chapter with three corollaries that show different applications of the characterizations of Σ-pure-injective modules given in Theorem 1.40.

Corollary 1.42 *Every pure submodule A_R of a Σ-pure-injective module B_R is a direct summand of B_R.*

PROOF. Let $G_0 \supseteq G_1 \supseteq G_2 \supseteq \ldots$ be a descending chain of finitely definable subgroups of A_R. For every $n \geq 0$ there exists a finitely definable subgroup H_n of B_R such that $G_n = H_n \cap A_R$ (Lemma 1.34). Then $\bigcap_{i=0}^{n} H_i$, $n \geq 0$, is a descending chain of finitely definable subgroups of B_R. Hence it is stationary. Thus the chain of the G_n is stationary. This proves that A_R is Σ-pure-injective. Since A_R is pure in B_R, A_R must be a direct summand of B_R. $\square$

Corollary 1.43 *Let R be a ring, let A_R be a Σ-pure-injective R-module and let r be an element of R such that right multiplication by r is an injective endomorphism of the abelian group A. Then right multiplication by r is an automorphism of the abelian group A.*

Proof. The descending chain $A \supseteq Ar \supseteq Ar^2 \supseteq Ar^3 \ldots$ of finitely definable subgroups of A_R is stationary, so that $Ar^n = Ar^{n+1}$ for some n. Since multiplication by r is an injective mapping $A \rightarrow A$, we have that $A = Ar$. $\square$

Corollary 1.44 *Let e be a non-zero idempotent in an arbitrary ring R. If A_R is a Σ-pure-injective R-module, then Ae is a Σ-pure-injective right eRe-module.*

Proof. Let $e \in R$ be a non-zero idempotent and let A_R be a Σ-pure-injective module. If G is a finitely definable subgroup of Ae_{eRe}, there exists a $n \times m$ matrix $M = (a_{ij})$ with entries in eRe such that if

$$S_M = \{ X \in (Ae)^n \mid XM = 0 \}$$

and $\pi_1 \colon (Ae)^n \rightarrow Ae$ is the canonical projection onto the first summand, then $\pi_1(S_M) = G$. Then $\{ X \in A^n \mid XM = 0 \} = S_M \oplus (A(1-e))^n$. Hence $G \oplus A(1-e)$ is a finitely definable subgroup of A_R. In particular, the d.c.c. on finitely definable subgroups of A_R implies the d.c.c. on finitely definable subgroups of Ae_{eRe}. $\square$

1.8   Notes on Chapter 1

E. Artin, E. Noether and J. H. M. Wedderburn were the first to consider the class of semisimple rings. [Noether] proved that conditions (a) and (c) of Theorem 1.2 are equivalent. The equivalence of condition (g) to the other conditions of Theorem 1.2 is the Wedderburn-Artin Theorem. The Jacobson radical was defined and characterized by Jacobson in 1945. He extended a previous characterization for finite-dimensional algebras over a field due to Perlis. In spite of the name, Nakayama's Lemma is due to [Azumaya 51]. The main ideas about the prime radical introduced in Section 1.1 are due to Levitzki, McCoy and Nagata.

Local rings appeared in commutative algebra first: Grell (1926) and Krull (1938) for commutative integral domains, Uzkov (1948) for arbitrary commutative rings. The systematic use of local rings and localization of a ring at a prime ideal begun with Krull in commutative algebra and with Chevalley and Zariski in algebraic geometry. Lemma 1.14 is essentially taken from [Warfield 75, Lemma 1.2]. Artinian serial rings were probably studied first in 1935 by Köthe and the term *uniserial* is due to him. The structure of these rings was then studied by [Asano], [Nakayama], who called "generalized uniserial ring" what we call artinian serial ring, [Fuller], [Eisenbud and Griffith 71], and by [Warfield 75], who defined serial rings in such a way that they need not be noetherian. Hence the definition of (right) serial rings we use in this monograph follows Warfield's terminology. The results on shrinkable uniserial modules presented in Section 1.3 are due to [Facchini and Salce]. The best reference on chain rings is probably the monograph [Bessenrodt, Brungs and Törner].

We have seen in Proposition 1.18 that every uniserial module over a commutative ring is unshrinkable. Much more can be said about uniserial modules over commutative rings. The following results are due to [Shores and Lewis]. Let M be a uniserial module over a commutative ring R. Then $\mathrm{End}_R(M)$ is a commutative local ring [Shores and Lewis, Corollary 3.4]. By passing from R to $R/\mathrm{ann}_R(M)$ we may assume M faithful without loss of generality. Hence let M be a faithful uniserial module over a commutative ring R. If M contains a finitely generated faithful submodule, M is said to be *finitely faithful*. If M is faithful but not finitely faithful, M is said to be *nonfinitely faithful*. The ring R has a finitely faithful uniserial module M if and only if R is a valuation ring. In this case M is isomorphic to an R-submodule of $Q(R)$ containing R (where $Q(R)$ denotes the classical quotient ring of R, that is, the total ring of fractions of R) and the endomorphism ring $\mathrm{End}_R(M)$ is isomorphic to the localization R_P of R at a suitable prime ideal P. In particular, $\mathrm{End}_R(M)$ is a valuation ring. If M is a nonfinitely faithful uniserial R-module, then $\mathrm{End}_R(M)$ is a valuation domain if and only if R is a domain, but there exist uniserial modules over valuation rings whose endomorphism ring is not a valuation ring. The proof of all these results can be found in [Shores and Lewis].

The history of pure submodules begins with [Prüfer] in 1923. He introduced the notion of pure subgroup of an abelian group. The definition for arbitrary modules is due to [Cohn], who gave it making use of characterization (c) in Theorem 1.27. Purity and algebraic compactness for general algebraic systems were defined and studied by Mycielski and Weglorz in the sixties. The notion of pure-injective module comes from abelian group theory. The class of pure-injective abelian groups was first considered by [Kaplansky 54], [Łoś], Balcerzyk and Maranda. See [Fuchs 70a, p. 178] for the history of the discovery of this class of abelian groups. The first systematic studies on purity in module theory go back to Fieldhouse, [Stenström] and [Warfield 69b], and most of the results presented in Section 1.4 are due to these three authors. We have mostly followed Warfield's presentation. [Warfield 69b] proved the existence of the pure-injective envelope of an arbitrary module M_R, that is, the smallest extension of M_R that is pure-injective and contains M_R as a pure submodule. We have not introduced pure-injective envelopes because we shall not need this notion in this book. At present, the best reference for purity in module theory is [Jensen and Lenzing, Chapters 6, 7 and 8 and Appendix B].

Finitely definable subgroups were introduced independently by [Gruson and Jensen 73] and [Zimmermann]. The equivalence of the first four conditions of Theorem 1.35 was proved by [Łoś] for abelian groups and by [Stenström] and [Warfield 69b] for modules over arbitrary rings. The last condition was discovered by [Zimmermann] and [Gruson and Jensen 73]. The content of Section 1.6, that is, the idea of using categories of additive functors to describe pure-injective modules as the injective objects of a suitable Grothendieck category, is due to [Gruson and Jensen 73 and 81]. Theorem 1.40 is due to [Zimmermann] and [Gruson and Jensen 76]. Modules satisfying both the a.c.c. and the d.c.c. on finitely definable subgroups have applications in representation theory ([Crawley-Boevey] and [Zimmermann-Huisgen and Zimmermann 90]). Example 1.41 is taken from [Zimmermann, p. 1096].

In this book we shall not consider the relationship between purity and model theory, for which we refer the reader to the books of [Jensen and Lenzing] and [Prest].

Chapter 2

The Krull-Schmidt-Remak-Azumaya Theorem

2.1 The exchange property

In the first two sections of this chapter we shall consider modules with the exchange property. Making use of the exchange property we shall study refinements of direct sum decompositions (Sections 2.3 and 2.10), prove the Krull-Schmidt-Remak-Azumaya Theorem (Section 2.4) and prove that every finitely presented module over a serial ring is serial (Section 3.5). If A, B, C are submodules of a module M and $C \leq A$, then $A \cap (B + C) = (A \cap B) + C$. This is called the *modular identity*. We begin with an immediate consequence of the modular identity that will be used repeatedly in the sequel.

Lemma 2.1 *If $A \subseteq B \subseteq A \oplus C$ are modules, then $B = A \oplus D$, where $D = B \cap C$.*

Proof. Application of the modular identity to the modules $A \subseteq B$ and C yields $B \cap (C + A) = (B \cap C) + A$, that is, $B = A + D$. This sum is direct because $A \cap D \subseteq A \cap C = 0$. $\qquad\square$

Given a cardinal $\aleph$, an R-module M is said to have the $\aleph$-*exchange property* if for any R-module G and any two direct sum decompositions

$$G = M' \oplus N = \oplus_{i \in I} A_i,$$

where $M' \cong M$ and $|I| \leq \aleph$, there are R-submodules B_i of A_i, $i \in I$, such that $G = M' \oplus (\oplus_{i \in I} B_i)$.

In this notation an application of Lemma 2.1 to the modules

$$B_i \subseteq A_i \subseteq B_i \oplus (M' \oplus (\oplus_{j \neq i} B_j))$$

yields $A_i = B_i \oplus D_i$, where $D_i = A_i \cap (M' \oplus (\oplus_{j \neq i} B_j))$. Hence the submodules B_i of A_i in the definition of module with the $\aleph$-exchange property are necessarily direct summands of A_i.

A module M has the *exchange property* if it has the $\aleph$-exchange property for every cardinal $\aleph$. A module M has the *finite exchange property* if it has the $\aleph$-exchange property for every finite cardinal $\aleph$.

A finitely generated module has the exchange property if and only if it has the finite exchange property.

Lemma 2.2 *If G, M', N, P, A_i $(i \in I)$, B_i $(i \in I)$ are modules, $B_i \subseteq A_i$ for every $i \in I$,*

$$G = M' \oplus N \oplus P = (\oplus_{i \in I} A_i) \oplus P \tag{2.1}$$

and

$$G/P = ((M' + P)/P) \oplus (\oplus_{i \in I}((B_i + P)/P)), \tag{2.2}$$

then

$$G = M' \oplus (\oplus_{i \in I} B_i) \oplus P.$$

Proof. From (2.2) it follows immediately that $G = M' + \left(\sum_{i \in I} B_i\right) + P$. In order to show that this sum is direct, suppose $m' + \left(\sum_{i \in I} b_i\right) + p = 0$ for some $m' \in M'$, $b_i \in B_i$ almost all zero, and $p \in P$. From (2.2) we have that $(m' + P) + \left(\sum_{i \in I}(b_i + P)\right) = 0$ in G/P, so that $m' \in P$ and $b_i \in P$ for every $i \in I$. Then by (2.1) we get $m' \in M' \cap P = 0$ and $b_i \in B_i \cap P \subseteq A_i \cap P = 0$. Therefore $p = 0$. $\qquad\qquad\square$

The proof of the next corollary follows immediately from the definitions and Lemma 2.2.

Corollary 2.3 *If G, M', N, P, A_i $(i \in I)$ are modules, $|I| \leq \aleph$,*

$$G = M' \oplus N \oplus P = (\oplus_{i \in I} A_i) \oplus P$$

and M' has the $\aleph$-exchange property, then for every $i \in I$ there exists a direct summand B_i of A_i such that

$$G = M' \oplus (\oplus_{i \in I} B_i) \oplus P. \qquad\qquad\square$$

The rest of this section is devoted to proving the first properties of modules with the exchange property. The following result shows that the class of modules with the $\aleph$-exchange property is closed under direct summands and finite direct sums.

Lemma 2.4 *Suppose $\aleph$ is a cardinal and $M = M_1 \oplus M_2$. The module M has the $\aleph$-exchange property if and only if both M_1 and M_2 have the $\aleph$-exchange property.*

Proof. Suppose $M = M_1 \oplus M_2$ has the $\aleph$-exchange property,

$$G = M_1' \oplus N = \oplus_{i \in I} A_i,$$

$M_1' \cong M_1$ and $|I| \le \aleph$. Then $G' = M_2 \oplus G = M' \oplus N = M_2 \oplus (\oplus_{i \in I} A_i)$, where $M' = M_1' \oplus M_2 \cong M$. Fix an element $k \in I$, and set $I' = I \setminus \{k\}$. Then $G' = M' \oplus N = (M_2 \oplus A_k) \oplus (\oplus_{i \in I'} A_i)$. Hence there exist submodules $B \subseteq M_2 \oplus A_k$ and $B_i \subseteq A_i$ for every $i \in I'$ such that

$$G' = M' \oplus B \oplus (\oplus_{i \in I'} B_i). \tag{2.3}$$

Since $M_2 \subseteq M_2 \oplus B \subseteq M_2 \oplus A_k$, it follows from Lemma 2.1 that

$$M_2 \oplus B = M_2 \oplus B_k,$$

where $B_k = (M_2 \oplus B) \cap A_k$. Thus $M' \oplus B = (M_1' \oplus M_2) \oplus B = M_1' \oplus M_2 \oplus B_k$. Substituting this into (2.3) we obtain

$$G' = M_1' \oplus M_2 \oplus (\oplus_{i \in I} B_i). \tag{2.4}$$

Application of the modular identity to the modules $M_1' \oplus (\oplus_{i \in I} B_i) \subseteq G$ and M_2 yields $G \cap (M_2 + (M_1' \oplus (\oplus_{i \in I} B_i))) = (G \cap M_2) + (M_1' \oplus (\oplus_{i \in I} B_i))$, that is, $G = M_1' \oplus (\oplus_{i \in I} B_i)$. Thus M_1 has the $\aleph$-exchange property.

Conversely, suppose M_1 and M_2 have the $\aleph$-exchange property and

$$G = M_1' \oplus M_2' \oplus N = \oplus_{i \in I} A_i,$$

where $M_1' \cong M_1$, $M_2' \cong M_2$ and $|I| \le \aleph$. Since M_1 has the $\aleph$-exchange property, there are submodules $A_i' \subseteq A_i$ such that $G = M_1' \oplus M_2' \oplus N = M_1' \oplus (\oplus_{i \in I} A_i')$. Since M_2 also has the $\aleph$-exchange property, from Corollary 2.3 it follows that for every $i \in I$ there exists a submodule $B_i \subseteq A_i'$ such that

$$G = M_2' \oplus (\oplus_{i \in I} B_i) \oplus M_1'.$$

Thus $M = M_1 \oplus M_2$ has the $\aleph$-exchange property. $\square$

Clearly, every module has the 1-exchange property. Modules the with 2-exchange property have the finite exchange property, as the next lemma shows.

Lemma 2.5 *If a module M has the 2-exchange property, then M has the finite exchange property.*

Proof. It is sufficient to show, for an arbitrary integer $n \ge 2$, that if M has the n-exchange property, then M has the $(n+1)$-exchange property. Let M be a module with the n-exchange property ($n \ge 2$) and suppose

$$G = M' \oplus N = A_1 \oplus A_2 \oplus \cdots \oplus A_{n+1},$$

where $M' \cong M$. Set $P = A_1 \oplus A_2 \oplus \cdots \oplus A_n$, so that $G = M' \oplus N = P \oplus A_{n+1}$. Since M has the 2-exchange property, there exist submodules $P' \subseteq P$ and $B_{n+1} \subseteq A_{n+1}$ such that $G = M' \oplus P' \oplus B_{n+1}$. An application of Lemma 2.1 to the modules $P' \subseteq P \subseteq P' \oplus (M' \oplus B_{n+1})$ and $B_{n+1} \subseteq A_{n+1} \subseteq B_{n+1} \oplus (M' \oplus P')$ yields $P = P' \oplus P''$ and $A_{n+1} = B_{n+1} \oplus A'_{n+1}$, where $P'' = P \cap (M' \oplus B_{n+1})$ and $A'_{n+1} = A_{n+1} \cap (M' \oplus P')$. From the decompositions

$$G = M' \oplus P' \oplus B_{n+1} = (P'' \oplus A'_{n+1}) \oplus (P' \oplus B_{n+1})$$

we infer that P'' is isomorphic to a direct summand of M'. Therefore P'' has the n-exchange property by Lemma 2.4. Since

$$P = P' \oplus P'' = A_1 \oplus A_2 \oplus \cdots \oplus A_n,$$

there exist submodules $B_i \subseteq A_i$ ($i = 1, 2, \ldots, n$) such that

$$P = P'' \oplus B_1 \oplus B_2 \oplus \cdots \oplus B_n.$$

Application of Lemma 2.1 to the modules

$$P'' \subseteq M' \oplus B_{n+1} \subseteq G = P'' \oplus (P' \oplus A_{n+1})$$

yields $M' \oplus B_{n+1} = P'' \oplus P'''$, where $P''' = (M' \oplus B_{n+1}) \cap (P' \oplus A_{n+1})$. Therefore

$$
\begin{aligned}
G &= M' \oplus P' \oplus B_{n+1} = P' \oplus P'' \oplus P''' = P \oplus P''' \\
&= B_1 \oplus \cdots \oplus B_n \oplus P'' \oplus P''' = B_1 \oplus \cdots \oplus B_n \oplus B_{n+1} \oplus M',
\end{aligned}
$$

that is, M has the $(n+1)$-exchange property. $\square$

2.2   Indecomposable modules with the exchange property

The aim of this section is to show that the indecomposable modules with the (finite) exchange property are exactly those with a local endomorphism ring. First we prove two elementary lemmas that will be used often in the sequel.

Lemma 2.6 *Let A be a module and let M_1, M_2, M' be submodules of A. Suppose $A = M_1 \oplus M_2$. Let $\pi_2 \colon A = M_1 \oplus M_2 \to M_2$ denote the canonical projection. Then $A = M_1 \oplus M'$ if and only if $\pi_2|_{M'} \colon M' \to M_2$ is an isomorphism. If these equivalent conditions hold, then the canonical projection $\pi_{M'} \colon A \to M'$ with respect to the decomposition $A = M_1 \oplus M'$ is $(\pi_2|_{M'})^{-1} \circ \pi_2$.*

Proof. The mapping $\pi_2|_{M'}$ is injective if and only if $M' \cap M_1 = 0$, and is surjective if and only if for every $x_2 \in M_2$ there exists $x' \in M'$ such that

$x' = x_1 + x_2$ for some $x_1 \in M_1$, that is, if and only if $M_2 \subseteq M' + M_1$, i.e., if and only if $M_1 + M_2 = M' + M_1$. This proves the first part of the statement. For the second part suppose that the equivalent conditions hold. Given an arbitrary element $a \in A$, one has $a = x_1 + \pi_{M'}(a)$ for a suitable element $x_1 \in M_1$. Hence $\pi_2(a) = \pi_2(x_1) + \pi_2|_{M'}(\pi_{M'}(a))$, from which $(\pi_2|_{M'})^{-1}\pi_2(a) = \pi_{M'}(a)$. $\square$

Lemma 2.7 *Let $M, N, P_1, \ldots, P_n$ be modules with $M \oplus N = P_1 \oplus \cdots \oplus P_n$. If M is an indecomposable module with the finite exchange property, then there is an index $j = 1, 2, \ldots, n$ and a direct sum decomposition $P_j = B \oplus C$ of P_j such that $M \oplus N = M \oplus B \oplus (\oplus_{i \neq j} P_i)$, $M \cong C$ and $N \cong B \oplus (\oplus_{i \neq j} P_i)$.*

Proof. Since M has the finite exchange property, for every $i = 1, 2, \ldots, n$ there exists a decomposition $P_i = B_i \oplus C_i$ of P_i such that

$$M \oplus B_1 \oplus \cdots \oplus B_n = P_1 \oplus \cdots \oplus P_n.$$

If we factorize modulo $B_1 \oplus \cdots \oplus B_n$ we find that $M \cong C_1 \oplus \cdots \oplus C_n$. But M is indecomposable. Hence there exists an index j such that $M \cong C_j$ and $C_i = 0$ for every $i \neq j$. Then $B_i = P_i$ for $i \neq j$, hence

$$M \oplus N = M \oplus B_1 \oplus \cdots \oplus B_n = M \oplus B_j \oplus (\oplus_{i \neq j} P_i).$$

In particular, $N \cong B_j \oplus (\oplus_{i \neq j} P_i)$. Thus $B = B_j$ and $C = C_j$ have the required properties. $\square$

Theorem 2.8 *The following conditions are equivalent for an indecomposable module M_R.*

(a) *The endomorphism ring of M_R is local.*

(b) *M_R has the finite exchange property.*

(c) *M_R has the exchange property.*

Proof. (a) $\Rightarrow$ (b). Let M_R be a module with local endomorphism ring $\mathrm{End}(M_R)$. By Lemma 2.5 in order to prove that the finite exchange property holds it suffices to show that M has the 2-exchange property. Let G, N, A_1, A_2 be modules such that $G = M \oplus N = A_1 \oplus A_2$. Let $\varepsilon_M, \varepsilon_{A_1}, \varepsilon_{A_2}, \pi_M, \pi_{A_1}, \pi_{A_2}$ be the embeddings of M, A_1, A_2 into G and the canonical projections of G onto M, A_1, A_2 with respect to these two decompositions. We must show that there are submodules $B_1 \subseteq A_1$ and $B_2 \subseteq A_2$ such that $G = M \oplus B_1 \oplus B_2$. Now

$$1_M = \pi_M \varepsilon_M = \pi_M(\varepsilon_{A_1}\pi_{A_1} + \varepsilon_{A_2}\pi_{A_2})\varepsilon_M = \pi_M \varepsilon_{A_1}\pi_{A_1}\varepsilon_M + \pi_M \varepsilon_{A_2}\pi_{A_2}\varepsilon_M.$$

Since $\mathrm{End}(M)$ is local, one of these two summands, say $\pi_M \varepsilon_{A_1}\pi_{A_1}\varepsilon_M$, must be an automorphism of M. Let H be the image of the monomorphism

$$\varepsilon_{A_1}\pi_{A_1}\varepsilon_M \colon M \to G,$$

so that $\varepsilon_{A_1}\pi_{A_1}\varepsilon_M$ induces an isomorphism $M \to H$ and $\pi_M|_H\colon H \to M$ is an isomorphism. From Lemma 2.6 we have that $G = N \oplus H$ and the projection $G \to H$ with respect to this decomposition is $(\pi_M|_H)^{-1}\pi_M$. Since

$$H = \varepsilon_{A_1}\pi_{A_1}\varepsilon_M(M) \subseteq A_1 \subseteq N \oplus H,$$

it follows from Lemma 2.1 that $A_1 = H \oplus B_1$, where $B_1 = A_1 \cap N$, and the projection $A_1 \to H$ with respect to this decomposition is $(\pi_M|_H)^{-1}\pi_M|_{A_1}$. Therefore $G = A_1 \oplus A_2 = H \oplus (B_1 \oplus A_2)$. With respect to this last decomposition of G the projection $G \to H$ is $(\pi_M|_H)^{-1}\pi_M|_{A_1}\pi_{A_1} = (\pi_M|_H)^{-1}\pi_M\varepsilon_{A_1}\pi_{A_1}$, and this mapping restricted to M is $(\pi_M|_H)^{-1}\pi_M\varepsilon_{A_1}\pi_{A_1}\varepsilon_M$. This is an isomorphism. Again by Lemma 2.6 we get that $G = M \oplus B_1 \oplus A_2$.

(b) $\Rightarrow$ (c). Let M_R be an indecomposable module with the finite exchange property and suppose $G = M \oplus N = \oplus_{i \in I}A_i$. Fix a non-zero element $x \in M$. There is a finite subset F of I such that $x \in \oplus_{i \in F}A_i$. Set $A' = \oplus_{i \in I\setminus F}A_i$, so that $G = M \oplus N = (\oplus_{i \in F}A_i) \oplus A'$. By Lemma 2.7 either there is an index $j \in F$ and a direct sum decomposition $A_j = B \oplus C$ of A_j such that

$$G = M \oplus B \oplus (\oplus_{i \in F,\ i \neq j}A_i) \oplus A',$$

or there is a direct sum decomposition $A' = B' \oplus C'$ of A' such that

$$G = M \oplus B' \oplus (\oplus_{i \in F}A_i).$$

The second possibility cannot occur because $M \cap (\oplus_{i \in F}A_i) \neq 0$. Hence there is an index $j \in F$ and a submodule B of A_j such that

$$G = M \oplus B \oplus (\oplus_{i \in F,\ i \neq j}A_i) \oplus A' = M \oplus B \oplus (\oplus_{i \in I,\ i \neq j}A_i).$$

(c) $\Rightarrow$ (a). Let M be an indecomposable module and suppose that $\mathrm{End}(M)$ is not a local ring. Then there exist two elements $\varphi, \psi \in \mathrm{End}(M)$ which are not automorphisms of M, such that $\varphi - \psi = 1_M$. Let $A = M_1 \oplus M_2$ be the external direct sum of two modules M_1, M_2 both equal to M, and let $\pi_i\colon A \to M_i$, $i = 1, 2$ be the canonical projections. The composite of the mappings

$$\begin{pmatrix} \varphi \\ \psi \end{pmatrix}\colon M \to M_1 \oplus M_2 \text{ and } (1_M\ -1_M)\colon M_1 \oplus M_2 \to M$$

is the identity mapping of M, so that if M' denotes the image of $\begin{pmatrix} \varphi \\ \psi \end{pmatrix}$ and K denotes the kernel of $(1_M\ -1_M)$, then $A = M' \oplus K$. If the exchange property were to hold for M, there would be direct summands B_1 of M_1 and B_2 of M_2 such that $A = M' \oplus K = M' \oplus B_1 \oplus B_2$. Since M_1 and M_2 are indecomposable, we would have either $A = M' \oplus M_1$ or $A = M' \oplus M_2$. If $A = M' \oplus M_1$, then $\pi_2|_{M'}\colon M' \to M_2$ is an isomorphism (Lemma 2.6). Then the composite mapping $\pi_2 \circ \begin{pmatrix} \varphi \\ \psi \end{pmatrix}\colon M \to M_2$ is an isomorphism. But $\pi_2 \circ \begin{pmatrix} \varphi \\ \psi \end{pmatrix} = \psi$, contradiction. Similarly if $A = M' \oplus M_2$. This shows that M does not have the exchange property. $\qquad\square$

2.3 Isomorphic refinements of finite direct sum decompositions

Let M be a module over a ring R. Suppose that $\{\, M_i \mid i \in I \,\}$ and $\{\, N_j \mid j \in J \,\}$ are two families of submodules of M such that $M = \oplus_{i \in I} M_i = \oplus_{j \in J} N_j$. Then these two decompositions are said to be *isomorphic* if there exists a one-to-one correspondence $\varphi \colon I \to J$ such that $M_i \cong N_{\varphi(i)}$ for every $i \in I$, and the second decomposition is a *refinement* of the first if there is a surjective map $\varphi \colon J \to I$ such that $N_j \subseteq M_{\varphi(j)}$ for every $j \in J$ (equivalently, if there is a surjective map $\varphi \colon J \to I$ such that $\oplus_{j \in \varphi^{-1}(i)} N_j = M_i$ for every $i \in I$). The first theorem of this section gives a criterion that assures the existence of isomorphic refinements of two direct sum decompositions.

Theorem 2.9 *Let $\aleph$ be a cardinal, let M be a module with the $\aleph$-exchange property, and let $M = \oplus_{i \in I} M_i = \oplus_{j \in J} N_j$ be two direct sum decompositions of M with I finite and $|J| \le \aleph$. Then these two direct sum decompositions of M have isomorphic refinements.*

Proof. We may assume $I = \{0, 1, 2, \ldots, n\}$. We shall construct a chain $N_j \supseteq N'_{0,j} \supseteq N'_{1,j} \supseteq \cdots \supseteq N'_{n-1,j} \supseteq N'_{n,j}$ for every $j \in J$ such that

$$M = \left(\oplus_{i=0}^{k} M_i\right) \oplus \left(\oplus_{j \in J} N'_{k,j}\right)$$

for every $k = 0, 1, 2, \ldots, n$. The construction of the $N'_{k,j}$ is by induction on k. For $k = 0$ the module M_0 has the $\aleph$-exchange property (Lemma 2.4). Hence there are submodules $N'_{0,j}$ of N_j such that $M = M_0 \oplus \left(\oplus_{j \in J} N'_{0,j}\right)$. Suppose $1 \le k \le n$ and that the modules $N'_{k-1,j}$ with $M = \left(\oplus_{i=0}^{k-1} M_i\right) \oplus \left(\oplus_{j \in J} N'_{k-1,j}\right)$ have been constructed. Apply Corollary 2.3 to the decompositions

$$M = M_k \oplus \left(\oplus_{i=k+1}^{n} M_i\right) \oplus \left(\oplus_{i=0}^{k-1} M_i\right) = \left(\oplus_{j \in J} N'_{k-1,j}\right) \oplus \left(\oplus_{i=0}^{k-1} M_i\right)$$

(note that M_k has the $\aleph$-exchange property by Lemma 2.4). Then there exist submodules $N'_{k,j}$ of $N'_{k-1,j}$ such that $M = M_k \oplus \left(\oplus_{j \in J} N'_{k,j}\right) \oplus \left(\oplus_{i=0}^{k-1} M_i\right)$, which is what we had to prove.

For $k = n$ we have that $M = (\oplus_{i=0}^{n} M_i) \oplus (\oplus_{j \in J} N'_{n,j})$, so that $N'_{n,j} = 0$ for every $j \in J$. Since the $N'_{k,j}$ are direct summands of M contained in $N'_{k-1,j}$, there is a direct sum decomposition $N'_{k-1,j} = N'_{k,j} \oplus N_{k,j}$ for every k and j (Lemma 2.1). Similarly, $N_j = N'_{0,j} \oplus N_{0,j}$. Hence $N_j = N_{0,j} \oplus N_{1,j} \oplus \cdots \oplus N_{n,j}$ for every $j \in J$, so that $M = \oplus_{j \in J} \oplus_{i=0}^{n} N_{i,j}$ is a refinement of the decomposition $M = \oplus_{j \in J} N_j$.

As $M = \left(\oplus_{i=0}^{k-1} M_i\right) \oplus \left(\oplus_{j \in J} N'_{k-1,j}\right) = \left(\oplus_{i=0}^{k} M_i\right) \oplus \left(\oplus_{j \in J} N'_{k,j}\right)$ for $k = 1, 2, \ldots, n$, factorizing modulo $\left(\oplus_{i=0}^{k-1} M_i\right) \oplus \left(\oplus_{j \in J} N'_{k,j}\right)$ we obtain that $\oplus_{j \in J} N_{k,j} \cong M_k$ for $k = 1, 2, \ldots, n$. Similarly $\oplus_{j \in J} N_{0,j} \cong M_0$. Hence for every

$i = 0, 1, 2, \ldots, n$ there is a decomposition $M_i = \oplus_{j \in J} N''_{i,j}$ of M_i with $N''_{i,j} \cong N_{i,j}$ for every i and j. Thus $\oplus^n_{i=0} \oplus_{j \in J} N''_{i,j}$ is a refinement of the decomposition $M = \oplus^n_{i=0} M_i$ isomorphic to the decomposition $M = \oplus_{j \in J} \oplus^n_{i=0} N_{i,j}$. $\square$

A second case in which it is possible to find isomorphic refinements is that of direct sum decompositions into countably many direct summands with the $\aleph_0$-exchange property. This is proved in the next theorem.

Theorem 2.10 *Let M be a module with two direct sum decompositions*

$$M = B_0 \oplus B_1 \oplus B_2 \oplus \ldots \tag{2.5}$$

$$M = C_0 \oplus C_1 \oplus C_2 \oplus \ldots \tag{2.6}$$

with countably many direct summands, where all the summands B_i and C_j have the $\aleph_0$-exchange property. Then the two direct sum decompositions (2.5) and (2.6) have isomorphic refinements.

Proof. Set $B'_{i,-1} = B_i$ and $C'_{i,-1} = C_i$ for every $i = 0, 1, 2, \ldots$. By induction on $j = 0, 1, 2, \ldots$ we shall construct direct summands $B_{i,j}, B'_{i,j}$ of B_i for every $i = j+1, j+2, j+3, \ldots$ and direct summands $C_{i,j}, C'_{i,j}$ of C_i for every $i = j, j+1, j+2, \ldots$ such that the following properties hold for every $j = 0, 1, 2, \ldots$:

(a) $B'_{i,j-1} = B_{i,j} \oplus B'_{i,j}$ for every $i = j+1, j+2, j+3, \ldots$;

(b) $C'_{i,j-1} = C_{i,j} \oplus C'_{i,j}$ for every $i = j, j+1, j+2, \ldots$;

(c) $M = (B'_{0,-1} \oplus C'_{0,0}) \oplus (B'_{1,0} \oplus C'_{1,1}) \oplus \cdots \oplus (B'_{j,j-1} \oplus C'_{j,j}) \oplus (\oplus^{\infty}_{k=j+1} C'_{k,j})$;

(d) $M = (B'_{0,-1} \oplus C'_{0,0}) \oplus (B'_{1,0} \oplus C'_{1,1}) \oplus \cdots \oplus (B'_{j,j-1} \oplus C'_{j,j}) \oplus (\oplus^{\infty}_{k=j+1} B'_{k,j})$;

(e) $C_{j,j} \oplus C_{j+1,j} \oplus C_{j+2,j} \oplus \cdots \cong B'_{j,j-1}$;

(f) $B_{j+1,j} \oplus B_{j+2,j} \oplus B_{j+3,j} \oplus \cdots \cong C'_{j,j}$.

Case $j = 0$. Since B_0 has the $\aleph_0$-exchange property, there exists a decomposition $C_i = C_{i,0} \oplus C'_{i,0}$ of C_i for every $i = 0, 1, 2, \ldots$ such that

$$M = B_0 \oplus C'_{0,0} \oplus C'_{1,0} \oplus C'_{2,0} \oplus \ldots . \tag{2.7}$$

From (2.6) and (2.7) it follows that

$$B_0 \cong C_{0,0} \oplus C_{1,0} \oplus C_{2,0} \oplus \ldots ,$$

that is, (e) holds.

The direct summand $C'_{0,0}$ of C_0 has the $\aleph_0$-exchange property by Lemma 2.4. Applying Corollary 2.3 to (2.5) and (2.7) we obtain direct sum decompositions $B_i = B_{i,0} \oplus B'_{i,0}$ of B_i for every $i = 1, 2, 3, \ldots$ such that

$$M = B_0 \oplus C'_{0,0} \oplus B'_{1,0} \oplus B'_{2,0} \oplus B'_{3,0} \oplus \ldots . \tag{2.8}$$

From (2.5) and (2.8) we get that $C'_{0,0} \cong B_{1,0} \oplus B_{2,0} \oplus B_{3,0} \oplus \ldots$, that is, (f) holds. This concludes the construction of the submodules $C_{i,0}, C'_{i,0}$ of C_i for every $i = 0, 1, 2, \ldots$ and the submodules $B_{i,0}, B'_{i,0}$ of B_i for every $i = 1, 2, 3, \ldots$. Now properties (a) and (b) hold because $B'_{i,-1} = B_i$ and $C'_{i,-1} = C_i$ for every $i = 0, 1, 2, \ldots$. Property (c) is given by equation (2.7), and (d) is given by equation (2.8). This concludes the first inductive step $j = 0$.

Now fix an integer $\ell > 0$ and suppose we have already constructed the direct summands $C_{i,j}, C'_{i,j}$ of C_i with the required properties for $j < \ell$ and $i \geq j$ and the direct summands $B_{i,j}, B'_{i,j}$ of B_i for $j < \ell$ and $i > j$. In particular we suppose that (d) holds for $j = \ell - 1$, that is, we suppose that

$$M = B'_{\ell,\ell-1} \oplus \left(\oplus_{k=\ell+1}^{\infty} B'_{k,\ell-1} \right) \oplus \left((B'_{0,-1} \oplus C'_{0,0}) \right.$$
$$\left. \oplus (B'_{1,0} \oplus C'_{1,1}) \oplus \cdots \oplus (B'_{\ell-1,\ell-2} \oplus C'_{\ell-1,\ell-1}) \right), \qquad (2.9)$$

and we suppose that (c) holds for $j = \ell - 1$, that is,

$$M = \left(\oplus_{k=\ell}^{\infty} C'_{k,\ell-1} \right) \oplus \left((B'_{0,-1} \oplus C'_{0,0}) \right.$$
$$\left. \oplus (B'_{1,0} \oplus C'_{1,1}) \oplus \cdots \oplus (B'_{\ell-1,\ell-2} \oplus C'_{\ell-1,\ell-1}) \right). \qquad (2.10)$$

Since $B'_{\ell,\ell-1}$ is a direct summand of B_ℓ, it has the $\aleph_0$-exchange property by Lemma 2.4. Hence we can apply Corollary 2.3 to the two decompositions (2.9) and (2.10), and obtain that there exist direct sum decompositions

$$C'_{k,\ell-1} = C_{k,\ell} \oplus C'_{k,\ell}$$

of $C'_{k,\ell-1}$ for every $k = \ell, \ell+1, \ell+2, \ldots$ such that

$$M = B'_{\ell,\ell-1} \oplus \left(\oplus_{k=\ell}^{\infty} C'_{k,\ell} \right) \oplus \left((B'_{0,-1} \oplus C'_{0,0}) \right.$$
$$\left. \oplus (B'_{1,0} \oplus C'_{1,1}) \oplus \cdots \oplus (B'_{\ell-1,\ell-2} \oplus C'_{\ell-1,\ell-1}) \right). \qquad (2.11)$$

This is property (c) for the integer ℓ. Now equality (2.10) can be rewritten as

$$M = \left(\oplus_{k=\ell}^{\infty} \left(C_{k,\ell} \oplus C'_{k,\ell} \right) \right) \oplus \left((B'_{0,-1} \oplus C'_{0,0}) \right.$$
$$\left. \oplus (B'_{1,0} \oplus C'_{1,1}) \oplus \cdots \oplus (B'_{\ell-1,\ell-2} \oplus C'_{\ell-1,\ell-1}) \right).$$

This and (2.11) yield

$$B'_{\ell,\ell-1} \cong \oplus_{k=\ell}^{\infty} C_{k,\ell},$$

that is, property (e) holds.

Now $C'_{\ell,\ell}$ is a direct summand of C_ℓ. Hence it has the $\aleph_0$-exchange property. Equality (2.11) can be rewritten as

$$M = C'_{\ell,\ell} \oplus \left(\oplus_{k=\ell+1}^{\infty} C'_{k,\ell} \right) \oplus \left((B'_{0,-1} \oplus C'_{0,0}) \right.$$
$$\left. \oplus (B'_{1,0} \oplus C'_{1,1}) \oplus \cdots \oplus (B'_{\ell-1,\ell-2} \oplus C'_{\ell-1,\ell-1}) \oplus B'_{\ell,\ell-1} \right) \qquad (2.12)$$

and (2.9) can be rewritten as

$$M = \left(\oplus_{k=\ell+1}^{\infty} B'_{k,\ell-1}\right) \oplus \left((B'_{0,-1} \oplus C'_{0,0})\right.$$
$$\left. \oplus (B'_{1,0} \oplus C'_{1,1}) \oplus \cdots \oplus (B'_{\ell-1,\ell-2} \oplus C'_{\ell-1,\ell-1}) \oplus B'_{\ell,\ell-1}\right). \quad (2.13)$$

Applying Corollary 2.3 to (2.12) and (2.13) we find that there exists a direct sum decomposition $B'_{k,\ell-1} = B_{k,\ell} \oplus B'_{k,\ell}$ for every $k = \ell+1, \ell+2, \dots$ such that

$$M = C'_{\ell,\ell} \oplus \left(\oplus_{k=\ell+1}^{\infty} B'_{k,\ell}\right) \oplus \left((B'_{0,-1} \oplus C'_{0,0})\right.$$
$$\left. \oplus (B'_{1,0} \oplus C'_{1,1}) \oplus \cdots \oplus (B'_{\ell-1,\ell-2} \oplus C'_{\ell-1,\ell-1}) \oplus B'_{\ell,\ell-1}\right). \quad (2.14)$$

This proves property (d) for the integer ℓ. From (2.13) and (2.14) it follows that $C'_{\ell,\ell} \cong \oplus_{k=\ell+1}^{\infty} B_{k,\ell}$. Hence (f) holds, and this concludes the construction by induction.

From (e) we infer that there exist modules $B_{i,j}$ for $i \leq j$ such that $B_{i,j} \cong C_{j,i}$ and

$$B_{j,j} \oplus B_{j,j+1} \oplus B_{j,j+2} \oplus \dots = B'_{j,j-1} \quad (2.15)$$

for every $j \geq 0$. From (a) we have that

$$B_i = B'_{i,-1} = B_{i,0} \oplus B'_{i,0} = B_{i,0} \oplus B_{i,1} \oplus B'_{i,1}$$
$$= \cdots = B_{i,0} \oplus B_{i,1} \oplus \cdots \oplus B_{i,i-1} \oplus B'_{i,i-1},$$

so that

$$B_i = \oplus_{k=0}^{\infty} B_{i,k} \quad (2.16)$$

by (2.15).

Similarly, from (f) we get that there exist modules $C_{i,j}$ for $i < j$ such that $C_{i,j} \cong B_{j,i}$ and

$$C_{j,j+1} \oplus C_{j,j+2} \oplus C_{j,j+3} \oplus \dots = C'_{j,j} \quad (2.17)$$

for every $j \geq 0$. From (b) it follows that $C_j = C_{j,0} \oplus C_{j,1} \oplus \cdots \oplus C_{j,j} \oplus C'_{j,j}$, so that

$$C_j = \oplus_{k=0}^{\infty} C_{j,k} \quad (2.18)$$

by (2.17).

Since $B_{i,j} \cong C_{j,i}$ for every $i,j = 0,1,2,\dots$, (2.16) and (2.18) yield the required isomorphic refinements of the decompositions (2.5) and (2.6). $\qquad \square$

2.4 The Krull-Schmidt-Remak-Azumaya Theorem

The Krull-Schmidt-Remak-Azumaya Theorem is one of the main topics of this volume. We shall obtain a proof of it using the exchange property. We begin with a lemma that is of independent interest.

Lemma 2.11 *If a module M is a direct sum of modules with local endomorphism rings, then every indecomposable direct summand of M has local endomorphism ring.*

Proof. Suppose $M = A \oplus B = \oplus_{i \in I} M_i$, where A is indecomposable and all the modules M_i have local endomorphism ring. Let F be a finite subset of I with $A \cap \oplus_{i \in F} M_i \neq 0$, and set $C = \oplus_{i \in F} M_i$. The module C has the exchange property (Lemma 2.4 and Theorem 2.8). Hence there exist direct sum decompositions $A = A' \oplus A''$ of A and $B = B' \oplus B''$ of B such that $M = C \oplus A' \oplus B'$. Note that A' is a proper submodule of A, because $A \cap C \neq 0$ and $A' \cap C = 0$. Since A is indecomposable, it follows that $A' = 0$. Thus $M = C \oplus B'$. From $M = C \oplus B' = A \oplus B' \oplus B''$ it follows that $C \cong A \oplus B''$. Thus A is isomorphic to a direct summand of C. Hence A has the exchange property by Lemma 2.4. Therefore A has local endomorphism ring by Theorem 2.8. $\square$

We are ready for the proof of the Krull-Schmidt-Remak-Azumaya Theorem.

Theorem 2.12 (Krull-Schmidt-Remak-Azumaya Theorem) *Let M be a module that is a direct sum of modules with local endomorphism rings. Then any two direct sum decompositions of M into indecomposable direct summands are isomorphic.*

Proof. Suppose that $M = \oplus_{i \in I} M_i = \oplus_{j \in J} N_j$, where all the M_i and N_j are indecomposable. By Lemma 2.11 all the modules M_i and N_j have local endomorphism rings. For $I' \subseteq I$ and $J' \subseteq J$ let

$$M(I') = \oplus_{i \in I'} M_i \quad \text{and} \quad N(J') = \oplus_{j \in J'} N_j.$$

By Lemma 2.4 and Theorem 2.8 the modules $M(I')$ and $N(J')$ have the exchange property whenever I' and J' are finite. Since the summands N_j are indecomposable, for every finite subset $I' \subseteq I$ there exists a subset $J' \subseteq J$ such that $M = M(I') \oplus N(J \setminus J')$. From $M = M(I') \oplus N(J \setminus J') = N(J') \oplus N(J \setminus J')$, we get that $M(I') \cong N(J')$. By Theorem 2.9 applied to the decompositions $M(I') \cong N(J')$, the two decompositions $M(I') = \oplus_{i \in I'} M_i$ and $N(J') = \oplus_{j \in J'} N_j$ have isomorphic refinements. From the indecomposability of the M_i and N_j, it follows that there is a one-to-one correspondence $\varphi \colon I' \to J'$ such that $M_i \cong N_{\varphi(i)}$ for every $i \in I'$. For every R-module A set

$$I_A = \{ i \in I \mid M_i \cong A \} \quad \text{and} \quad J_A = \{ j \in J \mid N_j \cong A \}.$$

From what we have just seen it follows that if I_A is finite, then $|I_A| \le |J_A|$, and if $I_A \ne \emptyset$, then $J_A \ne \emptyset$. By symmetry, if J_A is finite, then $|J_A| \le |I_A|$, and $J_A \ne \emptyset$ implies $I_A \ne \emptyset$. In order to prove the theorem it is sufficient to show that $|I_A| = |J_A|$ for every R-module A.

Suppose first that I_A is finite. In this case we argue by induction on $|I_A|$. If $|I_A| = 0$, then $|J_A| = 0$. If $|I_A| \ge 1$, fix an index $i_0 \in I_A$. Then there is an index $j_0 \in J$ such that $M = M(\{i_0\}) \oplus N(J \setminus \{j_0\})$. If we factorize the module $M(\{i_0\}) \oplus N(J \setminus \{j_0\}) = M(I)$ modulo $M(\{i_0\})$ we obtain that

$$N(J \setminus \{j_0\}) \cong M(I \setminus \{i_0\}).$$

From the induction hypothesis we get that $|I_A \setminus \{i_0\}| = |J_A \setminus \{j_0\}|$, so that $|I_A| = |J_A|$.

By symmetry we can conclude that if J_A is finite, then $|I_A| = |J_A|$ as well.

Hence we can suppose that both I_A and J_A are infinite sets. By symmetry it is sufficient to show that $|J_A| \le |I_A|$ for an arbitrary module A.

For each $i \in I_A$ set $J(i) = \{ j \in J \mid M = M_i \oplus N(J \setminus \{j\}) \}$. Obviously $J(i) \subseteq J_A$. If x is a non-zero element of M_i, then there is a finite subset J'' of J such that $x \in N(J'')$. Hence $M_i \cap N(K) \ne 0$ for every $K \subseteq J$ that contains J''. Thus $J(i) \subseteq J''$, so that $J(i)$ is finite.

We claim that $\bigcup_{i \in I_A} J(i) = J_A$. In order to prove the claim, fix $j \in J_A$. Then there exists a finite subset I' of I such that $N_j \cap M(I') \ne 0$. Hence there exists a finite subset $J' \subseteq J$ such that $M = M(I') \oplus N(J \setminus J')$. Note that $j \in J'$. Since $N(J' \setminus \{j\})$ has the exchange property, we can apply Corollary 2.3 to the decompositions $M = N(J' \setminus \{j\}) \oplus N_j \oplus N(J \setminus J') = (\oplus_{i \in I'} M_i) \oplus N(J \setminus J')$. Then for every $i \in I'$ there exists a direct summand M_i' of M_i such that $M = N(J' \setminus \{j\}) \oplus (\oplus_{i \in I'} M_i') \oplus N(J \setminus J')$. Then $N_j \cong \oplus_{i \in I'} M_i'$, so that there exists an index $k \in I'$ with $M_k' = M_k$ and $M_i' = 0$ for every $i \in I'$, $i \ne k$. Note that $M_k \cong N_j \cong A$, so that $k \in I_A$. Thus

$$M = N(J' \setminus \{j\}) \oplus M_k \oplus N(J \setminus J') = M_k \oplus N(J \setminus \{j\}),$$

that is, $j \in J(k)$. Hence $j \in \bigcup_{i \in I_A} J(i)$, which proves the claim.

It follows that

$$|J_A| = \left| \bigcup_{i \in I_A} J(i) \right| \le |I_A| \cdot \aleph_0 = |I_A|. \qquad \square$$

2.5 Applications

In this section we apply the Krull-Schmidt-Remak-Azumaya Theorem to some important classes of modules.

A first immediate application of the Krull-Schmidt-Remak-Azumaya Theorem 2.12 can be given to the class of semisimple modules.

Lemma 2.13 (Schur) *The endomorphism ring $\operatorname{End}(M)$ of a simple module M is a division ring.*

Proof. If M is a simple module and f is a non-zero endomorphism of M, then $\ker f$ and $f(M)$ must be submodules of M. Hence they are either 0 or M. If $\ker f = M$, then $f = 0$, contradiction. Therefore $\ker f = 0$ and f is injective. If $f(M) = 0$, then $f = 0$, contradiction. Therefore $f(M) = M$ and f is surjective. Thus f is an automorphism of M, that is, f is invertible in $\operatorname{End}(M)$. $\square$

Since division rings are local rings, we get the Krull-Schmidt-Remak-Azumaya Theorem for semisimple modules:

Theorem 2.14 *Any two direct sum decompositions of a semisimple module into simple direct summands are isomorphic.* $\square$

The next result describes the structure of the submodules and the homomorphic images of a semisimple module.

Proposition 2.15 *Let M be a semisimple R-module and $\{\, M_i \mid i \in I \,\}$ a family of simple submodules of M such that $M = \oplus_{i \in I} M_i$. Then for every submodule N of M there is a subset J of I such that $N \cong \oplus_{i \in J} M_i$ and $M/N \cong \oplus_{i \in I \setminus J} M_i$.*

Proof. By Proposition 1.1 the submodule N of the semisimple module M is a direct summand of M, so that $M = N \oplus N'$ for a submodule $N' \cong M/N$ of M. By Proposition 1.1 again, both N and N' are semisimple. Hence $N = \oplus_{\lambda \in \Lambda} N_\lambda$ and $N' = \oplus_{\mu \in \Lambda'} N'_\mu$ for suitable simple submodules N_λ, N'_μ. By Theorem 2.14 the two decompositions $\oplus_{i \in I} M_i = (\oplus_{\lambda \in \Lambda} N_\lambda) \oplus (\oplus_{\mu \in \Lambda'} N'_\mu)$ of M are isomorphic. Therefore there are a subset J of I and one-to-one correspondences $\varphi \colon J \to \Lambda$ and $\psi \colon I \setminus J \to \Lambda'$ such that $M_i \cong N_{\varphi(i)}$ for every $i \in J$ and $M_i \cong N'_{\psi(i)}$ for every $i \in I \setminus J$. The conclusion follows immediately. $\square$

As a second application of Theorem 2.12, we study the uniqueness of decomposition of some particular artinian or noetherian modules.

Lemma 2.16 *Let M be a module and f an endomorphism of M.*

(a) *If n is a positive integer such that $f^n(M) = f^{n+1}(M)$, then*

$$\ker(f^n) + f^n(M) = M.$$

(b) *If M is an artinian module, then f is an automorphism if and only if f is injective.*

Proof. (a) If n is such that $f^n(M) = f^{n+1}(M)$, then $f^t(M) = f^{t+1}(M)$ for every $t \geq n$, so that $f^n(M) = f^{2n}(M)$. Let us show that

$$\ker(f^n) + f^n(M) = M.$$

If $x \in M$, then $f^n(x) \in f^n(M) = f^{2n}(M)$, so that $f^n(x) = f^n(y)$ for some $y \in f^n(M)$. Therefore $z = x - y \in \ker(f^n)$, and $x = z + y \in \ker(f^n) + f^n(M)$.

(b) If f an injective endomorphism of the artinian module M, the descending chain

$$f(M) \supseteq f^2(M) \supseteq f^3(M) \supseteq \cdots$$

is stationary, so that $\ker(f^n) + f^n(M) = M$ for some positive integer n by part (a). As f^n is injective, $\ker(f^n) = 0$, and therefore $f^n(M) = M$. In particular, f is surjective. $\square$

Similarly it can be proved that

Lemma 2.17 *Let M be a module and f an endomorphism of M.*

(a) *If n is a positive integer such that $\ker f^n = \ker f^{n+1}$, then*

$$\ker(f^n) \cap f^n(M) = 0.$$

(b) *If M is a noetherian module, then f is an automorphism if and only if f is surjective.* $\square$

A submodule N of a module M_R is *fully invariant* if $\varphi(N) \subseteq N$ for every $\varphi \in \operatorname{End}(M_R)$, that is, if N is a submodule of the left $\operatorname{End}(M_R)$-module M. A submodule N of M_R is *essential* in M_R if $N \cap P \neq 0$ for every non-zero submodule P of M_R. The *socle* of a module M_R is the sum of all simple submodules of M_R. It is a semisimple fully invariant submodule of M_R and it will be denoted $\operatorname{soc}(M_R)$. Since every non-zero artinian module has a simple submodule, the socle is an essential submodule in every artinian module. If N_R is an artinian module and its socle $\operatorname{soc}(N_R)$ is a simple module, by Lemma 2.16(b) an endomorphism $f \in \operatorname{End}(N_R)$ is not an automorphism if and only $f(\operatorname{soc}(N_R)) = 0$. It follows that $\operatorname{End}(N_R)$ is a local ring with Jacobson radical $J(\operatorname{End}(N_R)) = \{\, f \in \operatorname{End}(N_R) \mid f(\operatorname{soc}(N_R)) = 0 \,\}$. Therefore Theorem 2.12 yields

Theorem 2.18 *Let M_R be an R-module that is a direct sum of artinian modules with simple socle. Then any two direct sum decompositions of M_R into indecomposable direct summands are isomorphic.* $\square$

The hypothesis that the artinian modules have simple socle is essential in Theorem 2.18 as we shall see in Section 8.2.

Recall that a module M is *local* if it has a greatest proper submodule. Hence noetherian local modules are the "duals" of artinian modules with a simple socle. The next result is the dual of Theorem 2.18 and is proved similarly.

Theorem 2.19 *Let M be an R-module that is a direct sum of noetherian local modules. Then any two direct sum decompositions of M into indecomposable direct summands are isomorphic.* □

Our third application of the Krull-Schmidt-Remak-Azumaya Theorem will be to the class of modules of finite composition length.

Lemma 2.20 (Fitting's Lemma) *If M is a module of finite length n and f is an endomorphism of M, then $M = \ker(f^n) \oplus f^n(M)$.*

Proof. Since M is of finite length n, both the chains

$$\ker f \subseteq \ker f^2 \subseteq \ker f^3 \subseteq \dots$$

and

$$f(M) \supseteq f^2(M) \supseteq f^3(M) \supseteq \dots$$

are stationary at the n-th step, so that $\ker(f^n) \oplus f^n(M) = M$ by Lemmas 2.16(a) and 2.17(a). □

We shall say that a module M_R is a *Fitting module* if for every $f \in \text{End}(M_R)$ there is a positive integer n such that $M = \ker(f^n) \oplus f^n(M)$. Thus by Lemma 2.20 every module of finite length is a Fitting module. It is easily seen that direct summands of Fitting modules are Fitting modules.

Lemma 2.21 *The endomorphism ring of an indecomposable Fitting module is a local ring.*

Proof. If M is a Fitting module and f is an endomorphism of M, there exists a positive integer n such that $M = \ker(f^n) \oplus f^n(M)$. If M is indecomposable, two cases may occur. In the first case $\ker(f^n) = 0$ and $f^n(M) = M$. Then f^n is an automorphism of M, so that f is an automorphism of M. In the second case $\ker(f^n) = M$, that is, f is nilpotent. Hence every endomorphism of M is either invertible or nilpotent.

In order to show that the endomorphism ring $\text{End}(M)$ of M is local, we must show that the sum of two non-invertible endomorphisms is non-invertible. Suppose that f and g are two non-invertible endomorphisms of M, but $f + g$ is invertible. If $h = (f+g)^{-1}$ is the inverse of $f+g$, then $fh + gh = 1$. Since f and g are not automorphisms, neither fh nor gh are automorphisms. Therefore, as we have just seen in the previous paragraph, there exists a positive integer n such that $(gh)^n = 0$. Since $1 = (1 - gh)(1 + gh + (gh)^2 + \dots + (gh)^{n-1})$, the endomorphism $1 - gh = fh$ is invertible. This contradiction proves the lemma. □

Theorem 2.12 and Lemma 2.21 yield

Theorem 2.22 *Let M be an R-module that is a direct sum of indecomposable Fitting modules. Then any two direct sum decompositions of M into indecomposable direct summands are isomorphic.*   □

In particular, from Lemma 2.20 it follows that

Corollary 2.23 (The Krull-Schmidt Theorem) *Let M be an R-module of finite length. Then M is the direct sum of a finite family of indecomposable modules, and any two direct sum decompositions of M as direct sums of indecomposable modules are isomorphic.*   □

A further class of modules to which the Krull-Schmidt-Remak-Azumaya Theorem can be applied immediately is the class of indecomposable injective modules. The proof of the following lemma is an easy exercise left to the reader.

Lemma 2.24 *Let $M \neq 0$ be an R-module. The following conditions are equivalent:*

(a) *The intersection of any two non-zero submodules of M is non-zero.*

(b) *The injective envelope of M is indecomposable.*

(c) *Every non-zero submodule of M is essential in M.*

(d) *Every non-zero submodule of M is indecomposable.*   □

An R-module $M \neq 0$ is said to be *uniform* if it satisfies the equivalent conditions of Lemma 2.24. For instance, an artinian module is uniform if and only if it has a simple socle.

We state the next lemma in the language of Grothendieck categories. The reader who is not used to this language may think of the case of an indecomposable injective R-module.

Lemma 2.25 *Let M be an indecomposable injective object of a Grothendieck category $\mathcal{C}$. Then*

(a) *An endomorphism of M is an automorphism if and only if it is a monomorphism.*

(b) *The endomorphism ring of M is a local ring.*

In particular, the endomorphism ring of every indecomposable injective module is a local ring.

Proof. (a) If $f: M \to M$ is a monomorphism, then $\text{im}(f)$ is a subobject of M isomorphic to M. In particular, $\text{im}(f)$ is a non-zero direct summand of M. Since M is indecomposable, $\text{im}(f) = M$ and f is an automorphism.

(b) Let $\text{End}_{\mathcal{C}}(M)$ denote the endomorphism ring of M. We must show that the sum of two non-invertible elements of $\text{End}_{\mathcal{C}}(M)$ is non-invertible. Suppose that f and g are two non-invertible endomorphisms of M. In (a) we have seen that f and g are not monomorphisms, that is, $\ker f \neq 0$ and $\ker g \neq 0$. Since M is coirreducible ($=$ uniform), we have $\ker f \cap \ker g \neq 0$. Now

$$\ker f \cap \ker g \subseteq \ker(f + g),$$

so that $\ker(f + g) \neq 0$. Therefore $f + g$ is not invertible in $\text{End}_{\mathcal{C}}(M)$. $\qquad\square$

As an immediate application of Lemma 2.25 to the category $\mathcal{C} = \text{Mod-}R$ and the Krull-Schmidt-Remak-Azumaya Theorem we have:

Theorem 2.26 *Let M be an R-module that is a direct sum of injective indecomposable modules. Then any two direct sum decompositions of M into indecomposable direct summands are isomorphic.* $\qquad\square$

As a second application of Lemma 2.25 to the category $\mathcal{C} = (_R\text{FP}, \text{Ab})$ we find:

Corollary 2.27 *The endomorphism ring of an indecomposable pure-injective module is a local ring.*

Proof. Let M_R be an indecomposable pure-injective module, let $_R\text{FP}$ be the full subcategory of R-Mod whose objects are the finitely presented left R-modules, let Ab be the category of abelian groups, and let $\mathcal{C} =(_R\text{FP}, \text{Ab})$ be the category of all additive functors from $_R\text{FP}$ to Ab. Then

$$M \otimes_R -: {}_R\text{FP} \to \text{Ab}$$

is an indecomposable injective object in the Grothendieck category $\mathcal{C}$ (Proposition 1.39). Since $\text{End}_R(M) \cong \text{End}_{\mathcal{C}}(M)$, the ring $\text{End}_R(M)$ is local by Lemma 2.25(b). $\qquad\square$

We conclude this section showing that the Krull-Schmidt Theorem holds for Σ-pure-injective modules. We need a preliminary proposition.

Proposition 2.28 *Let R be a ring and let $\mathcal{B} \subseteq \mathcal{C}$ be classes of non-zero right R-modules. Suppose that every module in $\mathcal{C}$ has a direct summand in $\mathcal{B}$ and that for every proper pure submodule P of any module $M \in \mathcal{C}$ there exists a submodule D of M such that $D \in \mathcal{B}$, $P \cap D = 0$ and $P + D = P \oplus D$ is pure in M. Then every module $M \in \mathcal{C}$ is a direct sum of modules belonging to $\mathcal{B}$.*

Proof. Suppose $M_R \in \mathcal{C}$. Let $\mathcal{I}$ be the set of submodules of M that belong to $\mathcal{B}$ and put $\mathcal{S} = \{ I \mid I \subseteq \mathcal{I}, \sum_{N \in I} N$ is a pure submodule of M and the sum $\sum_{N \in I} N$ is direct $\}$. Since every module in $\mathcal{C}$ has a direct summand in $\mathcal{B}$, the set $\mathcal{S}$ is non-empty. Partially order $\mathcal{S}$ by set inclusion. Then the union of a chain of elements of $\mathcal{S}$ is an element of $\mathcal{S}$ by Proposition 1.31(a). By Zorn's Lemma $\mathcal{S}$ has a maximal element J. Set $P = \sum_{N \in J} N = \oplus_{N \in J} N$, so that P is a pure submodule of M. If P is a proper submodule of M, then there is a submodule D of M such that $D \in \mathcal{B}$, $P \cap D = 0$ and $P + D$ is pure in M. Thus $J \cup \{D\} \in \mathcal{S}$, and $D \notin J$ because $P \cap D = 0$ and the module $D \in \mathcal{B}$ is non-zero. This contradicts the maximality of J. Hence $M = P$ is a direct sum of modules belonging to $\mathcal{B}$. $\qquad\square$

Theorem 2.29 *Let M_R be a Σ-pure-injective R-module. Then M_R is a direct sum of modules with local endomorphism ring, so that any two direct sum decompositions of M_R as direct sums of indecomposables are isomorphic.*

Proof. We shall apply Proposition 2.28 to the class $\mathcal{C}$ of Σ-pure-injective non-zero R-modules and the class $\mathcal{B}$ of indecomposable Σ-pure-injective R-modules. Firstly, we must show that every Σ-pure-injective non-zero module M_R has an indecomposable direct summand N. To see this, let x be a non-zero element of M_R. Let $\mathcal{P}$ be the set of all pure submodules of M_R that do not contain x. Then $\mathcal{P}$ is non-empty and the union of every chain in $\mathcal{P}$ is an element of $\mathcal{P}$ (Proposition 1.31(a)). By Zorn's Lemma $\mathcal{P}$ has a maximal element Q. By Corollary 1.42 there exists a submodule N of M such that $M = Q \oplus N$. If $N = N' \oplus N''$ with $N', N'' \neq 0$, then $Q \oplus N'$ and $Q \oplus N''$ are direct summands of M that do not belong to $\mathcal{P}$. Hence $x \in (Q \oplus N') \cap (Q \oplus N'') = Q$, a contradiction. The contradiction proves that N is indecomposable, as we wanted to prove.

Secondly, we must show that for every proper pure submodule P of a Σ-pure injective module M_R there exists an indecomposable Σ-pure-injective submodule D of M such that $P \cap D = 0$ with $P + D = P \oplus D$ pure in M. For such a pure submodule P we know that $M_R = P \oplus P'$ for a submodule P' (Corollary 1.42), and by the first part of the proof $P' = D \oplus P''$ for suitable submodules D and P'' with D indecomposable. The module D has the required properties.

Hence by Proposition 2.28 every Σ-pure-injective module M is a direct sum of indecomposable modules.

Finally, every indecomposable direct summand of M is pure-injective, hence every indecomposable direct summand of M has a local endomorphism ring (Corollary 2.27). $\qquad\square$

2.6   Goldie dimension of a modular lattice

The notion of Goldie dimension of a module concerns the lattice $\mathcal{L}(M)$ of all submodules of M, which is a modular lattice. Modular lattices seem to be the proper setting for the definition of Goldie dimension. Hence in this section we shall consider arbitrary modular lattices and their Goldie dimension.

Throughout this section $(L, \vee, \wedge)$ will denote a modular lattice with 0 and 1, that is, a lattice with a smallest element 0 and a greatest element 1 such that $a \wedge (b \vee c) = (a \wedge b) \vee c$ for every $a, b, c \in L$ with $c \le a$. If $a, b \in L$ and $a \le b$ let $[a, b] = \{ x \in L \mid a \le x \le b \}$ be the *interval* between a and b.

A finite subset $\{ a_i \mid i \in I \}$ of $L \setminus \{0\}$ is said to be *join-independent* if $a_i \wedge (\bigvee_{j \ne i} a_j) = 0$ for every $i \in I$. The empty subset of $L \setminus \{0\}$ is join-independent. An arbitrary subset A of $L \setminus \{0\}$ is *join-independent* if all its finite subsets are join-independent.

Lemma 2.30 *Let $A \subseteq L \setminus \{0\}$ be a join-independent subset of a modular lattice L. If B, C are finite subsets of A and $B \cap C = \emptyset$, then $(\bigvee_{b \in B} b) \wedge (\bigvee_{c \in C} c) = 0$.*

Proof. By induction on the cardinality $|B|$ of B. The case $|B| = 0$ is trivial. Suppose the lemma holds for subsets of cardinality $< |B|$. Fix an element $b \in B$ and set $B' = B \setminus \{b\}$ and $a = (\bigvee_{x \in B} x) \wedge (\bigvee_{y \in C} y)$. By the induction hypothesis

$$(b \vee a) \wedge \left(\bigvee_{x \in B'} x \right) \le \left(\bigvee_{y \in \{b\} \cup C} y \right) \wedge \left(\bigvee_{x \in B'} x \right) = 0, \qquad (2.19)$$

and by the definition of join-independent set

$$\left(\bigvee_{x \in B'} x \vee a \right) \wedge b \le \left(\bigvee_{x \in B' \cup C} x \right) \wedge b = 0. \qquad (2.20)$$

Then

$$
\begin{aligned}
a &\le (b \vee a) \wedge \left(\bigvee_{x \in B} x \right) \\
&= (b \vee a) \wedge \left(\left(\bigvee_{x \in B'} x \right) \vee b \right) \quad \text{(by the modular identity)} \\
&= \left((b \vee a) \wedge \left(\bigvee_{x \in B'} x \right) \right) \vee b \quad \text{(by (2.19))} \\
&= 0 \vee b = b,
\end{aligned}
\qquad (2.21)
$$

and

$$
\begin{aligned}
a &\le \left(\bigvee_{x \in B'} x \vee a \right) \wedge \left(\bigvee_{x \in B} x \right) \\
&= \left(\bigvee_{x \in B'} x \vee a \right) \wedge \left(b \vee \bigvee_{x \in B'} x \right) \quad \text{(by the modular identity)} \\
&= \left(\left(\bigvee_{x \in B'} x \vee a \right) \wedge b \right) \vee \left(\bigvee_{x \in B'} x \right) \quad \text{(by (2.20))} \\
&\le 0 \vee \bigvee_{x \in B'} x = \bigvee_{x \in B'} x.
\end{aligned}
\qquad (2.22)
$$

From (2.21) and (2.22) it follows that $a \le b \wedge \bigvee_{x \in B'} x = 0$, as desired. $\qquad \square$

Proposition 2.31 *Let $A \subseteq L \setminus \{0\}$ be a join-independent subset of a modular lattice L. Let $a \in L$ be a non-zero element such that $a \wedge (\bigvee_{b \in B} b) = 0$ for every finite subset B of A. Then $A \cup \{a\}$ is a join-independent subset of L.*

Proof. We must prove that every finite subset of $A \cup \{a\}$ is join-independent. This is obvious for finite subsets of A. Hence it suffices to show that if B is a finite subset of A, then $B \cup \{a\}$ is join-independent. Since $a \wedge (\bigvee_{b \in B} b) = 0$, we have to prove that $b \wedge (a \vee \bigvee_{x \in B \setminus \{b\}} x) = 0$ for each $b \in B$. Now

$$\left(\bigvee_{y \in B} y\right) \wedge \left(a \vee \left(\bigvee_{x \in B \setminus \{b\}} x\right)\right) \qquad \text{(by the modular identity)}$$
$$= \left(\left(\bigvee_{y \in B} y\right) \wedge a\right) \vee \left(\bigvee_{x \in B \setminus \{b\}} x\right) \quad \text{(by hypothesis)} \qquad (2.23)$$
$$= 0 \vee \left(\bigvee_{x \in B \setminus \{b\}} x\right) = \bigvee_{x \in B \setminus \{b\}} x,$$

so that

$$b \wedge (a \vee \bigvee_{x \in B \setminus \{b\}} x) \qquad\qquad \text{(since } b \le \bigvee_{y \in B} y)$$
$$= b \wedge \left(\bigvee_{y \in B} y\right) \wedge (a \vee \bigvee_{x \in B \setminus \{b\}} x) \quad \text{(by (2.23))} \qquad\qquad \square$$
$$= b \wedge \left(\bigvee_{x \in B \setminus \{b\}} x\right) = 0 \qquad\qquad \text{(because } B \text{ is join-independent)}.$$

By Zorn's Lemma every join-independent subset of $L \setminus \{0\}$ is contained in a maximal join-independent subset of $L \setminus \{0\}$.

An element $a \in L$ is *essential* in L if $a \wedge x \ne 0$ for every non-zero element $x \in L$. Thus 0 is essential in L if and only if $L = \{0\}$. If a, b are elements of L, $a \le b$ and a is essential in the lattice $[0, b]$, then a is said to be *essential* in b. In particular, 0 is essential in b if and only if $b = 0$.

Lemma 2.32 *Let a, b, c be elements of L. If a is essential in b and b is essential in c, then a is essential in c.*

Proof. Let x be a non-zero element of $[0, c]$. We must show that $a \wedge x \ne 0$. Now $b \wedge x \ne 0$ because b is essential in c, hence $a \wedge (b \wedge x) \ne 0$ because a is essential in b. But $a \wedge (b \wedge x) = a \wedge x$. $\qquad\square$

Lemma 2.33 *Let a, b, c, d be elements of L such that $b \wedge d = 0$. If a is essential in b and c is essential in d, then $a \vee c$ is essential in $b \vee d$.*

Proof. If any of the four elements a, b, c, d is zero, the statement of the lemma is trivial. Hence we shall assume that a, b, c, d are all non-zero.

We claim that if the hypotheses of the lemma hold for the four elements $a, b, c, d \in L \setminus \{0\}$, then $a \vee d$ is essential in $b \vee d$. Assume the contrary. Then there exists a non-zero element $x \in L$ such that $x \le b \vee d$ and

$$(a \vee d) \wedge x = 0.$$

Since $\{a, d\}$ is join-independent, the set $\{a, d, x\}$ is join-independent by Proposition 2.31. In particular, $a \wedge (d \vee x) = 0$, so that $a \wedge b \wedge (d \vee x) = 0$. This implies that $b \wedge (d \vee x) = 0$, because a is essential in b and $b \wedge (d \vee x) \leq b$. Now $\{d, x\} \subseteq \{a, d, x\}$ is join-independent, and thus $b \wedge (d \vee x) = 0$ forces that $\{b, d, x\}$ is join-independent (Proposition 2.31). In particular, $x \wedge (b \vee d) = 0$. But $x \leq b \vee d$, so that $x = 0$. This contradiction proves the claim.

If we apply the claim to the four elements c, d, a, a, we obtain that $c \vee a$ is essential in $d \vee a$, that is, $a \vee c$ is essential in $a \vee d$. The conclusion now follows from this, the claim and Lemma 2.32. $\qquad \square$

By an easy induction argument we obtain

Corollary 2.34 *Let $a_1, a_2, \ldots, a_n, b_1, b_2, \ldots, b_n$ be elements of L such that $\{b_1, b_2, \ldots, b_n\}$ is join-independent. If a_i is essential in b_i for every $i = 1, 2, \ldots, n$, then $a_1 \vee a_2 \vee \cdots \vee a_n$ is essential in $b_1 \vee b_2 \vee \cdots \vee b_n$.* $\qquad \square$

A lattice $L \neq \{0\}$ is *uniform* if all its non-zero elements are essential in L, that is, if $x, y \in L$ and $x \wedge y = 0$ imply $x = 0$ or $y = 0$. An element a of a modular lattice L is *uniform* if $a \neq 0$ and the lattice $[0, a]$ is uniform.

Lemma 2.35 *If a modular lattice L does not contain infinite join-independent subsets, then for every non-zero element $a \in L$ there exists a uniform element $b \in L$ such that $b \leq a$.*

Proof. Let $a \neq 0$ be an element of a modular lattice L such that every $b \leq a$ is not uniform. We shall define a sequence $a_1, a_2, a_3, \ldots$ of non-zero elements of $[0, a]$ such that for every $n \geq 1$ the set $\{a_1, a_2, \ldots, a_n\}$ is join-independent and $a_1 \vee \cdots \vee a_n$ is not essential in $[0, a]$. The construction of the elements a_n is by induction on n. For $n = 1$ note that a is not uniform, hence there exist non-zero elements $a_1, a_1' \in [0, a]$ such that $a_1 \wedge a_1' = 0$, i.e., a_1 has the required properties. Suppose $a_1, \ldots, a_{n-1}$ have been defined. Since $a_1 \vee \cdots \vee a_{n-1}$ is not essential in $[0, a]$, there exists a non-zero $b \in [0, a]$ such that $b \wedge (a_1 \vee \cdots \vee a_{n-1}) = 0$. The element b is not uniform. Hence there exist $a_n, a_n' \in [0, b]$, where a_n, a_n' are non-zero, such that $a_n \wedge a_n' = 0$. Then $a_n \wedge (a_1 \vee \cdots \vee a_{n-1}) = 0$, so that $\{a_1, a_2, \ldots, a_n\}$ is join-independent by Proposition 2.31. Moreover

$$
\begin{aligned}
&a_n' \wedge (a_1 \vee \; \cdots \vee a_n) &&\text{(since } a_n' \leq b) \\
&= a_n' \wedge b \wedge ((a_1 \vee \cdots \vee a_{n-1}) \vee a_n) &&\text{(by the modular identity)} \\
&= a_n' \wedge ((b \wedge (a_1 \vee \cdots \vee a_{n-1})) \vee a_n) \\
&= a_n' \wedge (0 \vee a_n) = 0.
\end{aligned}
$$

This completes the construction. Now $\{\, a_n \mid n \geq 1 \,\}$ is an infinite join-independent set. $\qquad \square$

Theorem 2.36 *The following conditions are equivalent for a modular lattice L with 0 and 1.*

(a) *L does not contain infinite join-independent subsets.*

(b) *L contains a finite join-independent subset $\{a_1, a_2, \ldots, a_n\}$ with a_i uniform for every $i = 1, 2, \ldots, n$ and $a_1 \vee a_2 \vee \cdots \vee a_n$ essential in L.*

(c) *The cardinality of the join-independent subsets of L is $\leq m$ for a nonnegative integer m.*

(d) *If $a_0 \leq a_1 \leq a_2 \leq \ldots$ is an ascending chain of elements of L, then there exists $i \geq 0$ such that a_i is essential in a_j for every $j \geq i$.*

Moreover, if these equivalent conditions hold and $\{a_1, a_2, \ldots, a_n\}$ is a finite join-independent subset of L with a_i uniform for every $i = 1, 2, \ldots, n$ and $a_1 \vee a_2 \vee \cdots \vee a_n$ essential in L, then any other join-independent subset of L has cardinality $\leq n$.

Proof. (a) $\Rightarrow$ (b). Let $\mathcal{F}$ be the family of all join-independent subsets of L consisting only of uniform elements. The family $\mathcal{F}$ is non-empty (Lemma 2.35). By Zorn's Lemma $\mathcal{F}$ has a maximal element X with respect to inclusion. By (a) the set X is finite, say $X = \{a_1, a_2, \ldots, a_n\}$. The element $a_1 \vee a_2 \vee \cdots \vee a_n$ is essential in L, otherwise there would exist a non-zero element $x \in L$ such that $(a_1 \vee a_2 \vee \cdots \vee a_n) \wedge x = 0$, and by Lemma 2.35 there would be a uniform element $b \in L$ such that $b \leq x$. Then $\{a_1, a_2, \ldots, a_n, b\}$ would be join-independent by Proposition 2.31, a contradiction.

(b) $\Rightarrow$ (c). Suppose that (b) holds, so that there exists a finite join-independent subset $\{a_1, a_2, \ldots, a_n\}$ of L with a_i uniform for every i and

$$a_1 \vee \cdots \vee a_n$$

essential in L. Assume that there exists a join-independent subset $\{b_1, b_2, \ldots, b_k\}$ of L of cardinality $k > n$. For every $t = 0, 1, \ldots, n$ we shall construct a subset X_t of $\{a_1, a_2, \ldots, a_n\}$ of cardinality t and a subset Y_t of $\{b_1, b_2, \ldots, b_k\}$ of cardinality $k - t$ such that $X_t \cap Y_t = \emptyset$ and $X_t \cup Y_t$ is join-independent. For $t = 0$ set $X_0 = \emptyset$ and $Y_0 = \{b_1, b_2, \ldots, b_k\}$. Suppose that X_t and Y_t have been constructed for some t, $0 \leq t < n$. We shall construct X_{t+1} and Y_{t+1}. Since $|Y_t| = k - t > n - t > 0$, there exists $j = 1, 2, \ldots, k$ with $b_j \in Y_t$. Set

$$c = \bigvee_{y \in (X_t \cup Y_t) \setminus \{b_j\}} y.$$

We claim that $c \wedge a_\ell = 0$ for some $\ell = 1, 2, \ldots, n$. Otherwise, if $c \wedge a_i \neq 0$ for every $i = 1, 2, \ldots, n$, then $c \wedge a_i$ is essential in a_i because a_i is uniform, so that $\bigvee_{i=1}^{n} c \wedge a_i$ is essential in $\bigvee_{i=1}^{n} a_i$ by Corollary 2.34. Since $\bigvee_{i=1}^{n} a_i$ is essential in 1, it follows that $\bigvee_{i=1}^{n} c \wedge a_i$ is essential in 1 (Lemma 2.32). Then $c \geq \bigvee_{i=1}^{n} c \wedge a_i$ is essential in 1, so that $c \wedge b_j \neq 0$. This contradicts the fact that $X_t \cup Y_t$ is join-independent and the contradiction proves the claim. From Proposition 2.31 and the claim it follows that $(X_t \cup \{a_\ell\}) \cup (Y_t \setminus \{b_j\})$ is join-independent, so that $X_{t+1} = X_t \cup \{a_\ell\}$ and $Y_{t+1} = Y_t \setminus \{b_j\}$ have the required properties. This completes the construction of the sets X_t and Y_t.

For $t = n$ we have a non-empty subset Y_n of $\{b_1, b_2, \ldots, b_k\}$ such that $\{a_1, a_2, \ldots, a_n\} \cup Y_n$ is a join-independent subset of cardinality k, so that

$$(a_1 \vee a_2 \vee \cdots \vee a_n) \wedge y = 0$$

for every $y \in Y_n$, and this contradicts the fact that $a_1 \vee a_2 \vee \cdots \vee a_n$ is essential in L. Hence every join-independent subset of L has cardinality $\leq n$.

(c) $\Rightarrow$ (d). If (d) does not hold, there is a chain $a_0 \leq a_1 \leq a_2 \leq \ldots$ of elements of L such that for every $i \geq 0$ there exists $j(i) > i$ with a_i not essential in $a_{j(i)}$. Set $j_0 = 0$ and $j_{n+1} = j(j_n)$ for every $n \geq 0$. Then for every $n \geq 0$ there exists a non-zero element $b_n \leq a_{j_{n+1}}$ such that $b_n \wedge a_{j_n} = 0$. The set $\{b_n \mid n \geq 0\}$ is join-independent by Proposition 2.31. Thus (c) does not hold.

(d) $\Rightarrow$ (a). If (a) is not satisfied, then L contains a countable infinite join-independent subset $\{b_i \mid i \geq 0\}$. Set $a_n = \bigvee_{i=0}^{n} b_i$. Then $a_0 \leq a_1 \leq a_2 \leq \ldots$, and for every $n \geq 0$ the element a_n is not essential in a_{n+1} because

$$a_n \wedge b_{n+1} = 0.$$

Hence (d) is not satisfied.

The last part of the statement has already been seen in the proof of (b) $\Rightarrow$ (c). $\qquad\square$

Thus, for a modular lattice L, either there is a finite join-independent subset $\{a_1, a_2, \ldots, a_n\}$ with a_i uniform for every $i = 1, 2, \ldots, n$ and

$$a_1 \vee a_2 \vee \cdots \vee a_n$$

essential in L, and in this case n is said to be the *Goldie dimension* dim L of L, or L contains infinite join-independent subsets, in which case L is said to have *infinite Goldie dimension*. The Goldie dimension of a lattice L is zero if and only if L has exactly one element.

2.7 Goldie dimension of a module

In this section we shall apply the Goldie dimension of modular lattices introduced in the previous section to the lattice $\mathcal{L}(M)$ of all submodules of a module M_R. If the lattice $\mathcal{L}(M)$ has finite Goldie dimension n, then n will be said to be the *Goldie dimension* dim M_R of the module M_R. Otherwise, if the lattice $\mathcal{L}(M)$ has infinite Goldie dimension, that is, if M_R contains an infinite direct sum of non-zero submodules, the module M_R will be said to have *infinite Goldie dimension* (dim $M_R = \infty$).

Since a module M is essential in its injective envelope $E(M)$,

$$\dim(M) = \dim(E(M)).$$

In Section 2.5 we had already defined uniform modules. Obviously, a module M is uniform if and only if the lattice $\mathcal{L}(M)$ is uniform. A module M has finite Goldie dimension n if and only if it contains an essential submodule that is the finite direct sum of n uniform submodules $U_1, \ldots, U_n$ (Theorem 2.36(b)). In this case $E(M) = E(U_1) \oplus E(U_2) \oplus \cdots \oplus E(U_n)$ is the finite direct sum of n indecomposable modules. Note that by Theorem 2.26 we already knew that if $E(M)$ is a finite direct sum of indecomposable modules, then the number of direct summands in any indecomposable decomposition of $E(M)$ does not depend on the decomposition. Hence a module M has finite Goldie dimension n if and only if its injective envelope $E(M)$ is the direct sum of n indecomposable modules.

In the next proposition we collect the most important arithmetical properties of the Goldie dimension of modules. Some of these properties have already been noticed. Their proof is elementary.

Proposition 2.37 *Let M be module.*

(a) $\dim(M) = 0$ *if and only if $M = 0$.*

(b) $\dim(M) = 1$ *if and only if M is uniform.*

(c) *If $N \le M$ and M has finite Goldie dimension, then N has finite Goldie dimension and* $\dim(N) \le \dim(M)$.

(d) *If $N \le M$ and M has finite Goldie dimension, then* $\dim(N) = \dim(M)$ *if and only if N is essential in M.*

(e) *If M and M' are modules of finite Goldie dimension, then $M \oplus M'$ is a module of finite Goldie dimension and* $\dim(M \oplus N) = \dim(M) + \dim(N)$.
□

Artinian modules and noetherian modules have finite Goldie dimension. For an artinian module M, the Goldie dimension of M is equal to the composition length of its socle $\mathrm{soc}(M)$. In particular, an artinian module M has Goldie dimension 1 if and only if it has a simple socle.

The next proposition contains a first application of the Goldie dimension of a ring.

Proposition 2.38 *Let R be a ring and suppose that R_R has finite Goldie dimension. Then every surjective endomorphism of a finitely generated projective right R-module P_R is an automorphism. In particular, every right or left invertible element of R is invertible.*

Proof. If P_R is a finitely generated projective R-module, then P_R has finite Goldie dimension. If φ is a surjective endomorphism of P_R, then

$$P_R \oplus \ker\varphi \cong P_R,$$

so that $\dim(\ker\varphi) = 0$, that is, φ is injective. For the second part of the statement we must show that if $x, y \in R$ and $xy = 1$, then $yx = 1$. Since $xy = 1$, left multiplication by x is a surjective endomorphism μ_x of R_R. From $xy = 1$ it follows that $yR \oplus \ker(\mu_x) = R$. Hence $yR = R$, i.e., y is also right invertible. Thus y is invertible and x is its two-sided inverse. □

2.8 Dual Goldie dimension of a module

We shall now apply the results on the Goldie dimension of modular lattices of Section 2.6 to the dual lattice of the lattice $\mathcal{L}(M)$ of all submodules of a module M. If $(L, \wedge, \vee)$ is a modular lattice, then its dual lattice $(L, \vee, \wedge)$ is also a modular lattice. In particular, the Duality Principle holds for modular lattices, that is, if a statement Φ expressed in terms of $\wedge$, $\vee$, $\leq$ and $\geq$ is true for all modular lattices, then the dual statement of Φ, obtained from Φ interchanging $\wedge$ with $\vee$ and $\leq$ with $\geq$, is also true for all modular lattices.

Since the dual of the lattice $\mathcal{L}(M)$ of all submodules of a module M is modular, all the results of Section 2.6 hold for this lattice. We now shall translate the results of Section 2.6 for the dual of the lattice $\mathcal{L}(M)$ to the language of modules.

Let M be a right R-module. A finite set $\{\, N_i \mid i \in I \,\}$ of proper submodules of M is said to be *coindependent* if $N_i + (\bigcap_{j \neq i} N_j) = M$ for every $i \in I$, or, equivalently, if the canonical injective mapping $M/\bigcap_{i \in I} N_i \to \oplus_{i \in I} M/N_i$ is bijective. An arbitrary set A of proper submodules of M is *coindependent* if its finite subsets are coindependent. If A is a coindependent set of proper submodules of M and N is a proper submodule of M such that $N + (\bigcap_{X \in B} X) = M$ for every finite subset B of A, then $A \cup \{N\}$ is a coindependent set of submodules of M (Proposition 2.31). By Zorn's Lemma, every coindependent set of submodules of M is contained in a maximal coindependent set.

The following lemma, which is dual to Lemma 2.24, has an elementary proof.

Lemma 2.39 *Let $M \neq 0$ be an R-module. The following conditions are equivalent:*

(a) *The sum of any two proper submodules of M is a proper submodule of M.*

(b) *Every proper submodule of M is superfluous in M.*

(c) *Every non-zero homomorphic image of M is indecomposable.* □

An R-module $M \neq 0$ is said to be *couniform*, or *hollow*, if it satisfies the equivalent conditions of the previous Lemma 2.39. Every local module is couniform, but not conversely. For instance, the $\mathbf{Z}$-module $\mathbf{Z}(p^\infty)$ (the Prüfer group) is couniform and is not local. Every proper submodule of a finitely generated module M is contained in a maximal submodule of M. Hence if M is a finitely generated module, M is couniform if and only if M is local. From Theorem 2.36 we obtain

Theorem 2.40 *The following conditions are equivalent for a right module M:*

(a) *There do not exist infinite coindependent sets of proper submodules of M.*

(b) *There exists a finite coindependent set $\{N_1, N_2, \ldots, N_n\}$ of proper submodules of M with M/N_i couniform for all i and $N_1 \cap N_2 \cap \cdots \cap N_n$ superfluous in M.*

(c) *The cardinality of the coindependent sets of proper submodules of M is $\leq m$ for a non-negative integer m.*

(d) *If $N_0 \supseteq N_1 \supseteq N_2 \supseteq \ldots$ is a descending chain of submodules of M, then there exists $i \geq 0$ such that N_i/N_j is superfluous in M/N_j for every $j \geq i$.*

Moreover, if these equivalent conditions hold and $\{N_1, N_2, \ldots, N_n\}$ is a finite coindependent set of proper submodules of M with M/N_i couniform for all i and $N_1 \cap N_2 \cap \cdots \cap N_n$ superfluous in M, then every other coindependent set of proper submodules of M has cardinality $\leq n$. $\qquad\square$

The *dual Goldie dimension* $\operatorname{codim}(M)$ of a right module M is the Goldie dimension of the dual lattice of the lattice $\mathcal{L}(M)$. Hence a module M has finite dual Goldie dimension n if and only if there exists a coindependent set $\{N_1, N_2, \ldots, N_n\}$ of proper submodules of M with M/N_i couniform for all i and $N_1 \cap N_2 \cap \cdots \cap N_n$ superfluous in M. And a module M has *infinite dual Goldie dimension* if there exist infinite coindependent sets of proper submodules of M. Note that if a module M has finite dual Goldie dimension, then for every proper submodule N of M there exists a proper submodule P of M containing N with M/P couniform (Lemma 2.35).

From Theorem 2.40(d) we obtain

Corollary 2.41 *Every artinian module has finite dual Goldie dimension.* $\qquad\square$

The proof of the next result is straightforward.

Proposition 2.42 *Let M be module.*

(a) $\operatorname{codim}(M) = 0$ *if and only if $M = 0$.*

(b) $\operatorname{codim}(M) = 1$ *if and only if M is couniform.*

(c) *If $N \leq M$ and M has finite dual Goldie dimension, then M/N has finite dual Goldie dimension and $\operatorname{codim}(M/N) \leq \operatorname{codim}(M)$.*

(d) *If M has finite dual Goldie dimension and $N \leq M$, then $\operatorname{codim}(M/N) = \operatorname{codim}(M)$ if and only if N is superfluous in M.*

(e) *If M and M' are modules of finite dual Goldie dimension, then $M \oplus M'$ is a module of finite dual Goldie dimension and*

$$\operatorname{codim}(M \oplus M') = \operatorname{codim}(M) + \operatorname{codim}(M'). \qquad \square$$

If a module M has finite dual Goldie dimension n, then there exists a set $\{N_1, N_2, \ldots, N_n\}$ of submodules of M such that $N = N_1 \cap N_2 \cap \cdots \cap N_n$ is superfluous in M and $M/N \cong \oplus_{i=1}^{n} M/N_i$ is a direct sum of n couniform modules. Note that there is no epimorphism of such a module M onto a direct sum of $n + 1$ non-zero modules. In Section 2.7 we saw that if a module M has the property that there are no monomorphisms from a direct sum of infinitely many non-zero modules into M, then M has finite Goldie dimension. This result cannot be dualized, that is, it is not true that if M is a module and there is no homomorphic image of M that is a direct product of infinitely many non-zero modules, then M has finite dual Goldie dimension. For instance, consider the **Z**-module **Z**, that is, the abelian group of integers. Then there is no homomorphic image of **Z** that is a direct product $\prod_{i \in I} G_i$ of infinitely many non-zero abelian groups G_i. But the set of all $p\mathbf{Z}$, p a prime number, is an infinite coindependent set of proper subgroups of **Z**, so that $\operatorname{codim}(\mathbf{Z}) = \infty$.

For a semisimple module the dual Goldie dimension coincides with the composition length of the module. Hence for a semisimple artinian ring R,

$$\dim(R_R) = \dim(_R R) = \operatorname{codim}(R_R) = \operatorname{codim}(_R R).$$

We shall denote this finite dimension $\dim(R)$.

Proposition 2.43 *The following conditions are equivalent for a ring R.*

(a) *The ring R is semilocal.*

(b) *The right R-module R_R has finite dual Goldie dimension.*

(c) *The left R-module $_R R$ has finite dual Goldie dimension.*

Moreover, if these equivalent conditions hold,

$$\operatorname{codim}(R_R) = \operatorname{codim}(_R R) = \dim(R/J(R)).$$

Proof. (a) $\Rightarrow$ (b). Suppose R_R has infinite dual Goldie dimension, and let $\{ I_n \mid n \geq 1 \}$ be an infinite coindependent set of proper right ideals of R. Then $R/ \bigcap_{n=1}^{k} I_n$ is a direct sum of k non-zero cyclic modules for every $k \geq 1$.

If C is a non-zero cyclic module, $C/CJ(R)$ is a non-zero module. Therefore $R/(J(R) + \bigcap_{n=1}^{k} I_n)$ is a direct sum of at least k non-zero modules for every $k \geq 1$. In particular $R/J(R)$ cannot have finite length, so that R cannot be semilocal.

(b) $\Rightarrow$ (a). Suppose that R_R has finite dual Goldie dimension. Let $\mathcal{I}$ be the set of all right ideals of R that are finite intersections of maximal right ideals. Note that if $I, J \in \mathcal{I}$ and $I \subset J$, then R/I and R/J are semisimple modules of finite length and

$$\operatorname{codim}(R/J) = \operatorname{length}(R/J) < \operatorname{length}(R/I) = \operatorname{codim}(R/I).$$

Since $\operatorname{codim}(R/I) \leq \operatorname{codim}(R_R)$ for every I, it follows that every descending chain in $\mathcal{I}$ is finite, i.e., the partially ordered set $\mathcal{I}$ is artinian. In particular $\mathcal{I}$ has a minimal element. Since any intersection of two elements of $\mathcal{I}$ belongs to $\mathcal{I}$, the set $\mathcal{I}$ has a least element, which is the Jacobson radical $J(R)$. Hence $J(R) \in \mathcal{I}$ is a finite intersection of maximal right ideals. Therefore $R/J(R)$ is a semisimple artinian right R-module, and R is semilocal.

Since (a) is right-left symmetric, (a), (b) and (c) are equivalent. Finally, $J(R)$ is a superfluous submodule of R_R (Nakayama's Lemma 1.4), so that if (b) holds, then $\operatorname{codim}(R_R) = \operatorname{codim}(R/J(R))$ by Proposition 2.42(d). $\square$

Corollary 2.44 *Let P_R be a finitely generated projective module over a semilocal ring R. Then every surjective endomorphism of P_R is an automorphism. In particular, every right or left invertible element of a semilocal ring is invertible.*

Proof. Since R is semilocal, R_R has finite dual Goldie dimension, so that P_R has finite dual Goldie dimension (Proposition 2.42). If $f\colon P_R \to P_R$ is surjective, then $\ker f$ is a direct summand of P_R, and $\ker f \oplus P_R \cong P_R$. Thus $\operatorname{codim}(\ker f) = 0$, i.e., $\ker f = 0$. The proof of the second part of the statement is analogous to the proof of the second part of the statement of Proposition 2.38. $\square$

We conclude this section with an example. A non-zero uniserial module is both uniform and couniform. Therefore a serial module has finite Goldie dimension if and only if it is the direct sum of a finite number of uniserial modules, if and only if it has finite dual Goldie dimension. More precisely, a serial module M has finite Goldie dimension n if and only if it is the direct sum of exactly n non-zero uniserial modules (so that the number n of direct summands of M that appear in any decomposition of M as a direct sum of non-zero uniserial modules does not depend on the decomposition), if and only if M has finite dual Goldie dimension n.

2.9 ℵ-small modules and ℵ-closed classes

Let R be an arbitrary ring. An R-module N_R is *small* if for every family

$$\{\, M_i \mid i \in I \,\}$$

of R-modules and any homomorphism $\varphi \colon N_R \to \oplus_{i \in I} M_i$, there is a finite subset $F \subseteq I$ such that $\pi_j \varphi = 0$ for every $j \in I \setminus F$. Here the $\pi_j \colon \oplus_{i \in I} M_i \to M_j$ are the canonical projections.

For instance, every finitely generated module is small. Another class of small modules is given by the class of uncountably generated uniserial modules, as the next proposition shows.

Proposition 2.45 *Every uniserial module that is not small can be generated by* $\aleph_0$ *elements.*

Proof. Let U be a uniserial module that is not small. Then there exist modules M_i, $i \in I$, and a homomorphism $\varphi \colon U \to \oplus_{i \in I} M_i$ such that if $\pi_j \colon \oplus_{i \in I} M_i \to M_j$ denotes the canonical projection for every $j \in I$, then $\pi_j \varphi \neq 0$ for infinitely many $j \in I$.

For every $x \in U$ set $\mathrm{supp}(x) = \{\, i \in I \mid \pi_i \varphi(x) \neq 0 \,\}$, so that $\mathrm{supp}(x)$ is a finite subset of I for every $x \in U$. Note that if $x, y \in U$ and $xR \subseteq yR$, then $\mathrm{supp}(x) \subseteq \mathrm{supp}(y)$. Define by induction a sequence of elements $x_n \in U$, $n \geq 0$, such that $\mathrm{supp}(x_0) \subset \mathrm{supp}(x_1) \subset \mathrm{supp}(x_2) \subset \dots$. Set $x_0 = 0$. If $x_n \in U$ has been defined, then $\mathrm{supp}(x_n)$ is finite, but $\pi_j \varphi \neq 0$ for infinitely many $j \in I$. Hence there exists $k \in I$ with $k \notin \mathrm{supp}(x_n)$ and $\pi_k \varphi \neq 0$. Let $x_{n+1} \in U$ be an element of U with $\pi_k \varphi(x_{n+1}) \neq 0$. Then $\mathrm{supp}(x_{n+1}) \not\subseteq \mathrm{supp}(x_n)$, so that $x_{n+1} R \not\subseteq x_n R$. Hence $x_n R \subseteq x_{n+1} R$, from which $\mathrm{supp}(x_n) \subset \mathrm{supp}(x_{n+1})$. This defines the sequence x_n.

If the elements x_n do not generate the module U, then there exists $v \in U$ such that $v \notin x_n R$ for every $n \geq 0$. Then $vR \supseteq x_n R$ for every n, so that $\mathrm{supp}(v) \supseteq \mathrm{supp}(x_n)$ for every n. This yields a contradiction, because $\mathrm{supp}(v)$ is finite and $\bigcup_{n \geq 0} \mathrm{supp}(x_n)$ is infinite. Hence the x_n generate U and U is countably generated.   □

Now we shall extend the definition of small module. Let ℵ be a cardinal number. An R-module N_R is ℵ-*small* if for every family $\{\, M_i \mid i \in I \,\}$ of R-modules and any homomorphism $\varphi \colon N_R \to \oplus_{i \in I} M_i$, the set $\{\, i \in I \mid \pi_i \varphi \neq 0 \,\}$ has cardinality $\leq$ ℵ.

For instance, every small module is $\aleph_0$-small, and every uniserial module is $\aleph_0$-small. It is easy to see that if ℵ is a finite cardinal number and N_R is ℵ-small, then $N_R = 0$.

Let R be a ring, $\mathcal{G}$ a non-empty class of right R-modules and let ℵ be a cardinal number. We say that $\mathcal{G}$ is ℵ-*closed* if:

(a) $\mathcal{G}$ is closed under homomorphic images, that is, if M_R, N_R are right R-modules, $f\colon M_R \to N_R$ is an epimorphism and $M_R \in \mathcal{G}$, then $N_R \in \mathcal{G}$;

(b) every module in $\mathcal{G}$ is $\aleph$-small;

(c) $\mathcal{G}$ is closed under direct sums of $\aleph$ modules, that is, if $M_i \in \mathcal{G}$ for every $i \in I$ and $|I| \leq \aleph$, then $\oplus_{i \in I} M_i \in \mathcal{G}$.

Examples 2.46 (1) For an infinite cardinal number $\aleph$ and a ring R, let $\mathcal{G}$ be the class of all $\aleph$-generated modules, that is, the right R-modules that are homomorphic images of $R^{(\aleph)}$. Then $\mathcal{G}$ is an $\aleph$-closed class.

(2) For a cardinal number $\aleph$ and a ring R, let $\mathcal{G}$ be the class of all $\aleph$-small right R-modules. Then $\mathcal{G}$ is an $\aleph$-closed class.

(3) Let R be a ring and let $\aleph$ be a finite cardinal number. We have already remarked that every $\aleph$-small right R-module is zero. Hence every $\aleph$-closed class of right R-modules consists of all zero R-modules.

(4) Let R be a ring and let $\mathcal{G}$ be the class of all σ-*small* R-modules, that is, the right R-modules that are countable ascending unions of small submodules. Then $\mathcal{G}$ is an $\aleph_0$-closed class. Note that by Proposition 2.45 every uniserial module is σ-small. □

The following theorem is essentially equivalent to an extension due to C. Walker of a theorem of [Kaplansky 58, Theorem 1]. Kaplansky proved it in the case in which $\aleph = \aleph_0$ and $\mathcal{G}$ is the class of $\aleph_0$-generated modules, and Walker extended it to the class of $\aleph$-generated modules for an arbitrary cardinal number $\aleph$. [Warfield 69c] remarked that the theorem holds for the classes of $\aleph$-small modules and σ-small modules, and that suitable versions for larger cardinals were also valid.

Theorem 2.47 *Let R be a ring, let $\aleph$ be a cardinal number and $\mathcal{G}$ an $\aleph$-closed class of right R-modules. If a module M_R is a direct sum of modules belonging to $\mathcal{G}$, then every direct summand of M_R is a direct sum of modules belonging to $\mathcal{G}$.*

Proof. Since the case of a finite cardinal number $\aleph$ is trivial (Example 2.46(3)), we may suppose $\aleph$ infinite. Let $M_R = \oplus_{i \in I} M_i$, where $M_i \in \mathcal{G}$ for every $i \in I$, and assume $M_R = N_R \oplus P_R$. Let $\mathcal{L}(N_R)$ and $\mathcal{L}(P_R)$ be the sets of all submodules of N_R and P_R, respectively, and let $\mathcal{T}$ be the set of all triples $(J, \mathcal{A}, \mathcal{B})$ such that

(1) $J \subseteq I$, $\mathcal{A} \subseteq \mathcal{L}(N_R) \cap \mathcal{G}$, $\mathcal{B} \subseteq \mathcal{L}(P_R) \cap \mathcal{G}$;

(2) the sum $\sum_{X \in \mathcal{A}} X$ is direct, that is, $\sum_{X \in \mathcal{A}} X = \oplus_{X \in \mathcal{A}} X$;

(3) the sum $\sum_{Y \in \mathcal{B}} Y$ is direct, that is, $\sum_{Y \in \mathcal{B}} Y = \oplus_{Y \in \mathcal{B}} Y$;

(4) $\oplus_{i \in J} M_i = (\oplus_{X \in \mathcal{A}} X) \oplus (\oplus_{Y \in \mathcal{B}} Y)$.

Note that $\mathcal{T}$ is non-empty, because $(\emptyset, \emptyset, \emptyset) \in \mathcal{T}$. Define a partial ordering on $\mathcal{T}$ by setting $(J, \mathcal{A}, \mathcal{B}) \leq (J', \mathcal{A}', \mathcal{B}')$ whenever $J \subseteq J'$, $\mathcal{A} \subseteq \mathcal{A}'$, and $\mathcal{B} \subseteq \mathcal{B}'$. It is easily seen that every chain in $\mathcal{T}$ has an upper bound in $\mathcal{T}$, so that by Zorn's Lemma $\mathcal{T}$ has a maximal element $(K, \mathcal{C}, \mathcal{D})$. Suppose $K \subset I$. Let $i \in I \setminus K$ and let ε be the idempotent endomorphisms of M_R with $\ker(\varepsilon) = P_R$ that is the identity on N_R.

Define an ascending chain $I_0 \subseteq I_1 \subseteq I_2 \subseteq \ldots$ of subsets of I of cardinality at most ℵ in the following way. Set $I_0 = \{i\}$. Suppose I_n has been defined. Since $\oplus_{j \in I_n} M_j \in \mathcal{G}$, its homomorphic images $\varepsilon(\oplus_{j \in I_n} M_j)$ and $(1 - \varepsilon)(\oplus_{j \in I_n} M_j)$ belong to $\mathcal{G}$, so that $\varepsilon(\oplus_{j \in I_n} M_j) + (1 - \varepsilon)(\oplus_{j \in I_n} M_j)$ is in $\mathcal{G}$. In particular, this module is ℵ-small, hence there exists a subset I_{n+1} of I of cardinality at most ℵ such that $\varepsilon(\oplus_{j \in I_n} M_j) + (1 - \varepsilon)(\oplus_{j \in I_n} M_j) \subseteq \oplus_{j \in I_{n+1}} M_j$. Note that $\oplus_{j \in I_n} M_j \subseteq \varepsilon(\oplus_{j \in I_n} M_j) + (1 - \varepsilon)(\oplus_{j \in I_n} M_j)$, so that $I_n \subseteq I_{n+1}$. This completes the construction of the subsets I_n by induction.

Let I' be the union of the countable ascending chain $I_0 \subseteq I_1 \subseteq I_2 \subseteq \ldots$, so that I' is a subset of I of cardinality at most ℵ. Hence $\oplus_{j \in I'} M_j \in \mathcal{G}$. Since $i \in I'$, it follows that $I' \cup K \supset K$. And $\varepsilon(\oplus_{j \in I_n} M_j) \subseteq \oplus_{j \in I_{n+1}} M_j$ for all n implies that $\varepsilon(\oplus_{j \in I'} M_j) \subseteq \oplus_{j \in I'} M_j$. Similarly, $(1 - \varepsilon)(\oplus_{j \in I'} M_j) \subseteq \oplus_{j \in I'} M_j$.

Now $\oplus_{j \in K} M_j = (\oplus_{X \in \mathcal{C}} X) \oplus (\oplus_{Y \in \mathcal{D}} Y)$ because $(K, \mathcal{C}, \mathcal{D}) \in \mathcal{T}$, and

$$\varepsilon(\oplus_{j \in I' \cup K} M_j) = \varepsilon(\oplus_{j \in I'} M_j + \oplus_{j \in K} M_j)$$
$$= \varepsilon(\oplus_{j \in I'} M_j + \oplus_{X \in \mathcal{C}} X + \oplus_{Y \in \mathcal{D}} Y) = \varepsilon(\oplus_{j \in I'} M_j) + \oplus_{X \in \mathcal{C}} X$$
$$\subseteq \oplus_{j \in I'} M_j + \oplus_{j \in K} M_j = \oplus_{j \in I' \cup K} M_j.$$

Hence the idempotent endomorphism ε of M_R induces an idempotent endomorphism on $\oplus_{j \in I' \cup K} M_j$, so that

$$\oplus_{j \in I' \cup K} M_j = \varepsilon(\oplus_{j \in I' \cup K} M_j) \oplus (1 - \varepsilon)(\oplus_{j \in I' \cup K} M_j).$$

The submodule $\oplus_{X \in \mathcal{C}} X$ is a direct summand of $\oplus_{j \in I' \cup K} M_j$ contained in $\varepsilon(\oplus_{j \in I' \cup K} M_j)$. Hence it is a direct summand of $\varepsilon(\oplus_{j \in I' \cup K} M_j)$, that is, there exists a submodule $\overline{X}$ of N_R such that

$$\varepsilon(\oplus_{j \in I' \cup K} M_j) = (\oplus_{X \in \mathcal{C}} X) \oplus \overline{X}.$$

Similarly, $\oplus_{Y \in \mathcal{D}} Y$ is a direct summand of $\oplus_{j \in I' \cup K} M_j$ contained in

$$(1 - \varepsilon)(\oplus_{j \in I' \cup K} M_j),$$

so that $(1 - \varepsilon)(\oplus_{j \in I' \cup K} M_j) = (\oplus_{Y \in \mathcal{D}} Y) \oplus \overline{Y}$ for some submodule $\overline{Y}$ of P_R. Therefore $\oplus_{j \in I' \cup K} M_j = (\oplus_{X \in \mathcal{C}} X) \oplus \overline{X} \oplus (\oplus_{Y \in \mathcal{D}} Y) \oplus \overline{Y}$. It follows that

$$\overline{X} \oplus \overline{Y} \cong \oplus_{j \in I' \cup K} M_j / \oplus_{j \in K} M_j \cong \oplus_{j \in I' \setminus K} M_j \in \mathcal{G},$$

so that both $\overline{X}$ and $\overline{Y}$ belong to $\mathcal{G}$. This shows that $(I' \cup K, \mathcal{C} \cup \{\overline{X}\}, \mathcal{D} \cup \{\overline{Y}\})$ is an element of $\mathcal{T}$ strictly greater than the maximal element $(K, \mathcal{C}, \mathcal{D})$. This contradiction proves that $K = I$. Thus $M_R = (\oplus_{X \in \mathcal{C}} X) \oplus (\oplus_{Y \in \mathcal{D}} Y)$. Hence $N_R = \oplus_{X \in \mathcal{C}} X$ and $P_R = \oplus_{Y \in \mathcal{D}} Y$. □

If we apply Theorem 2.47 to the $\aleph_0$-closed class of all countably generated
($= \aleph_0$-generated) modules we get a famous result of [Kaplansky 58]:

Corollary 2.48 (Kaplansky) *Any projective module is a direct sum of countably
generated modules.* $\square$

Corollary 2.49 *Let N be a direct summand of a serial module. Then there is a
decomposition $N = \oplus_{i \in I} N_i$, where each N_i is a direct summand of the direct
sum of a countable family $\{ U_{i,n} \mid n \in \mathbf{N} \}$ of uniserial modules.*

Proof. Let $\mathcal{G}$ be the class of all σ-small modules. This is an $\aleph_0$-closed class
that contains all uniserial modules (Example 2.46(4)). Apply Theorem 2.47.
Then N is a direct sum of modules N_i belonging to $\mathcal{G}$. Hence it suffices to
prove that a module N_i belonging to $\mathcal{G}$ that is a direct summand of a serial
module is a direct summand of the direct sum of a countable family of uniserial
modules. If N_i is a direct summand of a serial module $\oplus_{j \in J} U_j$, there are two
homomorphisms $\varphi \colon N_i \to \oplus_{j \in J} U_j$ and $\psi \colon \oplus_{j \in J} U_j \to N_i$ such that $\psi \varphi = 1_{N_i}$. If
$N_i \in \mathcal{G}$, N_i is $\aleph_0$-small, so that the set $C = \{ j \in J \mid \pi_j \varphi \neq 0 \}$ has cardinality
$\leq \aleph_0$. Now it is easily seen that N_i is a direct summand of the direct sum of
the countable family $\{ U_j \mid j \in C \}$. $\square$

We conclude the section with a proposition due to [Warfield 69c, Lemma
5], who proved it not only for modules, but for objects of more general abelian
categories. Here we consider the case of modules only.

Proposition 2.50 *Let R be a ring, $\aleph$ a cardinal number and $\mathcal{G}$ an $\aleph$-closed class
of right R-modules. If $M = \oplus_{i \in I} A_i = \oplus_{j \in J} B_j$, where A_i, B_j are non-zero
modules belonging to $\mathcal{G}$ for every $i \in I$ and every $j \in J$, then there exists a
partition $\{ I_\lambda \mid \lambda \in \Lambda \}$ of I and a partition $\{ J_\lambda \mid \lambda \in \Lambda \}$ of J with $|I_\lambda| \leq \aleph$,
$|J_\lambda| \leq \aleph$ and $\oplus_{i \in I_\lambda} A_i \cong \oplus_{j \in J_\lambda} B_j$ for every $\lambda \in \Lambda$.*

Proof. The case of a finite cardinal number $\aleph$ is trivial. Hence we may suppose
$\aleph$ infinite. We claim that if $i_0 \in I$, then there exist subsets $I' \subseteq I$ and $J' \subseteq J$
of cardinality $\leq \aleph$ such that $i_0 \in I'$ and $\oplus_{i \in I'} A_i = \oplus_{j \in J'} B_j$. In order to prove
the claim define sets I'_n and J'_n of cardinality $\leq \aleph$ for every integer $n \geq 0$ by
induction as follows. Set $I'_0 = \{i_0\}$ and $J'_0 = \emptyset$. Suppose I'_n and J'_n have been
defined. Then $(\oplus_{i \in I'_n} A_i) + (\oplus_{j \in J'_n} B_j) \in \mathcal{G}$, so that there exists a subset $J'_{n+1} \subseteq
J$ of cardinality at most $\aleph$ such that $(\oplus_{i \in I'_n} A_i) + (\oplus_{j \in J'_n} B_j) \subseteq \oplus_{j \in J'_{n+1}} B_j$. Since
$(\oplus_{i \in I'_n} A_i) + (\oplus_{j \in J'_{n+1}} B_j) \in \mathcal{G}$, there exists a subset $I'_{n+1} \subseteq I$ of cardinality at
most $\aleph$ such that $(\oplus_{i \in I'_n} A_i) + (\oplus_{j \in J'_{n+1}} B_j) \subseteq \oplus_{i \in I'_{n+1}} A_i$. It is now obvious that
$I' = \bigcup_{n \geq 0} I'_n$ and $J' = \bigcup_{n \geq 0} J'_n$ have the property required in the claim.

Define a chain of subsets $K_0 \subseteq K_1 \subseteq \cdots \subseteq K_\lambda \subseteq \ldots$ of I and a chain
of subsets $L_0 \subseteq L_1 \subseteq \cdots \subseteq L_\lambda \subseteq \ldots$ of J for each ordinal λ by transfinite
induction in the following way. Set $K_0 = L_0 = \emptyset$. If λ is a limit ordinal set
$K_\lambda = \bigcup_{\mu < \lambda} K_\mu$ and $L_\lambda = \bigcup_{\mu < \lambda} L_\mu$. For every ordinal μ such that $K_\mu = I$

set $K_{\mu+1} = K_\mu$ and $L_{\mu+1} = L_\mu$. Otherwise, if $K_\mu \subset I$, choose an element $i_0 \in I \setminus K_\mu$. By the claim there exist $I' \subseteq I$ and $J' \subseteq J$, both of cardinality $\leq \aleph$, such that $i_0 \in I'$ and $\oplus_{i \in I'} A_i = \oplus_{j \in J'} B_j$. In this case set $K_{\mu+1} = K_\mu \cup I'$ and $L_{\mu+1} = L_\mu \cup J'$.

Obviously $\oplus_{i \in K_\lambda} A_i = \oplus_{j \in L_\lambda} B_j$ for every λ, and there exists an ordinal $\overline{\lambda}$ such that $K_{\overline{\lambda}} = I$. Then $L_{\overline{\lambda}} = J$. Set $I_\lambda = K_{\lambda+1} \setminus K_\lambda$ and $J_\lambda = L_{\lambda+1} \setminus L_\lambda$ for every $\lambda < \overline{\lambda}$. Then

$$\oplus_{i \in K_{\lambda+1}} A_i = \oplus_{j \in L_{\lambda+1}} B_j \quad \text{and} \quad \oplus_{i \in K_\lambda} A_i = \oplus_{j \in L_\lambda} B_j$$

imply $\oplus_{i \in I_\lambda} A_i \cong \oplus_{j \in J_\lambda} B_j$. □

2.10 Direct sums of ℵ-small modules

In Section 2.3 we saw two cases in which there exist isomorphic refinements of two direct sum decompositions. The next theorem examines a third case.

Theorem 2.51 *Let M_R be a module that is a direct sum of $\aleph_0$-small submodules. Then any two direct sum decompositions of M into summands having the $\aleph_0$-exchange property have isomorphic refinements.*

Proof. We have already remarked that the class $\mathcal{G}$ of all $\aleph_0$-small R-modules is $\aleph_0$-closed (Example 2.46(2)). By Theorem 2.47 any decomposition of M_R refines into one in which the summands belong to $\mathcal{G}$. By Lemma 2.4 every refinement of a decomposition of M_R into summands with the $\aleph_0$-exchange property is a decomposition into summands with the $\aleph_0$-exchange property. Hence we may suppose that we have two direct sum decompositions of M into summands belonging to $\mathcal{G}$ and having the $\aleph_0$-exchange property. By Proposition 2.50 we may assume that the index sets are countable. In this case the result is given by Theorem 2.10. □

A fourth case in which isomorphic refinements exist is considered in the next important result, due to [Crawley and Jónsson, Theorem 7.1], who proved it for algebraic systems more general than modules. Here we shall present the proof given by [Warfield 69c, Theorem 7]. Also the proof given by Warfield holds in a context more general than ours, that is for suitable abelian categories, but we shall restrict our attention to the case we are interested in, that is, the case of modules. Recall that a module is σ-small if it is a countable ascending union of small submodules (Example 2.46(4)).

Theorem 2.52 *If a module M is a direct sum of σ-small modules each of which has the exchange property, then any two direct sum decompositions of M have isomorphic refinements.*

Proof. We claim that if $A = \oplus_{i=1}^{\infty} B_i$ is a direct sum of countably many σ-small modules B_i each of which has the exchange property and $A = C \oplus D$, then C is a direct sum of σ-small modules with the exchange property.

In order to prove the claim note that the direct summand C of A is σ-small, hence there is an ascending chain $0 = S_0 \subseteq S_1 \subseteq S_2 \subseteq \ldots$ of small submodules of C whose union is C itself. We shall construct submodules C_k, P_k of C for each $k \geq 0$ with the property that (1) $C = C_0 \oplus C_1 \oplus \cdots \oplus C_k \oplus P_k$, (2) $S_k \subseteq C_0 \oplus C_1 \oplus \cdots \oplus C_k$ and (3) C_k has the exchange property for every $k \geq 0$. Set $C_0 = 0$ and $P_0 = C$. Suppose $C_0, \ldots, C_{k-1}, P_0, \ldots, P_{k-1}$ with the required properties have been constructed. Then $C_0 \oplus C_1 \oplus \cdots \oplus C_{k-1}$ has the exchange property. Hence there exist direct summands B_i' of B_i such that $A = C_0 \oplus C_1 \oplus \cdots \oplus C_{k-1} \oplus (\oplus_{i=1}^{\infty} B_i')$. Since S_k is small, there exists a positive integer $n(k)$ such that $S_k \subseteq T_{n(k)}$, where

$$T_{n(k)} = C_0 \oplus C_1 \oplus \cdots \oplus C_{k-1} \oplus \left(\oplus_{i=1}^{n(k)} B_i' \right).$$

The module $T_{n(k)}$ has the exchange property, so that from

$$A = C \oplus D = T_{n(k)} \oplus \left(\oplus_{i=n(k)+1}^{\infty} B_i' \right),$$

we have that there exist $P_k \subseteq C$ and a direct sum decomposition $D_k \oplus D_k' = D$ such that $A = T_{n(k)} \oplus P_k \oplus D_k$. Set $C_k' = C \cap (T_{n(k)} \oplus D_k)$, so that $C = C_k' \oplus P_k$ by Lemma 2.1 and $S_k \subseteq C_k'$. Set $C_k = C_k' \cap P_{k-1}$. Then

$$C_0 \oplus C_1 \oplus \cdots \oplus C_{k-1} \subseteq C_k' \subseteq C = C_0 \oplus C_1 \oplus \cdots \oplus C_{k-1} \oplus P_{k-1}$$

forces $C_k' = C_0 \oplus C_1 \oplus \cdots \oplus C_{k-1} \oplus C_k$ (Lemma 2.1). Hence

$$C = C_0 \oplus C_1 \oplus \cdots \oplus C_k \oplus P_k$$

and $S_k \subseteq C_0 \oplus C_1 \oplus \cdots \oplus C_k$. Finally,

$$C_0 \oplus C_1 \oplus \cdots \oplus C_k \oplus P_k \oplus D_k \oplus D_k' = C \oplus D = A$$
$$= T_{n(k)} \oplus P_k \oplus D_k = C_0 \oplus C_1 \oplus \cdots \oplus C_{k-1} \oplus \left(\oplus_{i=1}^{n(k)} B_i' \right) \oplus P_k \oplus D_k$$

implies that $C_k \oplus D_k' \cong \oplus_{i=1}^{n(k)} B_i'$, so that C_k has the exchange property because it is isomorphic to a direct summand of $\oplus_{i=1}^{n(k)} B_i'$. This completes the construction by induction.

It is now obvious that $C = \oplus_{k=1}^{\infty} C_k$. Since B_i is σ-small, A itself is σ-small, so that each C_k is σ-small. This proves the claim.

In order to prove the theorem, suppose

$$M = \oplus_{i \in I} B_i \tag{2.24}$$

where, for each $i \in I$, B_i is σ-small and has the exchange property. Since every direct summand of B_i is σ-small and has the exchange property, it is enough to show that the decomposition (2.24) and any other decomposition

$$M = \oplus_{j \in J} M_j \tag{2.25}$$

have isomorphic refinements. By Theorem 2.47 the decomposition (2.25) has a refinement

$$M = \oplus_{k \in K} C_k \tag{2.26}$$

in which every C_k is σ-small. If we apply Proposition 2.50 to the decompositions (2.24) and (2.26) we see that we may assume I and K countable. By the claim the decomposition (2.26) has a refinement that is a direct sum of σ-small modules with the exchange property. Now Theorem 2.51 allows us to conclude. $\square$

From Theorem 2.52 we immediately obtain the following three corollaries:

Corollary 2.53 *If a module M is a direct sum of countably generated modules M_i, $i \in I$, each of which has the exchange property and N is a direct summand of M, then $N = \oplus_{i \in I} N_i$, where each N_i is isomorphic to a direct summand of M_i.* $\square$

Corollary 2.54 *If a module M is a direct sum of uniserial modules each of which has a local endomorphism ring, then any two direct sum decompositions of M have isomorphic refinements.* $\square$

Corollary 2.55 *If $M = \oplus_{i \in I} M_i$, where each M_i is a countably generated module with a local endomorphism ring, then any other direct sum decomposition of M can be refined to a decomposition isomorphic to the decomposition $M = \oplus_{i \in I} M_i$. In particular, any direct summand of M is isomorphic to $\oplus_{i \in J} M_i$ for a subset J of I.* $\square$

Corollary 2.55 is clearly a strengthened form of the Krull-Schmidt-Remak-Azumaya Theorem for direct sums of countably generated modules. It is apparently still an open question whether the hypothesis of being countably generated in Corollary 2.55 can be removed, that is, whether every direct summand of a direct sum of modules with local endomorphism rings is a direct sum of modules with local endomorphism rings. See [Elliger].

A ring R is said to be an *exchange ring* [Warfield 72] if R_R has the exchange property. For a ring R the right R-module R_R has the exchange property if and only if the left module $_R R$ has the exchange property [Warfield 72, Corollary 2]. We shall not need this fact, and its proof will be omitted.

Theorem 2.56 *If R is an exchange ring, then any projective right R-module is a direct sum of right ideals generated by idempotents.*

Proof. A projective R-module N_R is isomorphic to a direct summand of a free module M_R. Now apply Corollary 2.53 to M_R and N_R. $\qquad\square$

Every local ring is an exchange ring by Theorem 2.8. Hence from Theorem 2.56 we have that

Corollary 2.57 *Any projective right module over a local ring is free.* $\qquad\square$

2.11 The Loewy series

In this section, we introduce Loewy modules, which form a class containing all artinian modules. Let M be a module over an arbitrary ring R. Inductively define a well-ordered sequence of fully invariant submodules $\mathrm{soc}_\alpha(M)$ of M as follows:

$$\mathrm{soc}_0(M) = 0,$$
$$\mathrm{soc}_{\alpha+1}(M)/\mathrm{soc}_\alpha(M) = \mathrm{soc}(M/\mathrm{soc}_\alpha(M)) \text{ for every ordinal } \alpha,$$
$$\mathrm{soc}_\beta(M) = \bigcup_{\alpha<\beta} \mathrm{soc}_\alpha(M) \text{ for every limit ordinal } \beta.$$

The chain

$$\mathrm{soc}_0(M) \subseteq \mathrm{soc}_1(M) \subseteq \mathrm{soc}_2(M) \subseteq \cdots \subseteq \mathrm{soc}_\alpha(M) \subseteq \ldots$$

is called the *(ascending) Loewy series* of M. The module M is a *Loewy* module if there is an ordinal α such that $M = \mathrm{soc}_\alpha(M)$, and in this case the least ordinal α such that $M = \mathrm{soc}_\alpha(M)$ is called the *Loewy length* of M. Note that the Loewy series is always stationary, that is, for every module M there exists an ordinal α such that $\mathrm{soc}_\beta(M) = \mathrm{soc}_\alpha(M)$ for every $\beta \geq \alpha$ (for instance, it is sufficient to take any ordinal α whose cardinality is greater than the cardinality of M). For such an ordinal α, set $\delta(M) = \mathrm{soc}_\alpha(M)$. Then $\delta(M)$ is the largest Loewy submodule of M, and $M/\delta(M)$ has zero socle.

Lemma 2.58 *A module M is a Loewy module if and only if every non-zero homomorphic image of M has a non-zero socle.*

Proof. If M is a Loewy module and N is a proper submodule of M, consider the set of all the ordinal numbers α such that $\mathrm{soc}_\alpha(M) \subseteq N$. It is easily seen that this set has a greatest element β. Then M/N is a homomorphic image of $M/\mathrm{soc}_\beta(M)$, and the image of the socle $\mathrm{soc}_{\beta+1}(M)/\mathrm{soc}_\beta(M)$ of $M/\mathrm{soc}_\beta(M)$ in M/N is non-zero. Therefore the socle of M/N is non-zero.

Conversely, if M is not a Loewy module, then $M/\delta(M)$ is a non-zero homomorphic image of M with zero socle. $\qquad\square$

In particular, since every non-zero artinian module has a non-zero socle, every artinian module is Loewy. Every Loewy module is an essential extension of its socle.

Let $\mathcal{T}$ be the class of all Loewy right R-modules and let $\mathcal{F}$ be the class of all right R-modules with zero socle. Then $(\mathcal{T}, \mathcal{F})$ is a *torsion theory*, that is,

(a) $\mathrm{Hom}(T, F) = 0$ for all $T \in \mathcal{T}$, $F \in \mathcal{F}$.

(b) If M is a right R-module and $\mathrm{Hom}(M, F) = 0$ for all $F \in \mathcal{F}$, then $M \in \mathcal{T}$.

(c) If M is a right R-module and $\mathrm{Hom}(T, M) = 0$ for all $T \in \mathcal{T}$, then $M \in \mathcal{F}$.

The case of a right noetherian ring R is particularly interesting. If M is a Loewy right module over a right noetherian ring R and $x \in M$, then xR is a noetherian Loewy module, so that the ascending chain $\mathrm{soc}_n(xR)$, $n \geq 0$, must be stationary and $xR = \mathrm{soc}_m(xR)$ for some m. Since the modules $\mathrm{soc}_{n+1}(xR)/\mathrm{soc}_n(xR)$ are semisimple noetherian modules, it follows that xR is an R-module of finite composition length. Therefore:

Proposition 2.59 *If M is a Loewy right module over a right noetherian ring, then M is the sum of its submodules of finite composition length. In particular, M has Loewy length $\leq \omega$.* □

If x is an element in a right module M_R, the *annihilator*

$$\mathrm{ann}_R(x) = \{\, a \in R \mid xa = 0 \,\}$$

of x is always a right ideal of R. In particular, if $b \in R$, its *right annihilator* $\mathrm{r.ann}_R(b) = \{\, a \in R \mid ba = 0 \,\}$ is a right ideal of R. As a corollary of Proposition 2.59 we obtain

Corollary 2.60 *Let R be a right noetherian ring and let $\mathcal{G} = \{\, I \mid I \text{ is a right ideal of } R \text{ and } R/I \text{ is a right } R\text{-module of finite length} \,\}$. Then*

$$\delta(M_R) = \{\, x \in M \mid \mathrm{ann}_R(x) \in \mathcal{G} \,\}$$

for every R-module M_R.

Proof. Let x be an element of M. Then $\mathrm{ann}_R(x) \in \mathcal{G}$ if and only if xR is a right R-module of finite length, that is, if and only if $x \in \mathrm{soc}_n(M_R)$ for some positive integer n, i.e., if and only if $x \in \mathrm{soc}_\omega(M_R) = \delta(M_R)$. □

Proposition 2.59 can be adapted to commutative rings, as the next lemma shows.

Lemma 2.61 *The Loewy length of an artinian module over a commutative ring is $\leq \omega$.*

Proof. If $\operatorname{soc}_0(M) \subseteq \operatorname{soc}_1(M) \subseteq \operatorname{soc}_2(M) \subseteq \cdots \subseteq \operatorname{soc}_\alpha(M) \subseteq \ldots$ is the Loewy series of an artinian module M, then $\bigcup_{n \in \mathbf{N}} \operatorname{soc}_n(M) = M$, because if $x \in M$, then xR is an artinian module. Hence xR is a module of finite composition length, so that $xR \subseteq \operatorname{soc}_n(M)$ for some $n \in \mathbf{N}$. Therefore $M = \operatorname{soc}_\omega(M)$ has Loewy length $\leq \omega$.   □

Uniserial artinian modules of arbitrary Loewy length can be constructed over suitable non-commutative rings [Fuchs 70b, Facchini 84].

2.12   Artinian right modules over commutative or right noetherian rings

In this section we prove that the Krull-Schmidt Theorem holds for artinian right modules over rings which are either right noetherian or commutative. In Chapter 8 we shall see that it can fail for artinian modules over arbitrary non-commutative rings. Note that artinian modules are always finite direct sums of artinian indecomposable modules.

Lemma 2.62 *Let M_R be a module over an arbitrary ring R and let*

$$M_0 \subseteq M_1 \subseteq M_2 \subseteq \ldots$$

be an ascending chain of fully invariant submodules of M_R. Suppose that each M_i has finite composition length and $M = \bigcup_{i \geq 0} M_i$. Then

(a) *If $f \in \operatorname{End}(M_R)$, then $M = M' \oplus M''$, where $M' = \bigcup_{n \geq 0} \ker(f^n)$ and $M'' = \bigcup_{i \geq 0} \left(\bigcap_{n \geq 0} f^n(M_i) \right)$. Moreover, f restricts to an automorphism of M''.*

(b) *If M_R is indecomposable, then $\operatorname{End}(M_R)$ is a local ring.*

Proof. (a) For every $i \geq 0$ there is a positive integer n_i such that for every $j \geq n_i$

$$f^j(M_i) = f^{n_i}(M_i) \quad \text{and} \quad \ker(f^j) \cap M_i = \ker(f^{n_i}) \cap M_i.$$

By Lemmas 2.16(a) and 2.17(a)

$$M_i = \left(\ker(f^j) \cap M_i \right) \oplus f^j(M_i),$$

so that

$$M_i = \left(\bigcup_{n \geq 0} \ker(f^n) \cap M_i \right) \oplus \left(\bigcap_{n \geq 0} f^n(M_i) \right).$$

Further, f restricted to

$$M_i'' = \bigcap_{n \geq 0} f^n(M_i)$$

is a monomorphism, hence an automorphism of M_i''. Now (a) follows easily.

(b) If M_R is indecomposable, either $M_R = M'$ or $M_R = M''$. Hence for every $f \in \mathrm{End}(M_R)$, either $M_R = \bigcup_{n \geq 0} \ker(f^n)$ or f is an automorphism. This shows that if f is a non-invertible element of $\mathrm{End}(M_R)$, the restriction of f to any M_i is nilpotent. Now argue as in the second paragraph of the proof of Lemma 2.21 to show that the sum of two non-invertible elements of $\mathrm{End}(M_R)$ is non-invertible. Thus $\mathrm{End}(M_R)$ is local. $\square$

In the next proposition we prove that if the base ring R is either right noetherian or commutative, then all direct sum decompositions of a module that is a direct sum of artinian modules have an isomorphic common refinement. Hence the Krull-Schmidt Theorem holds for artinian modules over such rings.

Proposition 2.63 *Let R be a ring which is either right noetherian or commutative and let $M = \oplus_{i \in I} M_i$ be a right R-module which is the direct sum of indecomposable artinian modules M_i. Then any direct sum decomposition of M refines into a decomposition isomorphic to the decomposition $M = \oplus_{i \in I} M_i$ and any direct summand N of M is isomorphic to $\oplus_{i \in J} M_i$ for a subset $J \subseteq I$.*

Proof. Let A be an artinian right module over a ring R that is either right noetherian or commutative. By Proposition 2.59 and Lemma 2.61 the module A has Loewy length $\leq \omega$, so that $A = \mathrm{soc}_\omega(A) = \bigcup_{n \in \mathbb{N}} \mathrm{soc}_n(A)$. Every $\mathrm{soc}_n(A)$ is an artinian module of Loewy length $\leq n$. Since $\mathrm{soc}_{n+1}(A)/\mathrm{soc}_n(A)$ is a semisimple artinian module, every $\mathrm{soc}_n(A)$ is a module of finite composition length. By Lemma 2.62 every indecomposable artinian module A has a local endomorphism ring and is countably generated. Now apply Corollary 2.55. $\square$

2.13   Notes on Chapter 2

The exchange property was introduced by [Crawley and Jónsson]. Actually, Crawley and Jónsson's results were proved for a wide class of algebraic structures, namely for algebras in the sense of Jónsson-Tarski. Injective modules [Warfield 69c], quasi-injective modules [Fuchs 69], pure-injective modules [Zimmermann-Huisgen and Zimmermann 84], continuous modules (Mohamed and Müller, 1989), projective modules over perfect rings (Yamagata, 1974, and Harada-Ishii, 1975), and projective modules over Von Neumann regular rings [Stock] have the exchange property. It is not known whether the exchange property and the finite exchange property are equivalent for arbitrary modules. By Theorem 2.8 they are equivalent for indecomposable modules.

A module M is *continuous* if the following two conditions hold:

(C1) every submodule of M is essential in a direct summand of M;

(C2) if a submodule N of M is isomorphic to a direct summand of M, then N is a direct summand of M.

A module M is *quasi-continuous* if (C1) holds, and moreover

(C3) if N_1 and N_2 are direct summands of M such that $N_1 \cap N_2 = 0$, then $N_1 \oplus N_2$ is a direct summand of M.

Making use of ideas of [Oshiro and Rizvi], [Mohamed and Müller] have recently proved that the exchange property and the finite exchange property are equivalent for quasi-continuous modules. Note that there exist indecomposable quasi-continuous modules without the finite exchange property, for instance the abelian group **Z**.

The proofs of Lemma 2.2, Corollary 2.3, Lemmas 2.4 and 2.5 and Theorems 2.9 and 2.10 are taken from [Crawley and Jónsson]. In the proof of Theorem 2.8 the implication (b) ⇒ (c) is taken from [Crawley and Jónsson] and the remaining implications are due to [Warfield 69a, Proposition 1].

The history of the Krull-Schmidt-Remak-Azumaya Theorem begins with two papers of [Krull 25] and [Schmidt]. The present form of the theorem appeared for the first time in [Azumaya 50]. In that paper Azumaya proved the uniqueness of decomposition for infinite direct sums of modules with local endomorphism rings. In this book, the general result (Theorem 2.12) is referred to as the "Krull-Schmidt-Remak-Azumaya Theorem", whereas the "Krull-Schmidt Theorem" is the "classical" Krull-Schmidt Theorem, that is, the result concerning modules of finite length (Corollary 2.23). Krull himself used to term "Isomorphiesatz der direkte Zerlegung" (Isomorphism theorem of direct decomposition) for what we call the Krull-Schmidt Theorem. In [Krull 32] (last paragraph of the paper), Krull asked whether the "Isomorphiesatz der direkte Zerlegung" is independent of the descending chain condition, i.e., whether the Krull-Schmidt Theorem holds for artinian modules (cf. [Levy, p. 660]). The answer to this question appeared in [Facchini, Herbera, Levy and Vámos] and is the main topic of Section 8.2. The proofs of Lemma 2.11 and the Krull-Schmidt-Remak-Azumaya Theorem we have given here are taken from [Crawley and Jónsson].

Lemmas 2.20 and 2.21 are due to [Fitting, Satz 8], and Corollary 2.27 is due to [Zimmermann and Zimmermann-Huisgen 78, Theorem 9]. The proof of Proposition 2.28 is based on an argument of [Eisenbud and Griffith, Proof of Proposition 1.1].

The Goldie dimension for modules and rings was introduced by [Goldie 60], who called it "dimension". The Goldie dimension of a module is also called the *uniform dimension*, or the *uniform rank*, or simply the *rank* of the module. Concepts such as having finite Goldie dimension or uniform submodules and their basic properties go back to [Goldie 58, 60]. Goldie dimension for arbitrary modular lattices was introduced by [Grzeszczuk and Puczyłowski]. For the proof of Lemma 2.30 and Proposition 2.31 we have followed [Năstăsescu and Van Oystaeyen]. The rest of the material in Section 2.6 is taken from [Grzeszczuk and Puczyłowski].

The notion of dual Goldie dimension is due to [Varadarajan], who used the term *corank* for what we call dual Goldie dimension of a module. There are a number of different ways that one could attempt to dualize the notion of Goldie dimension; for instance, [Fleury] considers the spanning dimension of a module, a possible different dualization of the Goldie dimension. The *spanning dimension* of a module M is defined as the least integer k such that M is a sum $N_1 + \cdots + N_k$ (not necessarily direct) of k couniform submodules N_i of M. In our presentation of dual Goldie dimension we have followed [Grzeszczuk and Puczyłowski].

Proposition 2.45 is essentially taken from [Fuchs and Salce, Lemma 24].

Theorem 2.51, Corollary 2.55, the definition of exchange ring and Theorem 2.56 are due to [Warfield 69c, 69a, 72]. He also proved that a right module M_R has the finite exchange property if and only if its endomorphism ring $\operatorname{End}(M_R)$ is an exchange ring [Warfield 72, Theorem 2]. From Lemma 2.4 it follows immediately that if e is an idempotent in a ring R, then R is an exchange ring if and only if eRe and $(1-e)R(1-e)$ are exchange rings. There are further characterizations of exchange rings. For instance, [Monk] proved that a ring R is an exchange ring if and only if for every $a \in R$ there exist $b, c \in R$ such that $bab = b$ and $c(1-a)(1-ba) = 1 - ba$. Goodearl ([Goodearl and Warfield 76, p. 167]) and [Nicholson] independently proved that a ring R is an exchange ring if and only if for every $x \in R$ there exists an idempotent $e \in R$ such that $e \in xR$ and $1 - e \in (1-x)R$. This characterization has allowed the notion of exchange ring to be extended to rings without unit [Ara 97]. [Nicholson] also proved that R is an exchange ring if and only if $R/J(R)$ is an exchange ring and idempotents lift modulo $J(R)$.

Corollary 2.57 is a famous result of [Kaplansky 58, Theorem 2].

Loewy started using Loewy series in 1905 in the study of representations of matrix groups. Later, in 1926, Krull defined the term "Loewy series" and in [Krull 28] he observed that transfinite Loewy series could be defined. The results in Section 2.12 (i.e., that the Krull-Schmidt Theorem holds for artinian modules over rings which are either right noetherian or commutative) are due to [Warfield 69a]. The most important case of artinian module over a commutative noetherian ring was discovered by [Matlis]. He proved that if R is a noetherian commutative ring, S is a simple R-module and $E(S)$ is the injective envelope of S, then $E(S)$ is an artinian R-module whose endomorphism ring is a local noetherian complete commutative ring [Matlis, Theorems 3.7 and 4.2]. Conversely, if M_R is an artinian module with simple socle over a commutative ring R, then $E = \operatorname{End}(M_R)$ is a local noetherian complete commutative ring and $_E M$ is the injective envelope of the unique simple E-module [Facchini 81,

Theorem 2.8]. The following results hold for Loewy modules over commutative rings. Let M be a Loewy module over a commutative ring R. For each ordinal α the α-*th Loewy factor* of M is the semisimple module $\text{soc}_{\alpha+1}(M)/\text{soc}_\alpha(M)$, and the composition length of $\text{soc}_{\alpha+1}(M)/\text{soc}_\alpha(M)$ is the α-*th Loewy invariant* of M, denoted by $d_\alpha(M)$. The *support* of M is the set of all maximal ideals P of R such that $(0 :_M P) \neq 0$. If R is a commutative ring and M is a Loewy R-module with finite support $\{P_1, P_2, \ldots, P_n\}$, then $M = \oplus_{i=1}^n M_i$, where for each $i = 1, 2, \ldots, n$, M_i is a Loewy module whose Loewy factors are all direct sums of copies of R/P_i. If M is a Loewy module over a commutative ring, α is an ordinal and r is a positive integer such that both $d_\alpha(M)$ and $d_{\alpha+r}(M)$ are finite, then $d_\beta(M)$ is finite for every $\beta > \alpha + r$ and $M = \text{soc}_{\alpha+\omega}(M)$ [Shores, Theorem 4.2]. From this result we again obtain that every artinian module over a commutative ring has Loewy length $\leq \omega$. A module M over a commutative ring R is artinian if and only if it is a Loewy module with finite Loewy invariants [Facchini 81, Theorem 2.7]. Let M be an artinian module over a commutative ring such that $d_1(M) \leq n$. Then $d_r(M) \leq \begin{pmatrix} n + r - 1 \\ r \end{pmatrix}$ for every $r \geq 1$ ([Shores, Theorem 4.4] and [Facchini 81, Theorem 3.1]). Now let t be an indeterminate over the ring $\mathbf{Z}$ of integers. If M is an artinian module over a commutative ring, define $P(M, t) = \sum_{n=0}^{\infty} d_n(M)t^n \in \mathbf{Z}[[t]]$. Then $P(M, t)$ is a rational function in t of the form $f(t)/(1 - t)^s$, where $f(t) \in \mathbf{Z}[t]$ and $s = d_0(M)d_1(M)$. If d is the order of the pole of $P(M, t)$ at $t = 1$, then, for all sufficiently large n, $d_n(M)$ and the composition length $l(\text{soc}_n(M))$ of $\text{soc}_n(M)$ are polynomials in n with rational coefficients of degree $d - 1$ and d respectively [Facchini 81, Theorem 3.2].

A ring R is *right semiartinian* if R_R is a Loewy module. If R is right semiartinian, every right R-module is a Loewy module. Right semiartinian rings are exchange rings [Baccella].

Chapter 3

Semiperfect Rings

3.1 Projective covers and lifting idempotents

A *projective cover* of a module M_R is an epimorphism $\varphi\colon P_R \to M_R$ from a projective module P_R onto M_R whose kernel $\ker\varphi$ is a superfluous submodule of P_R.

The notion of projective cover is dual to that of injective hull. Although each module has an injective hull, projective covers seldom exist. For instance, let R be a ring with $J(R) = 0$. Then $\mathrm{rad}(F) = 0$ for every free R-module F, so that $\mathrm{rad}(P) = 0$ for every projective R-module P. Since the radical is the sum of the superfluous submodules, every projective R-module does not contain non-zero superfluous submodules. Hence an R-module has a projective cover if and only if it is projective. Thus a **Z**-module, that is, an abelian group, has a projective cover if and only if it is free.

Proposition 3.1 *Let $\varphi\colon P_R \to M_R$ be a projective cover of a module M_R and let $\psi\colon Q_R \to M_R$ be an epimorphism with Q_R projective. Then*

(a) *There is a direct sum decomposition $Q = Q' \oplus Q''$ of Q with $Q' \cong P$, $\psi|_{Q'}\colon Q' \to M_R$ a projective cover, and $\psi|_{Q''}\colon Q'' \to M_R$ the zero homomorphism.*

(b) *If $\psi\colon Q_R \to M_R$ is a projective cover, then there exists an isomorphism $f\colon Q_R \to P_R$ such that $\varphi \circ f = \psi$.*

Proof. Since Q is a projective module and $\varphi\colon P \to M$ is surjective, there is a mapping $f\colon Q \to P$ such that $\varphi \circ f = \psi$. But ψ is surjective, so that $\varphi(f(Q)) = \psi(Q) = M$. It follows that $f(Q) + \ker\varphi = P$. Now $\ker\varphi$ is superfluous in P, and therefore $f(Q) = P$. This shows that $f\colon Q \to P$ is surjective. Since P is projective, $\ker f$ is a direct summand of Q, that is, there is a direct sum decomposition $Q = Q' \oplus \ker(f)$ and $f|_{Q'}\colon Q' \to P$ an isomorphism. Using the fact that $\varphi\colon P_R \to M_R$ is a projective cover and $f|_{Q'}\colon Q' \to P$ is an isomorphism, it is obvious that $\psi|_{Q'} = \varphi \circ f|_{Q'}\colon Q' \to M_R$ is a projective cover. Finally $\psi(\ker f) = \varphi(f(\ker f)) = 0$. This proves (a).

Now suppose that ψ is a projective cover. Then $\ker(\psi)$ is superfluous in Q. Since $Q = Q' + \ker(f) \subseteq Q' + \ker(\psi)$, we have that $Q' = Q$. Therefore $f\colon Q \to P$ is the required isomorphism. $\square$

The proof of the next proposition is left to the reader.

Proposition 3.2 *If $\varphi_i\colon P_i \to M_i$ is the projective cover of an R-module M_i for every $i = 1, 2, \ldots, n$, then $\varphi_1 \oplus \cdots \oplus \varphi_n\colon P_1 \oplus \cdots \oplus P_n \to M_1 \oplus \cdots \oplus M_n$ is a projective cover of $M_1 \oplus \cdots \oplus M_n$.* $\square$

We shall apply the previous proposition to the study of the direct summands of R_R, that is, to the right ideals eR, e an idempotent. If $e \in R$ is an idempotent and $J(R)$ is the Jacobson radical of R, the canonical projection $\pi\colon eR \to eR/eJ(R)$ is a projective cover, because by Nakayama's Lemma (Corollary 1.4) the right ideal $eJ(R)$ is a superfluous submodule of eR.

Proposition 3.3 *Let e and f be two idempotents in a ring R and let $J(R)$ be the Jacobson radical of R. Then*

(a) *The right R-modules eR and fR are isomorphic if and only if the the left R-modules Re and Rf are isomorphic.*

(b) *The right R-modules eR and fR are isomorphic if and only if the the right R-modules $eR/eJ(R)$ and $fR/fJ(R)$ are isomorphic.*

Proof. (a) If the right R-modules eR and fR are isomorphic, their duals $\mathrm{Hom}_R(eR, R_R)$ and $\mathrm{Hom}_R(fR, R_R)$ are isomorphic left R-modules. But $\mathrm{Hom}_R(eR, R_R) \cong Re$ and $\mathrm{Hom}_R(fR, R_R) \cong Rf$.

(b) It is clear that every isomorphism $eR \to fR$ induces an isomorphism $eR/eJ(R) \to fR/fJ(R)$. Conversely, if $\alpha\colon eR/eJ(R) \to fR/fJ(R)$ is an isomorphism and $\pi_1\colon eR \to eR/eJ(R)$, $\pi_2\colon fR \to fR/fJ(R)$ are the canonical projections, then $\alpha \circ \pi_1\colon eR \to fR/fJ(R)$ and $\pi_2\colon fR \to fR/fJ(R)$ are two projective covers of $fR/fJ(R)$. Proposition 3.1(b) yields an isomorphism $eR \cong fR$. $\square$

Let I be an ideal in a ring R. We say that *idempotents can be lifted modulo I* if for every idempotent $e \in R/I$ there is an idempotent $f \in R$ such that $e = f + I$. Herstein proved that if I is a nil ideal then idempotents can be lifted modulo I. For a proof of this result and the following proposition see [Anderson and Fuller, §27].

Proposition 3.4 *Let $I \subseteq J(R)$ be an ideal of a ring R. Assume idempotents can be lifted modulo I. If $\{e_1, \ldots, e_n\}$ is a set of orthogonal idempotents of R/I, then there exists a set $\{f_1, \ldots, f_n\}$ of orthogonal idempotents of R such that $e_i = f_i + I$ for every i.* $\square$

As an application of lifting idempotents we present the following theorem due to [Warfield 72, Theorem 3].

Theorem 3.5 *Let $J(R)$ be the Jacobson radical of a ring R. Suppose that $R/J(R)$ is von Neumann regular and idempotents can be lifted modulo $J(R)$. Then R is an exchange ring.*

Proof. Since R_R is finitely generated, it suffices to prove that R_R has the finite exchange property. By Lemma 2.5 it is enough to show that R_R has the 2-exchange property. Suppose that $G = F \oplus N = A \oplus B$, where $F \cong R_R$. Fix an isomorphism $F \to R_R$, and let g be the generator of F corresponding to the generator 1 of R_R. Let $\pi_F: G \to F$ and $\iota_F: F \to G$ be the projection and the inclusion relative to the decomposition $G = F \oplus N$, and let

$$\varepsilon_A, \varepsilon_B \in \mathrm{End}(G_R)$$

be the orthogonal idempotent endomorphisms of G_R relative to the decomposition $G = A \oplus B$, so that $\varepsilon_A + \varepsilon_B = 1_G$, $\varepsilon_A G = A$ and $\varepsilon_B G = B$. Identify R and $\mathrm{End}(F_R)$ and use the same symbol t for an element $t \in R$ and the endomorphism of F given by $gx \mapsto gtx$. Since $\pi_F \varepsilon_A \iota_F \in \mathrm{End}(F_R) \cong R$, there exists $a \in R$ such that $\pi_F \varepsilon_A \iota_F = a$, that is, $\pi_F \varepsilon_A \iota_F(gx) = gax$ for every $x \in R$. Then $\pi_F \varepsilon_B \iota_F(gx) = \pi_F(1_G - \varepsilon_A)\iota_F(gx) = gx - gax = g(1-a)x$ for every $x \in R$, i.e., $\pi_F \varepsilon_B \iota_F = 1 - a$. Let $\overline{R}$ denote $R/J(R)$, and for every $x \in R$ let $\overline{x}$ denote $x + J(R)$. As $\overline{R}$ is von Neumann regular, the principal left ideal $\overline{R}\overline{a}$ of $\overline{R}$ is a direct summand of $\overline{R}$ [Anderson and Fuller, p. 175]. But idempotents can be lifted modulo $J(R)$, hence there exists an idempotent $e \in R$ such that $\overline{R}\overline{a} = \overline{R}\overline{e}$. In particular, there exists $r \in R$ and $j \in J(R)$ such that $e = ra + j$.

Since $j \in J(R)$, the element $1 - eje = 1 - e + erae$ is invertible in R. Let u be its inverse and $w = euer$, so that $e = eu(1 - e + erae)e = euerae = wae$ and $ew = w$. Now $\overline{R}\overline{a} = \overline{R}\overline{e}$, so that right multiplication by $\overline{e}$ is the identity on $\overline{R}\overline{a}$, in particular $\overline{a}\overline{e} = \overline{a}$, hence $\overline{e} = \overline{w}\overline{a}$. Then

$$\overline{(1-a)(1-e)} = \overline{(1-a) - e + ae} = \overline{1 - a - e + a} = \overline{1 - e},$$

so that

$$\overline{(1-e)(1-a)(1-e) + e} = \overline{1}.$$

Thus $(1-e)(1-a)(1-e) + e$ is invertible in R. Let $v \in R$ be its inverse. Then $1 - e = (1-e)v((1-e)(1-a)(1-e)+e)(1-e) = (1-e)v(1-e)(1-a)(1-e)$. An elementary direct calculation shows that $1 + (1-e) - (1-e)v(1-e)(1-a)$ is a two-sided inverse of $e + (1-e)v(1-e)(1-a)$, so that $e + (1-e)v(1-e)(1-a)$ is invertible in R. Therefore

$$\overline{e + (1-e)v(1-e)(1-a)} = \overline{wa + (1-e)v(1-e)(1-a)}$$

is invertible in $\overline{R}$, from which we deduce that $wa + (1-e)v(1-e)(1-a)$ is invertible in R.

Define two endomorphisms

$$\sigma = \varepsilon_A \iota_F w \pi_F \varepsilon_A \quad \text{and} \quad \tau = \varepsilon_B \iota_F (1-e) v (1-e) \pi_F \varepsilon_B$$

of G_R. Then σ is an idempotent endomorphism of G_R because

$$\sigma^2 = \varepsilon_A \iota_F w \pi_F \varepsilon_A \iota_F w \pi_F \varepsilon_A = \varepsilon_A \iota_F w a w \pi_F \varepsilon_A$$
$$= \varepsilon_A \iota_F w a e w \pi_F \varepsilon_A = \varepsilon_A \iota_F w \pi_F \varepsilon_A = \sigma.$$

Similarly, τ is idempotent.

Note that $\sigma(G) \subseteq A$, so that $A = \sigma(G) \oplus A'$ where $A' = A \cap \ker \sigma$ (Lemma 2.1). Similarly, $B = \tau(G) \oplus B'$ where $B' = B \cap \ker \tau$, hence

$$G = \sigma(G) \oplus \tau(G) \oplus A' \oplus B',$$

and with respect to this decomposition it is easily seen that the canonical projection $\pi_2 \colon G \to \sigma(G) \oplus \tau(G)$ is defined by $\pi_2(g) = \sigma(g) + \tau(g)$ for every $g \in G$. In order to conclude the proof it is sufficient to show that $G = F \oplus A' \oplus B'$, and for this it suffices to prove that $\pi_2 \iota_F \colon F \to \sigma(G) \oplus \tau(G)$ is an isomorphism (Lemma 2.6).

Let $f \colon \sigma(G) \oplus \tau(G) \to F$ be the homomorphism defined by

$$f = (w \pi_F \sigma + (1-e) v (1-e) \pi_F \tau)|_{\sigma(G) \oplus \tau(G)}.$$

Then

$$f \pi_2 \iota_F = w \pi_F \sigma \iota_F + (1-e) v (1-e) \pi_F \tau \iota_F$$
$$= w \pi_F \varepsilon_A \iota_F w \pi_F \varepsilon_A \iota_F$$
$$+ (1-e) v (1-e) \pi_F \varepsilon_B \iota_F (1-e) v (1-e) \pi_F \varepsilon_B \iota_F$$
$$= w a w a + (1-e) v (1-e)(1-a)(1-e) v (1-e)(1-a)$$
$$= w a + (1-e) v (1-e)(1-a),$$

which is invertible in R as we have already seen. Thus $f \pi_2 \iota_F$ is an automorphism of F. In particular $f \colon \sigma(G) \oplus \tau(G) \to F$ is surjective.

We shall now show that f is injective. Let $g, g' \in G$ be such that

$$f(\sigma(g) + \tau(g')) = 0.$$

Then $\sigma(g) = \varepsilon_A \iota_F w \pi_F \varepsilon_A(g) = \varepsilon_A(x)$ for $x = w \pi_F \varepsilon_A(g) \in F$. Hence $x = ex$. Similarly, $\tau(g') = \varepsilon_B(x')$ for some $x' = (1-e)x' \in F$. Then

$$0 = f(\sigma(g) + \tau(g'))$$
$$= (w \pi_F \sigma + (1-e) v (1-e) \pi_F \tau)|_{\sigma(G) \oplus \tau(G)} (\sigma(g) + \tau(g'))$$
$$= w \pi_F \sigma(g) + (1-e) v (1-e) \pi_F \tau(g')$$
$$= w \pi_F \varepsilon_A(x) + (1-e) v (1-e) \pi_F \varepsilon_B(x')$$
$$= w \pi_F \varepsilon_A \iota_F(x) + (1-e) v (1-e) \pi_F \varepsilon_B \iota_F(x')$$
$$= w a x + (1-e) v (1-e)(1-a) x'$$
$$= w a e x + (1-e) v (1-e)(1-a)(1-e) x' = e x + (1-e) x',$$

so that $x = ex = 0$ and $x' = (1 - e)x' = 0$. Thus $\sigma(g) = \varepsilon_A(x) = 0$ and $\tau(g') = \varepsilon_B(x') = 0$. This proves that f is injective.

Since f and $f\pi_2\iota_F$ are isomorphisms, it follows that $\pi_2\iota_F$ is an isomorphism, which concludes the proof. □

In particular, every von Neumann regular ring is an exchange ring.

If R is a ring satisfying the hypotheses of Theorem 3.5, then any projective right R-module is a direct sum of right ideals generated by idempotents (Theorem 2.56). As a special case, we get that every projective right module over a von Neumann regular ring is a direct sum of principal right ideals.

3.2 Semiperfect rings

The aim of this section is to introduce the class of semiperfect rings. We prove the following theorem.

Theorem 3.6 *Let R be a ring. The following conditions are equivalent:*

(a) *The ring R is semilocal and idempotents can be lifted modulo $J(R)$.*

(b) *The right R-module R_R is a direct sum of local modules.*

(c) *The ring R has a complete set $\{e_1, \ldots, e_n\}$ of orthogonal idempotents for which every $e_i R e_i$ is a local ring.*

(d) *Every simple right R-module has a projective cover.*

(e) *Every finitely generated right R-module has a projective cover.*

Proof. (a) $\Rightarrow$ (b). Let R be a semilocal ring such that idempotents can be lifted modulo $J(R)$. Since $R/J(R)$ is a semisimple ring, there exists a complete set $\{f_1, \ldots, f_n\}$ of primitive orthogonal idempotents of $R/J(R)$. By Proposition 3.4 there exists a set $\{e_1, \ldots, e_n\}$ of orthogonal idempotents of R such that $f_i = e_i + J(R)$ for every i. Then $e_1 + \cdots + e_n + J(R) = 1 + J(R)$ and $e_1 + \cdots + e_n$ is an idempotent of R, so that $1 - e_1 - \cdots - e_n$ is an idempotent of R that belongs to the Jacobson radical $J(R)$. Since the Jacobson radical never contains non-zero idempotents, it follows that $e_1 + \cdots + e_n = 1$. Hence $\{e_1, \ldots, e_n\}$ is a complete set of orthogonal idempotents in R, i.e.,

$$R_R = e_1 R \oplus \cdots \oplus e_n R.$$

The canonical projection $e_i R \rightarrow f_i(R/J(R))$ is an epimorphism of the projective cyclic module $e_i R$ onto the simple module $f_i(R/J)$ and its kernel is $e_i R \cap J(R) = e_i J(R) = \mathrm{rad}(e_i R)$. Therefore $\mathrm{rad}(e_i R)$ is a maximal submodule of $e_i R$. Hence $\mathrm{rad}(e_i R)$ is the greatest proper submodule of $e_i R$ and $e_i R$ is local.

(b) $\Rightarrow$ (c). This follows at once from Proposition 1.11.

(c) $\Rightarrow$ (d). Suppose that R has a complete set $\{e_1, \ldots, e_n\}$ of orthogonal idempotents such that each $e_i R e_i$ is a local ring. By Proposition 1.11 each $e_i R$ is a local module, so that every proper submodule of $e_i R$ is superfluous. If M_R is a simple module, then there is an epimorphism $R_R = \oplus_{i=1}^n e_i R \to M_R$. Hence there exists an index $j = 1, 2, \ldots, n$ and a non-zero homomorphism $\varphi \colon e_j R \to M_R$. Thus $\ker \varphi$ is superfluous in $e_j R$, and φ is surjective because M_R is simple. This shows that φ is a projective cover of M_R.

(d) $\Rightarrow$ (e). Let M_R be a finitely generated R-module. Let $\mathcal{S}$ be a set of representatives of all simple right R-modules up to isomorphism. For each $S \in \mathcal{S}$ let $\varphi_S \colon P_S \to S$ be a projective cover of S. Then $\ker \varphi_S$ is a small maximal submodule of P_S, hence each P_S is a local projective module. Set $P = \oplus_{S \in \mathcal{S}} P_S$. Let $Tr_M(P)$ be the *trace* of P in M, that is, $Tr_M(P) = \sum_{f \in \mathrm{Hom}_R(P,M)} f(P)$. Then $Tr_M(P) = M$, because if $Tr_M(P) \neq M$, there is a maximal submodule N of the finitely generated module M containing $Tr_M(P)$. Since M/N is simple, there is an epimorphism $P \to M/N$. But P is projective, and so there is a homomorphism $P \to M$ whose image is not contained in N. This contradiction proves that $Tr_M(P) = M$. Since M is finitely generated, there are a finite number $P_1, \ldots, P_n$ of local projective modules and an epimorphism $P_1 \oplus \cdots \oplus P_n \to M$. This induces an epimorphism $P_1/P_1 J(R) \oplus \cdots \oplus P_n/P_n J(R) \to M/MJ(R)$. For a local projective module P_i the maximal submodule is $\mathrm{rad}(P_i) = P_i J(R)$, so that $P_i/P_i J(R)$ is a simple module. Therefore $P_1/P_1 J(R) \oplus \cdots \oplus P_n/P_n J(R)$ is a semisimple module of finite length, and therefore there exists a finite subfamily of $P_1, \ldots, P_n$ (without loss of generality we may suppose that it is $P_1, \ldots, P_m$ with $m \leq n$) and a homomorphism $\psi \colon P_1 \oplus \cdots \oplus P_m \to M$ that induces an isomorphism $P_1/P_1 J(R) \oplus \cdots \oplus P_m/P_m J(R) \to M/MJ(R)$. By Nakayama's Lemma 1.4 the homomorphism ψ is surjective. The kernel of ψ is contained in $P_1 J(R) \oplus \cdots \oplus P_m J(R) = (P_1 \oplus \cdots \oplus P_m)J(R)$, and therefore it is superfluous again by Nakayama's Lemma. Therefore ψ is a projective cover of M.

(e) $\Rightarrow$ (a). Let e be an idempotent of $R/J(R)$. Then the cyclic right R-modules $e(R/J(R))$ and $(1-e)(R/J(R))$ have projective covers

$$\varphi_1 \colon P_1 \to e(R/J(R)) \quad \text{and} \quad \varphi_2 \colon P_2 \to (1-e)(R/J(R)).$$

By Proposition 3.2 the homomorphism

$$\varphi_1 \oplus \varphi_2 \colon P_1 \oplus P_2 \to e(R/J(R)) \oplus (1-e)(R/J(R))$$

is a projective cover. Since $J(R)$ is superfluous in R_R, the canonical projection $\pi \colon R_R \to R/J(R)$ is also a projective cover. By Proposition 3.1(b) there is an idempotent $f \in R$ such that

$$\pi(fR) = e(R/J(R)) \quad \text{and} \quad \pi((1-f)R) = (1-e)(R/J(R)).$$

Then

$$(f + J(R))(R/J(R)) = e(R/J(R))$$

and

$$(1 - f + J(R))(R/J(R)) = (1 - e)(R/J(R)),$$

from which $f + J(R) = e$. This shows that idempotents can be lifted modulo $J(R)$.

In order to prove that R is semilocal, we must show that every submodule $I/J(R)$ of $R/J(R)$ is a direct summand, where I is an arbitrary right ideal of R containing $J(R)$. Since the cyclic module R/I has a projective cover, if

$$\pi \colon R \to R/I$$

is the canonical projection, then there is a direct sum decomposition

$$R_R = eR \oplus (1 - e)R$$

such that $\pi|_{eR} \colon eR \to R/I$ is a projective cover (Proposition 3.1), where e is an idempotent element of R. Let us prove that $\ker(\pi|_{eR}) = eJ(R)$. One the one hand, $\pi|_{eR}(eJ(R)) = eJ(R) + I/I = 0$ because $I \supseteq J(R)$. On the other hand, $\ker(\pi|_{eR})$ is a superfluous submodule of eR, so that

$$\ker(\pi|_{eR}) \subseteq \operatorname{rad}(eR) = eJ(R).$$

Thus $\ker(\pi|_{eR}) = eJ(R)$ and $R/I \cong eR/eJ(R) \cong (e + J(R))(R/J(R))$ is a projective $R/J(R)$-module. Therefore the kernel $I/J(R)$ of the canonical epimorphism $R/J(R) \to R/I$ is a direct summand of $R/J(R)$. $\quad\square$

Note that conditions (a) and (c) of Theorem 3.6 are left-right symmetric, so that "right" can be replaced by "left" everywhere in the conditions of the statement of the theorem.

A ring R is a *right perfect ring* if every right R-module has a projective cover. A ring R is *semiperfect* if every finitely generated right (or left) R-module has a projective cover, that is, if it satisfies the equivalent conditions of Theorem 3.6. For example, local rings, right artinian rings and right serial rings are semiperfect. Semiperfect rings are exchange rings by Theorem 3.5.

Example 3.7 *If R is a semiperfect ring, then $\mathbf{M}_n(R)$ is a semiperfect ring for every n.*

This follows immediately from condition (c) in Theorem 3.6. $\quad\square$

Example 3.8 Let $D_1, \ldots, D_n$ be division rings and let V_{ij} be a D_i-D_j-bimodule for every $i, j = 1, \ldots, n$, $i \neq j$. Let R be the ring of $n \times n$ matrices

$$\begin{pmatrix} D_1 & V_{12} & \cdots & V_{1n} \\ V_{21} & D_2 & \cdots & V_{2n} \\ \vdots & & \ddots & \vdots \\ V_{n1} & V_{n2} & \cdots & D_n \end{pmatrix},$$

where $V_{ij}V_{jk} = 0$ for every $i \neq j$ and $j \neq k$. Then R is a semiperfect ring by Theorem 3.6(c). Note that

$$J(R) = \begin{pmatrix} 0 & V_{12} & \dots & V_{1n} \\ V_{21} & 0 & \dots & V_{2n} \\ \vdots & & \ddots & \vdots \\ V_{n1} & V_{n2} & \dots & 0 \end{pmatrix},$$

so that $J(R)^2 = 0$. $\qquad\qquad\square$

3.3   Modules over semiperfect rings

Let R be a semiperfect ring. The ring R has a complete set $\{e_1, \dots, e_n\}$ of orthogonal idempotents such that each $e_i R e_i$ is a local ring. Equivalently, the projective right R-modules $e_i R$ are local (Proposition 1.11). Therefore their maximal submodule is their radical $\mathrm{rad}(e_i R) = e_i J(R)$, the module $e_i R / e_i J(R)$ is simple for every i, and

$$R/J(R) = e_1 R / e_1 J(R) \oplus \cdots \oplus e_n R / e_n J(R).$$

Given an arbitrary simple R-module S_R, S_R is a simple $R/J(R)$-module in a natural way, so that there exists an index i such that $S_R \cong e_i R / e_i J(R)$ (Proposition 2.15). Hence every simple right R-module is isomorphic to one of the form $e_i R / e_i J(R)$. In the paragraph before Proposition 3.3 we have already remarked that the canonical projection $\pi \colon e_i R \to e_i R / e_i J(R)$ is the projective cover of $e_i R / e_i J(R)$. This notation will be used in the proof of the following proposition.

Proposition 3.9 *Let* $\varphi \colon P_R \to M_R$ *be a projective cover of a finitely generated module* M_R *over a semiperfect ring* R. *Then* φ *induces an isomorphism* $P/PJ(R) \to M/MJ(R)$.

Proof. By Lemma 1.14 the module $M/MJ(R)$ is a semisimple R-module of finite length, that is, it is isomorphic to a finite direct sum of modules $e_i R / e_i J(R)$. Let P'_R be the finite direct sum of the corresponding modules $e_i R$. Since P'_R is projective, the isomorphism $\oplus_i e_i R / e_i J(R) \to M/MJ(R)$ is induced by a homomorphism $\varphi \colon P'_R \to M$. The projective cover is uniquely determined up to isomorphism (Proposition 3.1(b)), so the proposition will be proved if we show that $\varphi \colon P'_R \to M$ is a projective cover. Since φ induces an isomorphism $P'_R / P' J(R) \to M/MJ(R)$, it follows that $\ker \varphi \subseteq P' J(R)$ and $\varphi(P') + MJ(R) = M$. But $P' J(R)$ is superfluous in P', so that $\ker \varphi$ is superfluous in P'. And $\varphi(P') + MJ(R) = M$ implies $\varphi(P') = M$ by Nakayama's Lemma, and therefore φ is a surjective mapping. This proves that $\varphi \colon P'_R \to M$ is a projective cover. $\qquad\square$

If R is a semiperfect ring, a module M_R is said to be *primitive* if there is a primitive idempotent $e \in R$ such that $M_R \cong eR$. A set $\{e_1, \ldots, e_m\}$ of idempotents of R is *basic* in case the e_i are pairwise orthogonal and $e_1 R, \ldots, e_m R$ is a set of representatives of the primitive right R-modules up to isomorphism. Note that a set $\{e_1, \ldots, e_m\}$ of idempotents of R is a basic set of primitive idempotents of R if and only if $\{e_1 + J(R), \ldots, e_m + J(R)\}$ is a basic set of primitive idempotents of $R/J(R)$, if and only if $\{e_1 R/e_1 J(R), \ldots, e_m R/e_m J(R)\}$ is a set of representatives of the simple right R-modules up to isomorphism. In particular, every semiperfect ring has a basic set of primitive idempotents.

Theorem 3.10 *Let $\{e_1, \ldots, e_m\}$ be a basic set of primitive idempotents of a semiperfect ring R. Then*

(a) *$\{e_1 R, \ldots, e_m R\}$ is a set of representatives of the indecomposable projective right R-modules up to isomorphism.*

(b) *For every projective R-module P_R there exist sets $A_1, \ldots, A_m$, uniquely determined up to cardinality, such that $P_R \cong e_1 R^{(A_1)} \oplus \cdots \oplus e_m R^{(A_m)}$.*

Proof. The semiperfect ring R has a complete set of orthogonal idempotents $\{f_1, \ldots, f_n\}$ for which every $f_i R f_i$ is a local ring (Theorem 3.6). The right ideals $f_i R$ are primitive. Hence for every f_i there exists an e_j with $f_i R \cong e_j R$. Thus R_R is isomorphic to the direct sum of a family of principal right ideals $e_j R$, each of which has a local endomorphism ring isomorphic to $e_j R e_j$. The same property holds for any free right R-module. Thus (b) follows from Corollary 2.55 and the Krull-Schmidt-Remak-Azumaya Theorem 2.12. Finally, (a) follows immediately from (b). □

Example 3.11 *Let R be a right serial ring. A right R-module is an indecomposable projective R-module if and only if it is primitive, and all such modules are uniserial.* □

In this example we have given a description of all indecomposable projective right modules over a particular ring, in this case a right serial ring. Whenever we have a description of the finitely generated projective right modules or any result about the finitely generated projective right modules over a ring, it is possible to obtain a dual result about the finitely generated projective left modules over that ring. The technique is the following one.

Given a ring S, let Mod-S denote the category of all right S-modules, proj-S the full subcategory of Mod-S whose objects are all finitely generated projective right S-modules, S-Mod the category of all left S-modules, and S-proj the full subcategory of S-Mod whose objects are all finitely generated projective left S-modules.

Proposition 3.12 *The contravariant functors*

$$\mathrm{Hom}_S(-, S)\colon \mathrm{Mod}\text{-}S \to S\text{-Mod} \quad and \quad \mathrm{Hom}_S(-, S)\colon S\text{-Mod} \to \mathrm{Mod}\text{-}S$$

induce a duality between the full subcategory proj-S *of* Mod-S *and the full subcategory* S-proj *of* S-Mod. *Hence the category* S-proj *is equivalent to the the opposite category of* proj-S.   □

The proof of the proposition follows immediately from the fact that the functors $\mathrm{Hom}_S(-, S)$ are additive, hence they preserve finite direct sums. Note that if $e \in S$ is idempotent, the object eS of proj-S corresponds to the the object Se of S-proj.

For instance, if S is a ring such that every finitely generated projective right S-module is free, then every finitely generated projective left S-module also is free. From 3.11 it follows that if R is a right serial ring, a finitely generated left R-module is an indecomposable projective R-module if and only if it is primitive, and every finitely generated projective left R-module is a direct sum of primitive left R-modules.

Example 3.13 *An artinian right serial ring that is not left serial.*

Let F, K be fields, where K is an extension of F, and consider the ring

$$R = \begin{pmatrix} F & K \\ 0 & K \end{pmatrix}.$$

Then e_{11}, e_{22} is a complete set of orthogonal primitive idempotents of R, so that $R_R = e_{11}R \oplus e_{22}R$. It is easy to check that the unique proper non-zero submodule of

$$e_{11}R = \begin{pmatrix} F & K \\ 0 & 0 \end{pmatrix}$$

is

$$\begin{pmatrix} 0 & K \\ 0 & 0 \end{pmatrix},$$

so that $e_{11}R$ is a uniserial module of composition length 2. The module

$$e_{22}R = \begin{pmatrix} 0 & 0 \\ 0 & K \end{pmatrix}$$

is simple. Hence R is a right artinian right serial ring.

If $K = F$, then R is the ring of 2×2 upper triangular matrices over F, so that R is a serial ring (Example 1.21).

Suppose $K \supset F$. If R is left serial, then the primitive projective left R-module

$$Re_{22} = \begin{pmatrix} 0 & K \\ 0 & K \end{pmatrix}$$

must be serial by the left-right dual of 3.11. But if A is any vector subspace of the F-vector space K, it is easily seen that

$$\begin{pmatrix} 0 & A \\ 0 & 0 \end{pmatrix}$$

is a left ideal of R, so that Re_{22} is not uniserial. This proves that R is not left serial whenever $K \supset F$.

Finally, suppose that K is a proper extension of finite degree of F. Then R is a finite dimensional F-algebra, hence R is artinian. Thus R is the required example in this case.

Note that if K is an extension of F of infinite degree, then Re_{22} is a left ideal of R of infinite Goldie dimension. □

There are two main reasons why we have introduced the class of semiperfect rings. On the one hand this class of rings contains the class of right (or left) serial rings and is contained in the class of semilocal rings. On the other hand semiperfect rings are exactly the endomorphism rings of the modules that are finite direct sums of modules with local endomorphism rings, as the following proposition shows.

Proposition 3.14 *Let M_R be a module and let $E = \text{End}(M_R)$ be its endomorphism ring. The following conditions are equivalent:*

(a) *M_R is a finite direct sum of R-modules with local endomorphism rings.*

(b) *E_E is a finite direct sum of E-modules with local endomorphism rings.*

(c) *E is semiperfect.*

Proof. (a) $\Leftrightarrow$ (c) A module M_R is a finite direct sum of R-modules with local endomorphism rings if and only if there exists a complete set $\{e_1, \dots, e_n\}$ of orthogonal idempotents in $E = \text{End}(M_R)$ with $e_i E e_i$ a local ring for every $i = 1, \dots, n$. This is one of the equivalent conditions of Theorem 3.6.

(b) $\Leftrightarrow$ (c) is exactly (a) $\Leftrightarrow$ (c) applied to the module $M_R = E_E$. □

We conclude this section showing that a finitely generated module M over a semiperfect ring R decomposes uniquely as the direct sum of a projective part and a submodule with no non-zero projective direct summands. In the proof we shall use an invariant of such a module M. We have already remarked that if M is a finitely generated module over a semilocal ring R, then $M/MJ(R)$ is a semisimple module of finite length (Lemma 1.14). Define $\text{Gen}(M)$ to be the composition length of $M/MJ(R)$, that is, the number of summands in a decomposition of $M/MJ(R)$ as a direct sum of simple modules.

Theorem 3.15 *Let M be a finitely generated module over a semiperfect ring R. Then $M = N \oplus P$, where P is projective and N has no non-zero projective*

*summands. This decomposition is unique up to isomorphism, in the sense that
if $M = N' \oplus P'$ is another decomposition with P' projective and N' without
non-zero projective summands, then $N \cong N'$ and $P \cong P'$.*

Proof. We use induction on $\mathrm{Gen}(M)$, the case $\mathrm{Gen}(M) = 0$ being trivial. Suppose $\mathrm{Gen}(M) > 0$. Let $\{e_1, \ldots, e_n\}$ be a complete set of orthogonal idempotents
of R such that each $e_i R e_i$ is a local ring (Theorem 3.6). If M has no non-zero
projective summands, there is nothing to prove. If M has a non-zero projective
summand, then M has a decomposition $M = M' \oplus Q$ with $Q \cong e_i R$ for some
$i = 1, 2, \ldots, n$ (Theorem 3.10(b)). Since $\mathrm{Gen}(M') < \mathrm{Gen}(M)$, the induction
hypothesis applied to M', yields $M' = N \oplus P_1$, where P_1 is projective and N
has no non-zero projective summands. Then

$$M = N \oplus (P_1 \oplus Q)$$

is a decomposition of M of the required type. Let $M = N' \oplus P'$ be another decomposition with P' projective and N' without non-zero projective summands.
Since $\mathrm{End}(Q_R) \cong e_i R e_i$ is a local ring, we can apply Lemma 2.7 to the two decompositions $M = N \oplus P_1 \oplus Q = N' \oplus P'$ The module Q cannot be isomorphic
to a direct summand of N', so that $P' = D \oplus E$, $Q \cong E$ and $N \oplus P_1 \cong N' \oplus D$.
By the induction hypothesis applied to $M' = N \oplus P_1$ it follows that $N \cong N'$
and $P_1 \cong D$. Then $P_1 \oplus Q \cong D \oplus E = P'$. $\square$

3.4 Finitely presented and Fitting modules

The classical Krull-Schmidt Theorem (Corollary 2.23) says that every module
of finite length is the direct sum of a finite number of indecomposable modules
in an essentially unique way. We proved this result in two steps: firstly, we
showed that every module of finite length is a Fitting module (Lemma 2.20);
secondly, we proved that the endomorphism ring of an indecomposable Fitting
module is a local ring (Lemma 2.21), so that it is possible to apply the Krull-
Schmidt-Remak-Azumaya Theorem 2.12.

   If we look for an analogous result for finitely presented modules, it is
natural to try to follow the same idea and ask over which rings every finitely
presented module is a Fitting module and is a direct sum of indecomposable
modules. These rings will be characterized in Theorem 3.22. For the characterization we need the notion of strongly π-regular ring, which we introduce in
the first part of this section.

Theorem 3.16 *The following conditions on a ring R are equivalent:*

  (a) *R satisfies the d.c.c. on chains of the form $rR \supseteq r^2 R \supseteq r^3 R \supseteq \ldots$, that
      is, for every $r \in R$ there exist $n \geq 1$ and $s \in R$ such that $r^n = r^{n+1} s$;*

(b) R satisfies the d.c.c. on chains of the form $Rr \supseteq Rr^2 \supseteq Rr^3 \supseteq \ldots$, that is, for every $r \in R$ there exist $m \geq 1$ and $u \in R$ such that $r^m = ur^{m+1}$.

Proof. It suffices to show that (a) implies (b). Suppose that (a) holds. If $r \in R$, there exists $n \geq 1$ and $s \in R$ such that $r^n = r^{n+1}s$. Hence

$$r^n = r^{n+i}s^i \qquad (3.1)$$

for every $i \geq 1$. Since the chain $sR \supseteq s^2R \supseteq s^3R \supseteq \ldots$ is stationary, there exists $m \geq n$ such that $s^m R = s^{m+1}R$, i.e., $s^m = s^{m+1}t$ for some $t \in R$. Then $s^m = s^{m+j}t^j$ for every $j \geq 1$. In particular, if $b = s^m$ and $c = t^m$, then $b = b^2c$. From (3.1) we get that $r^n = r^{n+m}b$, and multiplying this by r^{m-n} on the left we obtain $r^m = r^{2m}b$. Set $a = r^m$, so that $a = a^2b$. Applying (a) again, there exists $k \geq 1$ and $d \in R$ such that $(c - a)^k = (c - a)^{k+1}d$.

Now

$$ac = a^2bc = a^3b^2c = a^3b = a^2 \quad \text{and} \quad abc = (a^2b)bc = a^2b = a, \qquad (3.2)$$

so that

$$(c - a)^2 = c^2 - ca - ac + a^2 = c^2 - ca = c(c - a). \qquad (3.3)$$

From (3.2) and (3.3) we obtain

$$ab(c - a)^2 = abc(c - a) = a(c - a) = ac - a^2 = 0. \qquad (3.4)$$

From (3.3) we have

$$b^2(c - a)^2 = b^2c(c - a) = b(c - a), \qquad (3.5)$$

from which $b^k(c - a)^k = b(c - a)$. Therefore

$$b(c - a)^2d = b^k(c - a)^{k+1}d = b^k(c - a)^k = b(c - a). \qquad (3.6)$$

From (3.4) and (3.6) it follows that

$$0 = ab(c - a)^2d = ab(c - a), \qquad (3.7)$$

so that $aba = abc = a$ by (3.2). Now (3.5), (3.6) and (3.7) imply that

$$0 = ab(c - a)d = ab^2(c - a)^2d = ab^2(c - a).$$

Hence $a = aba = ab^2ca = ab^2a^2$, that is, $r^m = (ab^2r^{m-1})r^{m+1}$. □

A ring R satisfying the equivalent conditions of Theorem 3.16 is called a *strongly π-regular ring*.

Lemma 3.17 *The Jacobson radical $J(R)$ is nil for every strongly π-regular ring R.*

Proof. Let r be an element in the Jacobson radical $J(R)$ of a strongly π-regular ring R. Then $r^n = r^{n+1}s$ for some positive integer n and some $s \in R$, i.e., $r^n(1 - rs) = 0$. Since $r \in J(R)$, the element $1 - rs$ is invertible, and thus $r^n = 0$. □

The relation between strongly π-regular rings and Fitting modules, which we introduced in Section 2.5, is given by the following proposition.

Proposition 3.18 *A module M_R over an arbitrary ring R is a Fitting module if and only if its endomorphism ring $\mathrm{End}(M_R)$ is strongly π-regular.*

Proof. Let M_R be a Fitting module. In order to show that $\mathrm{End}(M_R)$ is strongly π-regular we must show that for every $h \in \mathrm{End}(M_R)$ there exists a positive integer m such that $h^m \in \mathrm{End}(M_R)h^{m+1}$. Since M_R is a Fitting module, there is a positive integer m such that $M = \ker(h^m) \oplus h^m(M)$. Set $f = h^m$, so that $M = \ker(f) \oplus f(M)$. Then $f(M) = f(\ker(f) \oplus f(M)) = f^2(M)$. The equality $f(M) = f^2(M)$ implies that f induces a surjective endomorphism of $f(M)$. And since $\ker(f) \cap f(M) = 0$, f induces an automorphism of $f(M)$. As $f(M)$ is a direct summand of M, the inverse mapping of this automorphism of $f(M)$ can be extended to an endomorphism g of M. Then $gf(y) = y$ for every $y \in f(M)$. Thus $gf^2(x) = f(x)$ for every $x \in M$, that is, $gf^2 = f$. Therefore $h^m = gh^{2m} \in \mathrm{End}(M_R)h^{m+1}$.

For the converse, let M_R be a module with $\mathrm{End}(M_R)$ strongly π-regular. If $f \in \mathrm{End}(M_R)$, there exist $g, h \in \mathrm{End}(M_R)$ and a positive integer n such that $f^n = gf^{2n} = f^{2n}h$. From the equality $f^n = f^{2n}h$ it follows that $f^n(M) = f^{2n}(M)$, so that $M = \ker(f^n) + f^n(M)$ by Lemma 2.16(a). From the equality $f^n = gf^{2n}$ it follows that $\ker(f^n) \cap f^n(M) = 0$. Thus $M = \ker(f^n) \oplus f^n(M)$. $\square$

We are now ready to begin the study of the rings for which every finitely presented module is a Fitting module and is a direct sum of indecomposable modules. If R is such a ring, then the module R_R itself is a direct sum of indecomposable Fitting modules, hence a direct sum of modules with local endomorphism rings (Lemma 2.21). Therefore $R \cong \mathrm{End}(R_R)$ is a semiperfect ring (Proposition 3.14). Moreover, the direct sum R_R^n must be a Fitting R-module, so that $\mathbf{M}_n(R) \cong \mathrm{End}(R_R^n)$ must be a strongly π-regular ring for all n (Proposition 3.18). Thus we shall concentrate on the study of finitely presented modules over the semiperfect rings R whose matrix rings $\mathbf{M}_n(R)$ are strongly π-regular for all n.

We split the proof of Theorem 3.22 into three preparatory lemmas.

Lemma 3.19 *Let R be a semiperfect ring with $J(R)$ nil. Let r be an element of R such that $rR = r^2R$. Then rR is a direct summand of R_R.*

Proof. Let $^-$ denote the image in $R/J(R)$, so that $\overline{rR}$ is a semisimple R-module of finite length. There exists $s \in rR$ with $r = rs$, so that left multiplication by r is a surjective endomorphism $\psi \colon \overline{rR} \to \overline{rR}$. Thus ψ is an automorphism of $\overline{rR}$. Now $rs = rs^2$, so that $\psi(\overline{s - s^2}) = \overline{0}$, from which $\overline{s - s^2} = \overline{0}$. Hence $s - s^2 \in J(R)$ is nilpotent. Let k be a positive integer such that $(s - s^2)^k = 0$. Consider the polynomial $(1 - \lambda)^k \in \mathbf{Z}[\lambda]$. Then $(1 - \lambda)^k$ can be written in the form $1 - \lambda p(\lambda)$ for a suitable polynomial $p(\lambda) \in \mathbf{Z}[\lambda]$ such that $p(1) = 1$.

The equality $(s - s^2)^k = 0$ can now be written $s^k(1 - sp(s)) = 0$. From the equality $s^k = s^{k+1}p(s)$ it is easily seen that the element $e = (sp(s))^k$ of R is idempotent. Hence in order to conclude the proof it is sufficient to show that $rR = eR$. Note that $\overline{s - s^2} = \overline{0}$, i.e., $\overline{s}$ is idempotent. Hence $s^k(1 - sp(s)) = 0$ forces $\overline{s} = \overline{sp(s)}$, from which $\overline{e} = \overline{(sp(s))^k} = \overline{s^k} = \overline{s}$. Since the value $p(1)$ of the polynomial $p(\lambda)$ in $\lambda = 1$ is 1, it follows that the value of the polynomial $(\lambda p(\lambda))^k$ in $\lambda = 1$ is 1, that is, the sum of the coefficients of $(\lambda p(\lambda))^k$ is 1. Moreover $r = rs$ implies that $r = rs^i$ for all $i \geq 0$. Thus $r(sp(s))^k = r$, that is, $re = r$.

Since $e \in sR \subseteq rR$, there exists an element $t \in R$ with $e = rt$. Now

$$r(ete) = (re)(te) = r(te) = e^2 = e,$$

hence replacing t by ete we may suppose that $t \in eRe$. Since

$$\text{length } \overline{eR} \geq \text{length } \overline{reR} = \text{length } \overline{rR}$$

and $eR \subseteq rR$, we have $\overline{eR} = \overline{rR}$. Now

$$\text{length } \overline{tR} \geq \text{length } \overline{rtR} = \text{length } \overline{eR}$$

and $\overline{tR} \subseteq \overline{eR}$ because $t \in eRe$, so that $\overline{tR} = \overline{eR}$. If $u \in R$ is such that $\overline{tu} = \overline{e}$, then $\overline{(t + 1 - e)(eu + (1 - e))} = \overline{tu + (1 - e)} = \overline{1}$. Thus $\overline{t + 1 - e}$ is right invertible in $\overline{R}$, so that $t + 1 - e$ is right invertible in R. Then

$$rR = r(t + 1 - e)R = (e + r - re)R = eR. \qquad \square$$

Lemma 3.20 *Let R be a semiperfect ring with $\mathbf{M}_n(R)$ strongly π-regular for all n. Let M be a finitely generated right R-module and let f be an endomorphism of M. Then*

(a) *If M is projective and $f(M) = f^2(M)$, then $M = \ker(f) \oplus f(M)$.*

(b) *There exists a positive integer t such that $f^t(M) = f^{t+i}(M)$ for every $i \geq 0$.*

Proof. (a) Let M_R be a finitely generated projective R-module and let f be an endomorphism of M_R such that $f(M) = f^2(M)$. Then

$$M = \ker(f) + f(M)$$

by Lemma 2.16(a). Since M is projective, M is isomorphic to a direct summand of a finitely generated free module R^n, $R^n = M \oplus P$, say. The endomorphism f of M can be extended to an endomorphism g of R^n by putting $g(x) = x$ for every $x \in P$. Note that $g(R^n) = g^2(R^n)$. If we regard the elements of R^n as column vectors, g is left multiplication by a matrix $a \in \mathbf{M}_n(R)$, and $a\mathbf{M}_n(R) = a^2\mathbf{M}_n(R)$. Since $\mathbf{M}_n(R)$ is a semiperfect ring (Example 3.7) and

$J(\mathbf{M}_n(R))$ is nil (Lemma 3.17), it is possible to apply Lemma 3.19, and obtain that $a\mathbf{M}_n(R)$ is a direct summand of $\mathbf{M}_n(R)_{\mathbf{M}_n(R)}$. Thus $a\mathbf{M}_n(R) = e\mathbf{M}_n(R)$ for an idempotent matrix $e \in \mathbf{M}_n(R)$. Since the left $\mathbf{M}_n(R)$-modules $\mathbf{M}_n(R)$ and $R^n \oplus \cdots \oplus R^n$ (n summands) are isomorphic, the equality

$$a\mathbf{M}_n(R) = e\mathbf{M}_n(R)$$

can be rewritten as $aR^n \oplus \cdots \oplus aR^n = eR^n \oplus \cdots \oplus eR^n$, from which $aR^n = eR^n$. As e is an idempotent matrix, $g(R^n) = aR^n = eR^n$ is a direct summand of R^n, that is, $g(R^n)$ is a projective R-module. The mapping

$$g|_{g(R^n)} \colon g(R^n) \to g^2(R^n) = g(R^n)$$

is obviously surjective. By Corollary 2.44 the endomorphism $g|_{g(R^n)}$ of $g(R^n)$ is an automorphism. In particular $\ker g \cap g(R^n) = 0$. Thus $\ker f \cap f(M) = 0$.

(b) Let $x_1, \ldots, x_n$ be generators of M_R, so that $M = x_1 R + \cdots + x_n R$. Then the direct sum M^n is a cyclic right $\mathbf{M}_n(R)$-module generated by $(x_1, \ldots, x_n)$, and $(f, \ldots, f) \colon M^n \to M^n$ is an endomorphism of $M^n_{\mathbf{M}_n(R)}$. As $M^n_{\mathbf{M}_n(R)}$ is cyclic generated by $(x_1, \ldots, x_n)$, there exists $a \in \mathbf{M}_n(R)$ such that

$$(f, \ldots, f)(x_1, \ldots, x_n) = (x_1, \ldots, x_n)a.$$

Then $(f^m(x_1), \ldots, f^m(x_n)) = (x_1, \ldots, x_n)a^m$ for every positive integer m. The ring $\mathbf{M}_n(R)$ is strongly π-regular, so that there exists a positive integer t such that $a^t \mathbf{M}_n(R) = a^{t+i}\mathbf{M}_n(R)$ for every $i \geq 0$. Thus

$$(f^t(x_1), \ldots, f^t(x_n))\mathbf{M}_n(R) = (f^{t+i}(x_1), \ldots, f^{t+i}(x_n))\mathbf{M}_n(R),$$

from which $f^t(M) = f^{t+i}(M)$. $\square$

Lemma 3.21 *Let R be a semiperfect ring with $\mathbf{M}_n(R)$ strongly π-regular for all n. Then every finitely presented R-module is a Fitting module.*

Proof. Let M_R be a finitely presented R-module and suppose that f is an endomorphism of M_R such that $f(M) = f^2(M)$. Let $\varphi \colon P \to M$ be a projective cover of M. Then $\varphi g = f\varphi$ for an endomorphism g of P. Let K be the kernel of φ, so that K is a finitely generated R-module. Then

$$\varphi g(K) = f\varphi(K) = 0,$$

so that $g(K) \subseteq K$. By Lemma 3.20(b) there is a positive integer t such that $g^t(K) = g^{t+i}(K)$ and $g^t(P) = g^{t+i}(P)$ for every $i \geq 0$. From Lemma 3.20(a) applied to the module P and its endomorphism g^t we obtain that

$$P = \ker(g^t) \oplus g^t(P),$$

hence g^t restricts to an automorphism of $g^t(P)$.

The mapping φ induces an epimorphism $\varphi' \colon g^t(P) \to \varphi g^t(P)$ by restriction, and $\ker \varphi' = g^t(P) \cap \ker \varphi = g^t(P) \cap K$. The submodule $\ker \varphi'$ of $g^t(P)$ is superfluous, because if N is a submodule of $g^t(P)$ and $N + \ker \varphi' = g^t(P)$, then $\ker(g^t) + N + \ker \varphi' = \ker(g^t) + g^t(P) = P$, so that $\ker(g^t) + N + \ker \varphi = P$, from which $\ker(g^t) + N = P = \ker(g^t) \oplus g^t(P)$, and thus $N = g^t(P)$. Hence the mapping $\varphi' \colon g^t(P) \to \varphi g^t(P)$ is an epimorphism from a projective module and has a superfluous kernel, i.e., it is a projective cover.

Now $g^t(K) \subseteq g^t(P) \cap K = \ker \varphi'$, so that

$$g^{2t}(K) \subseteq g^t(\ker \varphi') \subseteq g^t(K) = g^{2t}(K),$$

from which $g^t(\ker \varphi') = g^{2t}(K)$. Since g^t restricts to an automorphism of $g^t(P)$ and $\ker \varphi', g^t(K)$ are submodules of $g^t(P)$, it follows that

$$\ker \varphi' = g^t(K). \tag{3.8}$$

In particular, $\ker \varphi'$ is finitely generated, so that $\varphi g^t(P)$ is finitely presented. From $\varphi g = f \varphi$, it follows that $\varphi g^t = f^t \varphi$ by iteration, hence

$$\varphi g^t(P) = f^t \varphi(P) = f^t(M) = f(M).$$

Therefore $f(M)$ is finitely presented.

Now we are in a situation very similar to the original one, that is, we have a finitely presented R-module $f(M)$ and the restriction of f to $f(M)$ is a surjective endomorphism of $f(M)$. Hence we may replace M by $f(M)$ and the endomorphism f of M by its restriction to $f(M)$. In other words we may assume that f is a surjective map, so that g is a surjective map, from which $\varphi' = \varphi$. By Corollary 2.44 the mapping $g \colon P \to P$ is an isomorphism. In particular, we may take $t = 1$. Then equality (3.8) becomes $K = g(K)$. Now g an isomorphism and $g(K) = K$ imply that f is an isomorphism. If we go back to the original notation, we see that we have found that the restriction of f to $f(M)$ is an automorphism of $f(M)$, so that $\ker(f) \cap f(M) = 0$. Thus $M = \ker(f) \oplus f(M)$ by Lemma 2.16(a).

Now let h be an arbitrary endomorphism of the finitely presented module M. By Lemma 3.20(b) there is a positive integer t such that

$$h^t(M) = h^{t+i}(M)$$

for every $i \geq 0$. In particular $h^t(M) = h^{2t}(M)$, so that $M = \ker(h^t) \oplus h^t(M)$ by the previous case. Thus M_R is a Fitting module. $\square$

We are now ready to characterize the rings over which every finitely presented right module is a Fitting module and is a direct sum of indecomposables.

Theorem 3.22 *The following conditions are equivalent for a ring R:*

(a) *Every finitely presented right R-module is a Fitting module and is a direct sum of indecomposable modules.*

(b) *The ring R is semiperfect and the rings* $\mathbf{M}_n(R)$ *are strongly* π*-regular for all* n.

Proof. The implication (a) $\Rightarrow$ (b) has already been proved immediately before the statement of Lemma 3.19. The implication (b) $\Rightarrow$ (a) is proved in Lemmas 3.21 and 1.14(c). $\square$

From this theorem and Theorem 2.22 we obtain

Corollary 3.23 *Let R be a semiperfect ring with* $\mathbf{M}_n(R)$ *strongly* π*-regular for all* n *and let M be a finitely presented R-module. Then M has a decomposition* $M = M_1 \oplus \cdots \oplus M_n$ *into indecomposable direct summands, and any two such direct sum decompositions are isomorphic.* $\square$

[Cedó and Rowen] have constructed an example of a local ring R whose Jacobson radical $J(R)$ is nil, but $\mathbf{M}_2(R)$ is not strongly regular (a ring S is *strongly regular* if $a \in a^2 S$ for every $a \in S$). In particular this ring R is semiperfect and strongly π-regular, but it does not satisfy condition (b) in Theorem 3.22. Therefore condition (b) in Theorem 3.22 cannot be relaxed to "R is semiperfect and strongly π-regular".

3.5   Finitely presented modules over serial rings

In this section we prove that every finitely presented module over a serial ring is serial, an important result due to [Warfield 75]. The presentation of the proof we give here is due to Nguyen Viet Dung. First we recall a definition.

Let M and N be modules over an arbitrary ring R. The module N is *M-injective* if every homomorphism $A \rightarrow N$ for any submodule A of M can be extended to a homomorphism $M \rightarrow N$ [Anderson and Fuller]. We need a weaker notion, obtained by modifying a definition due to [Baba] and [Harada].

Definition 3.24 *Let M and N be modules over an arbitrary ring R. We say that N is* almost-*M*-injective *if every homomorphism* $A \rightarrow N$ *for any cyclic submodule A of M can be extended to a homomorphism* $g: B \rightarrow N$, *where B is a submodule of M containing A, and either* $B = M$ *or g is an isomorphism.*

The reason why we have introduced such a definition is that if P, Q are indecomposable projective modules over a serial ring, then P is almost-Q-injective, as the following lemma shows.

Lemma 3.25 *Let R be a serial ring and let* $\{e_1, \ldots, e_n\}$ *be a complete set of orthogonal primitive idempotents of R. Then* $e_i R$ *is almost-*$e_j R$*-injective for all* $i, j = 1, 2, \ldots, n$.

Proof. Fix a cyclic submodule A of $e_j R$ and a homomorphism $f: A \to e_i R$. If $f = 0$, the zero homomorphism $e_j R \to e_i R$ extends f and has the required properties. Hence we may suppose $f \neq 0$. Since A is a cyclic uniserial R-module, there exists an epimorphism $p: e_k R \to A$ for some $k = 1, 2, \dots, n$. Let $\varepsilon_A: A \to R$ and $\varepsilon_i: e_i R \to R$ be the inclusions. Then

$$\varepsilon_A p, \varepsilon_i f p \in \operatorname{Hom}(e_k R, R_R) \cong R e_k,$$

which is a uniserial left R-module. Hence there exists $r \in R$ such that either $\varepsilon_A p = r \varepsilon_i f p$ or $r \varepsilon_A p = \varepsilon_i f p$. Since p is an epimorphism, either $\varepsilon_A = r \varepsilon_i f$ or $r \varepsilon_A = \varepsilon_i f$.

Suppose $\varepsilon_A = r \varepsilon_i f$. Note that for every $a \in A$ we have that $a \in e_j R$ and $f(a) \in e_i R$, so that $e_j a = a$ and $e_i f(a) = f(a)$. Thus

$$a = e_j a = e_j \varepsilon_A(a) = e_j r \varepsilon_i f(a) = e_j r f(a) = e_j r e_i f(a).$$

Consider the epimorphism $e_i R \to e_j r e_i R$ given by left multiplication by $e_j r e_i$. We show that it is an isomorphism. If it is not injective, its kernel has a non-zero intersection with the non-zero submodule $f(A)$ of $e_i R$. In other words, there exists $a \in A$ with $f(a) \neq 0$ and $e_j r e_i f(a) = 0$. But $a = e_j r e_i f(a)$, and we get a contradiction. Hence left multiplication by $e_j r e_i$ is an isomorphism $e_i R \to e_j r e_i R$. Let $g: e_j r e_i R \to e_i R$ be its inverse. Then $e_j r e_i R$ is a submodule of $e_j R$ that contains A, because $a = e_j r e_i f(a) \in e_j r e_i R$ for every $a \in A$. Also g extends f, because for every $a \in A$ one has $f(a) = g(e_j r e_i f(a)) = g(a)$.

Finally, suppose $r \varepsilon_A = \varepsilon_i f$. Then $ra = f(a)$ for every $a \in A$, so that $e_i r a = e_i f(a) = f(a)$. Therefore left multiplication by $e_i r$ is an homomorphism $R_R \to e_i R$ that extends f. □

In order to prove the main results of this section (Theorem 3.29 and Corollary 3.30) we need some preliminary results.

Lemma 3.26 *Let $P = P_1 \oplus \dots \oplus P_n$ be a direct sum of uniserial modules P_i over an arbitrary ring R. If P_i is almost-P_j-injective for every $i \neq j$, then every cyclic uniserial submodule of P is contained in a uniserial direct summand of P.*

Proof. Let C be a cyclic uniserial submodule of P. For every $i = 1, 2, \dots, n$ let $\pi_i: P \to P_i$ denote the canonical projection. Then

$$\bigcap_{i=1}^{n} \ker \pi_i \cap C = 0.$$

Since C is uniserial, there exists an index k such that $\ker \pi_k \cap C = 0$. Set $A = \pi_k(C)$, so that if we restrict the domain of $\pi_k: P \to P_k$ to C and its codomain to A we obtain an isomorphism $f: C \to A$. For every $i = 1, 2, \dots, n$ consider the mapping $f_i = \pi_i|_C \circ f^{-1}: A \to P_i$. For each $i \neq k$ the module P_i is almost-P_k-injective and $f_i: A \to P_i$ is a mapping whose domain A is a cyclic submodule of P_k. Hence each f_i can be extended to a homomorphism $g_i\colon B_i \to P_i$, where B_i is a submodule of P_k containing A, and either $B_i = P_k$ or g_i is an isomorphism.

We now choose an index $m = 1, 2, \ldots, n$ as follows. If B_i is a proper submodule of P_k for some $i \neq k$, let $m = 1, 2, \ldots, n$, $m \neq k$ be the index such that $B_m \subseteq B_i$ for every $i \neq k$ (it exists because P_k is uniserial). In this case $g_m\colon B_m \to P_m$ is an isomorphism. If $B_i = P_k$ for every $i \neq k$, set $m = k$ and $B_m = P_k$.

Let $g\colon B_m \to P = P_1 \oplus \cdots \oplus P_n$ be the mapping defined by

$$g(x) = (g_1(x), \ldots, g_{k-1}(x), x, g_{k+1}(x), \ldots, g_n(x))$$

for every $x \in B_m$. Since g is a monomorphism, if we set $D = g(B_m)$, then $D \cong B_m$, so that D is uniserial. We claim that D is the module with the required properties, that is, D is a direct summand of P that contains C.

In order to show that $C \subseteq D$, fix an element $x \in C$. Then $f(x) \in A$, so that $g_i(f(x)) = f_i(f(x)) = (\pi_i|_C \circ f^{-1})(f(x)) = \pi_i(x)$ for every $i \neq k$, and thus

$$
\begin{aligned}
x &= (\pi_1(x), \ldots, \pi_{k-1}(x), \pi_k(x), \pi_{k+1}(x), \ldots, \pi_n(x)) \\
&= (g_1(f(x)), \ldots, g_{k-1}(f(x)), f(x), g_{k+1}(f(x)), \ldots, g_n(f(x))) \\
&= g(f(x)) \in g(A) \subseteq g(B_m) = D.
\end{aligned}
$$

It remains to prove that D is a direct summand of P. Apply Lemma 2.6 to the decomposition $P = M \oplus P_m$, where $M = \oplus_{i \neq m} P_i$. In order to show that $P = M \oplus D$ it is sufficient to prove that $\pi_m|_D\colon D \to P_m$ is an isomorphism. We must consider the two cases $m \neq k$ and $m = k$.

If $m \neq k$, then $g_m\colon B_m \to P_m$ is an isomorphism. Since $\pi_k|_D$ induces an isomorphism between D and B_m, it follows that $\pi_m|_D = g_m \circ \pi_k|_D$ is an isomorphism of D onto P_m.

If $m = k$, then $B_m = P_k$, so that $\pi_k|_D\colon D \to P_k$ is obviously an isomorphism. □

From Theorem 3.10 and Lemmas 3.25 and 3.26 we have

Corollary 3.27 *Let R be a serial ring and let P be a finitely generated projective R-module. Then every cyclic uniserial submodule of P is contained in a uniserial direct summand of P.* □

The proof of the next result is elementary and left to the reader.

Lemma 3.28 *Suppose $P = P_1 \oplus P_2$. Let $\pi_1\colon P \to P_1$ be the canonical projection onto the first direct summand, and let A be a submodule of P. Then*

$$A = (A \cap P_1) \oplus (A \cap P_2)$$

if and only if $\pi_1(A) = A \cap P_1$. □

We are ready to prove the main result of this section.

Theorem 3.29 *Let R be a serial ring, let P be a finitely generated projective module, and let M be a finitely generated submodule of P. Then there is a decomposition $P = P_1 \oplus P_2 \oplus \cdots \oplus P_n$ of P into uniserial projective modules such that $M = (M \cap P_1) \oplus (M \cap P_2) \oplus \cdots \oplus (M \cap P_n)$.*

Proof. Since R is a finite direct sum of cyclic uniserial modules, every finitely generated R-module is a finite sum of cyclic uniserial submodules. Suppose $M = U_1 + U_2 + \cdots + U_m$, where the submodules U_i are cyclic and uniserial. The proof is by induction on $m + \dim(P)$, where $\dim(P)$ denotes the Goldie dimension of P. Note that for $m = 0$ the result follows immediately from Theorem 3.10.

Assume $M = U_1 + U_2 + \cdots + U_m$, $m \geq 1$ and $\dim(P) = n$. Then

$$M = M' + U_m,$$

where $M' = U_1 + U_2 + \cdots + U_{m-1}$. By the induction hypothesis we may assume $U_m \neq 0$. Again by the induction hypothesis there is a decomposition

$$P = P_1 \oplus \cdots \oplus P_n$$

of P into non-zero uniserial projective modules such that

$$M' = (M' \cap P_1) \oplus \cdots \oplus (M' \cap P_n).$$

Since U_m is cyclic and uniserial, U_m is contained in a uniserial direct summand D of P (Corollary 3.27). The module D is a finitely generated uniserial projective module, so that D is isomorphic to a uniserial direct summand of R_R, hence the endomorphism ring of D is a right chain ring (Lemma 1.19). Thus D has the exchange property (Theorem 2.8). By Lemma 2.7 there exist an index $j = 1, 2, \ldots, n$ and a direct summand P_j' of P_j such that $P = D \oplus P_j' \oplus (\oplus_{i \neq j} P_i)$. But P_j is indecomposable, so that either $P_j' = P_j$ or $P_j' = 0$. Since $\dim(D) = 1$ and $\dim(P_i) = 1$ for every i, it follows that $P_j' = 0$, and $P = D \oplus (\oplus_{i \neq j} P_i)$. Rearrange the indices, so that without loss of generality we may suppose $j = 1$, i.e., $P = D \oplus P_2 \oplus \cdots \oplus P_n$. Let $\pi_1 \colon P_1 \oplus \cdots \oplus P_n \to P_1$ denote the canonical projection.

Since the uniserial module P_1 contains both

$$\pi_1(U_m) \quad \text{and} \quad \pi_1(M') = M' \cap P_1,$$

there are two possible cases:

First case: $\pi_1(U_m) \subseteq \pi_1(M')$. Then

$$\pi_1(M) = \pi_1(M' + U_m) = \pi_1(M') + \pi_1(U_m) = \pi_1(M') = M' \cap P_1 \subseteq M \cap P_1.$$

Since obviously $\pi_1(M) \supseteq M \cap P_1$, we get that $\pi_1(M) = M \cap P_1$. Thus by Lemma 3.28 we have that $M = (M \cap P_1) \oplus (M \cap (P_2 \oplus \cdots \oplus P_n))$.

Second case: $\pi_1(U_m) \supseteq \pi_1(M')$. Since

$$P = P_1 \oplus \cdots \oplus P_n = D \oplus P_2 \oplus \cdots \oplus P_n,$$

by Lemma 2.6 the restriction $\pi_1|_D \colon D \to P_1$ is an isomorphism and the canonical projection $p \colon D \oplus P_2 \oplus \cdots \oplus P_n \to D$ is $p = (\pi_1|_D)^{-1}\pi_1$. Therefore

$$p(U_m) = (\pi_1|_D)^{-1}(\pi_1(U_m)) \supseteq (\pi_1|_D)^{-1}(\pi_1(M')) = p(M').$$

It follows that $p(M) = p(M') + p(U_m) = p(U_m) = U_m \subseteq M \cap D$. The other inclusion $M \cap D \subseteq p(M)$ is trivial, so that $p(M) = M \cap D$. From Lemma 3.28 we obtain $M = (M \cap D) \oplus (M \cap (P_2 \oplus \cdots \oplus P_n))$.

In both cases the module $N = M \cap (P_2 \oplus \cdots \oplus P_n)$ is a direct summand of M, so that N itself is a sum of m cyclic uniserial modules. Since the Goldie dimension of the projective module $P' = P_2 \oplus \cdots \oplus P_n$ is $n - 1$, the induction hypothesis can be applied to the submodule N of P'. Hence there is a decomposition $P_2' \oplus \cdots \oplus P_n'$ of P' into uniserial projective modules such that $N = (N \cap P_2') \oplus \cdots \oplus (N \cap P_n')$. Then $P = P_1 \oplus P_2' \oplus \cdots \oplus P_n'$ is the required decomposition of P in the first case, whereas $P = D \oplus P_2' \oplus \cdots \oplus P_n'$ is the required decomposition of P in the second case. $\square$

Corollary 3.30 *Every finitely presented module over a serial ring is serial.* $\square$

3.6 Notes on Chapter 3

One easily sees that every module is a homomorphic image of a free, hence projective, module. Dually, [Eckmann and Schopf] proved that not only can every module be embedded in an injective module, but also there is a "minimal" embedding of a module in an injective module, that is, they proved the existence of the injective envelope. Dualizing the notion of an injective envelope, [Bass 60] introduced projective covers, perfect rings and semiperfect rings ("minimal epimorphisms" had already been studied by [Eilenberg and Nakayama] and [Eilenberg]). Theorem 3.6 is mainly due to [Bass 60, Theorem 2.1] (the equivalence of conditions (c) and (d) to the other conditions was proved by [Müller 70, Theorem 1 and p. 465]). Example 3.8 is due to [Müller 70, p. 465], Theorem 3.10 is due to [Klatt] and [Müller 70, Theorem 3] independently, and Theorem 3.15 is due to [Warfield 75, Theorem 1.4].

In Theorem 3.5 we have encountered an interesting class of rings, that is, *F-semiperfect* rings. A ring R is *F-semiperfect* if $R/J(R)$ is von Neumann regular and idempotents can be lifted modulo $J(R)$. A ring R is F-semiperfect if and only if every finitely presented right (or left) R-module has a projective cover.

Every semiperfect ring is F-semiperfect. Endomorphism rings of injective modules are F-semiperfect. More generally, endomorphism rings of injective objects in an arbitrary Grothendieck category and endomorphism rings of quasinjective modules are F-semiperfect. Most of these results about F-semiperfect rings were proved by [Oberst and Schneider].

The rings R that satisfy the d.c.c. on chains of the form

$$rR \supseteq r^2R \supseteq r^3R \supseteq \ldots$$

are often called *right π-regular* in the literature. Since this condition is left-right symmetric (Theorem 3.16), we prefer to call them *strongly π-regular*. These rings were introduced by [Kaplansky 50]. Theorem 3.16 is due to [Dischinger], and the proof of Theorem 3.16 we have given here was found by Zöschinger. Every strongly π-regular ring is an exchange ring [Stock, Example 2.3]. Proposition 3.18 is due to [Armendariz, Fisher and Snider, Proposition 2.3]. Lemmas 3.19, 3.20, 3.21, Theorem 3.22 and Corollary 3.23 are due to [Rowen 86].

In Theorem 3.22 we characterized the rings over which finitely presented right modules are Fitting modules and are direct sums of indecomposable modules. A more natural class of rings is the class of *Krull-Schmidt rings,* that is, the rings R for which every finitely presented right R-module is a direct sum of modules with local endomorphism rings (the condition is left-right symmetric, that is, it holds for finitely presented right R-modules if and only if it holds for finitely presented left R-modules). These rings have been defined and characterized in terms of the category $({}_R\text{FP}, \text{Ab})$ by [Herzog]. Every ring that satisfies the equivalent conditions of Theorem 3.22 is a Krull-Schmidt ring (because every indecomposable Fitting modules has a local endomorphism ring, Lemma 2.21). Every Krull-Schmidt ring is semiperfect (Proposition 3.14). By Corollary 2.55 every pure-projective module over a Krull-Schmidt ring is a direct sum of finitely presented modules with local endomorphism rings. This property was noticed by [Gómez Pardo and Guil Asensio].

Theorem 3.29 and Corollary 3.30 (finitely presented modules over serial rings are serial) are due to [Warfield 75, Theorem 3.3 and Corollary 3.4]. This had been proved previously in the commutative case by [Kaplansky 49] (Kaplansky proved that every finitely presented module over a commutative valuation ring, that is, a commutative chain ring, is a direct sum of cyclic modules; note that a commutative ring is serial if and only if it is a finite direct product of valuation rings) and by [Roux] in the case of chain rings. A different proof of Corollary 3.30 was found independently by [Drozd]. The uniqueness of the decomposition of a finitely presented module over a serial ring as a direct sum of uniserial modules will be one of the main topics of Chapter 9.

Chapter 4

Semilocal Rings

4.1 The Camps-Dicks Theorem

We have already defined semilocal rings as the rings R for which $R/J(R)$ is a semisimple artinian ring (Section 1.2). We have seen that they are the rings of finite dual Goldie dimension (Proposition 2.43). More precisely, if R is a semilocal ring, $\operatorname{codim}(R_R) = \operatorname{codim}(_RR)$ is equal to the number of summands in a direct sum decomposition of the semisimple right module $R/J(R)$ as a direct sum of simple right modules. In this section we shall prove a theorem containing two further characterizations of semilocal rings (Theorem 4.2). We begin with a lemma whose proof is taken from [Herbera and Shamsuddin, proof of Theorem 3].

Lemma 4.1 *Let M be a right module over an arbitrary ring R and let f, g be two endomorphisms of M. Then*

(a) $\ker(f - fgf) = \ker(f) \oplus \ker(1 - gf)$ *and* $\ker(1 - gf) \cong \ker(1 - fg)$;

(b)
$$\operatorname{coker}(f - fgf) \cong \operatorname{coker}(f) \oplus \operatorname{coker}(1 - fg)$$

and

$$\operatorname{coker}(1 - gf) \cong \operatorname{coker}(1 - fg).$$

Proof. (a) It is clear that $\ker(f) + \ker(1 - gf) \subseteq \ker(f - fgf)$. Conversely, if $x \in \ker(f - fgf)$, then $(1 - gf)(x) \in \ker(f)$, $gf(x) \in \ker(1 - gf)$ and $x = (1 - gf)(x) + gf(x)$, so that $\ker(f) + \ker(1 - gf) = \ker(f - fgf)$. It is easily verified that $\ker(f) \cap \ker(1 - gf) = 0$.

The homomorphism $f \colon M \to M$ induces a homomorphism

$$\ker(1 - gf) \to \ker(1 - fg)$$

by restriction. By symmetry, g induces a homomorphism

$$\ker(1 - fg) \to \ker(1 - gf).$$

It is immediately verified that the compositions of these two mappings are the identity mappings of $\ker(1 - gf)$ and $\ker(1 - fg)$ respectively, so that $\ker(1 - gf) \cong \ker(1 - fg)$.

(b) Consider the mapping $\varphi \colon M \to \operatorname{coker}(f) \oplus \operatorname{coker}(1 - fg)$ defined by $\varphi(x) = (x + f(M), x + (1 - fg)(M))$ for every $x \in M$. We show that φ is a surjective mapping. Note that $M = fg(M) + (1 - fg)(M) \subseteq f(M) + (1 - fg)(M)$. Therefore for any $y, z \in M$, there exists $v \in f(M)$ and $w \in (1 - fg)(M)$ such that $y - z = v + w$. Set $x = y - v = z + w$. Then

$$\varphi(x) = (y + f(M), z + (1 - fg)(M)).$$

This shows that φ is surjective. The kernel of φ is $f(M) \cap (1 - fg)(M)$, and thus we must show that $f(M) \cap (1 - fg)(M) = (f - fgf)(M)$. Now if $f(x) = (1 - fg)(y)$, $x, y \in M$, then $y = f(x) + fg(y)$, so that

$$f(x) = (1 - fg)(y) = (1 - fg)(f(x + g(y))) = (f - fgf)(x + g(y)).$$

This proves that $f(M) \cap (1 - fg)(M) \subseteq (f - fgf)(M)$. The opposite inclusion is easily verified.

Define a mapping

$$\operatorname{coker}(1 - gf) \to \operatorname{coker}(1 - fg)$$

by

$$x + (1 - gf)(M) \mapsto f(x) + (1 - fg)(M)$$

for every $x \in M$. This is a well-defined homomorphism. Similarly, there is a well-defined homomorphism

$$\operatorname{coker}(1 - fg) \to \operatorname{coker}(1 - gf)$$

given by

$$x + (1 - fg)(M) \mapsto g(x) + (1 - gf)(M).$$

The compositions of these two homomorphisms are the identity mappings of $\operatorname{coker}(1 - gf)$ and $\operatorname{coker}(1 - fg)$ respectively, so that

$$\operatorname{coker}(1 - gf) \cong \operatorname{coker}(1 - fg). \qquad \square$$

For a ring R we denote the multiplicative group of invertible elements of R by $U(R)$.

Theorem 4.2 (Camps and Dicks) *The following conditions are equivalent for a ring R.*

(a) *R is semilocal.*

(b) *There exists an integer $n \geq 0$ and a function $d\colon R \to \{0, 1, \ldots, n\}$ such that*

    (i) *for every $a, b \in R$, $d(1 - ab) + d(a) = d(a - aba)$;*

    (ii) *if $a \in R$ and $d(a) = 0$, then $a \in U(R)$.*

(c) *There exists a partial order $\leq$ on the set R such that*

    (iii) *$(R, \leq)$ is an artinian poset;*

    (iv) *if $a, b \in R$ and $1 - ab \notin U(R)$, then $a - aba < a$.*

Moreover, if these equivalent conditions hold, then $\mathrm{codim}(R) \leq n$ for any integer n that satisfies condition (b).

Proof.

    (a) $\Rightarrow$ (b). If R is a semilocal ring, then R_R has finite dual Goldie dimension (Proposition 2.43). Let $n = \mathrm{codim}(R_R)$ and let $d\colon R \to \{0, 1, \ldots, n\}$ be defined by $d(a) = \mathrm{codim}(R/aR)$ for every $a \in R$.

    In order to prove that d has property (i), fix $a, b \in R$ and apply Lemma 4.1(b) to the two endomorphisms of the module R_R given by left multiplication by a and b respectively. Then $R/(a - aba)R \cong R/aR \oplus R/R(1 - ab)$, so that $d(a - aba) = d(a) + d(1 - ab)$. Hence (i) holds.

    If $a \in R$ and $d(a) = 0$, then $R/aR = 0$, so that $a \in U(R)$ by Proposition 2.38.

    (b) $\Rightarrow$ (c). If (b) holds, define a partial order $\leq$ on R via

$$a \leq b \quad \text{if } a = b \text{ or } d(a) > d(b).$$

Then (c) is easily verified.

    (c) $\Rightarrow$ (a). Let $\overline{R}$ denote $R/J(R)$, and for every $r \in R$ let $\overline{r}$ denote $r + J(R)$. Suppose that (c) holds. Set

$$\mathcal{F} = \{\, r \in R \mid \overline{r}^2 = \overline{r} \text{ and } \overline{(1 - r)}\overline{R} \text{ is a right ideal of finite length of } \overline{R} \,\}.$$

Note that $\mathcal{F} \neq \emptyset$, because $1 \in \mathcal{F}$. Since $(R, \leq)$ is artinian, there exists an element $a \in \mathcal{F}$ minimal with respect to the order $\leq$.

    Suppose $\overline{a} \neq \overline{0}$. Then $a \notin J(R)$, so that $aR \setminus J(R) \neq \emptyset$, and we can choose an element $ab \in aR \setminus J(R)$ that is minimal with respect to the order $\leq$. Since $ab \notin J(R)$, there exists $c \in R$ such that $1 - abc \notin U(R)$. Then by (iv) we get $a - abca < a$. Set $a' = a - abca$, so that $a' < a$. We show that $a' \in \mathcal{F}$.

    We claim that if $x \in R$ and $1 - abx \notin U(R)$, then $\overline{abxab} = \overline{ab}$. In order to prove the claim fix $x \in R$ with $1 - abx \notin U(R)$. From property (iv) we get that $ab - abxab < ab$. Since ab is minimal in $aR \setminus J(R)$, it follows that $ab - abxab \notin aR \setminus J(R)$, that is, $ab - abxab \in J(R)$. Hence $\overline{ab} = \overline{abxab}$. This proves the claim.

Now apply the claim to $x = c$. Then $\overline{abcab} = \overline{ab}$, so that $\overline{abc}$ is idempotent. Then $\overline{a'}$ also is idempotent, because

$$\left(\overline{a'}\right)^2 = \left(\overline{a - abca}\right)^2 = \overline{a} - \overline{abca} - \overline{abca} + \overline{abcabca} = \overline{a} - \overline{abca} = \overline{a'}.$$

Note that $\overline{a - a'} = \overline{abca}$ also is idempotent.

It is easily verified that $\{\overline{1 - a}, \overline{a - a'}, \overline{a'}\}$ is a complete set of orthogonal idempotents in $\overline{R}$. Therefore $\overline{R}_{\overline{R}} = \overline{(1 - a)}\overline{R} \oplus \overline{(a - a')}\overline{R} \oplus \overline{a'}\overline{R}$. We show that $\overline{(a - a')}\overline{R}$ is a simple $\overline{R}$-module. Since $\overline{a - a'} = \overline{abca}$, we get that

$$\overline{(a - a')}\overline{R} = \overline{abca}\overline{R} = \overline{ab}\overline{R},$$

because $\overline{abcab} = \overline{ab}$ as we have already seen. Moreover $\overline{ab}\overline{R} \neq \overline{0}$, otherwise $ab \in J(R)$, which is not true. Now consider any $abd \in abR \setminus J(R)$. Since $abd \notin J(R)$, there exists $e \in R$ such that $1 - abde \notin U(R)$. Applying the claim with $x = de$, we see that $\overline{abdeab} = \overline{ab}$, so that $\overline{abd}\overline{R} = \overline{ab}\overline{R}$. This shows that $\overline{(a - a')}\overline{R}$ is a simple $\overline{R}$-module.

Now $\overline{(1 - a)}\overline{R}$ is a module of finite length, so that

$$\overline{(1 - a)}\overline{R} \oplus \overline{(a - a')}\overline{R} = \overline{(1 - a')}\overline{R}$$

has finite length. Thus $a' \in \mathcal{F}$.

But $a' < a$ and a was minimal in $\mathcal{F}$. This contradiction shows that $\overline{a} = \overline{0}$. Therefore $\overline{(1 - a)}\overline{R} = \overline{R}_{\overline{R}}$ has finite length, that is, $\overline{R}$ is right artinian and the proof of (a) is concluded.

Now suppose the equivalent conditions of the statement hold. We want to show that if $m = \dim \overline{R} = \operatorname{codim} R$ and n is any integer that satisfies condition (b), then $m \leq n$. As $m = \dim(\overline{R})$, there are elements $e_1, \ldots, e_m \in R$ such that $\{\overline{e_1}, \ldots, \overline{e_m}\}$ is a complete set of non-zero orthogonal idempotents of $\overline{R}$. Define $a_0, a_1, \ldots, a_m \in R$ by induction as follows: $a_0 = 1$ and $a_i = a_{i-1} - a_{i-1}e_i a_{i-1}$ for $i = 1, 2, \ldots, m$. Note that $\overline{a_i} = \overline{e_{i+1}} + \overline{e_{i+2}} + \cdots + \overline{e_m}$ for every $i = 0, 1, \ldots, m$, so that $\overline{1 - a_{i-1}e_i} = \overline{1 - e_i} \notin U(\overline{R})$. It follows that $1 - a_{i-1}e_i \notin U(R)$. Applying property (ii) we see that $d(1 - a_{i-1}e_i) > 0$, and applying property (i) we get $d(1 - a_{i-1}e_i) + d(a_{i-1}) = d(a_i)$, so that $d(a_{i-1}) < d(a_i)$ for every $i = 1, \ldots, m$. From $d(a_0) < d(a_1) < \cdots < d(a_m)$ we obtain that $m \leq n$. $\square$

4.2 Modules with semilocal endomorphism ring

Theorem 4.2 is very useful in testing whether the endomorphism ring of a module is semilocal.

Theorem 4.3 (Herbera and Shamsuddin) *Let M_R be a right module over a ring R.*

(a) *If M_R has finite Goldie dimension and every injective endomorphism of M_R is bijective, then the endomorphism ring $\mathrm{End}(M_R)$ is semilocal and*

$$\mathrm{codim}(\mathrm{End}(M_R)) \leq \dim(M_R).$$

(b) *If M_R has finite dual Goldie dimension and every surjective endomorphism of M_R is bijective, then the endomorphism ring $\mathrm{End}(M_R)$ is semilocal and*

$$\mathrm{codim}(\mathrm{End}(M_R)) \leq \mathrm{codim}(M_R).$$

(c) *If M_R has finite Goldie dimension and finite dual Goldie dimension, then the endomorphism ring $\mathrm{End}(M_R)$ is semilocal and*

$$\mathrm{codim}(\mathrm{End}(M_R)) \leq \dim(M_R) + \mathrm{codim}(M_R).$$

Proof. If $\dim(M_R)$ is finite, set $n = \dim(M_R)$ and define

$$d_1 \colon \mathrm{End}(M_R) \to \{0, 1, 2, \ldots, n\}$$

by $d_1(f) = \dim(\ker(f))$. By Lemma 4.1(a), the mapping d_1 satisfies conditions (i) and (ii) of Theorem 4.2. This proves (a)

If $\mathrm{codim}(M_R)$ is finite, set $m = \mathrm{codim}(M)$ and define

$$d_2 \colon \mathrm{End}(M_R) \to \{0, 1, 2, \ldots, m\}$$

by $d_2(f) = \mathrm{codim}(\mathrm{coker}(f))$. By Lemma 4.1(b), the mapping d_2 satisfies conditions (i) and (ii) of Theorem 4.2. This proves (b).

If $\dim(M_R)$ and $\mathrm{codim}(M_R)$ are both finite, set

$$d = d_1 + d_2 \colon \mathrm{End}(M) \to \{0, 1, 2, \ldots, n + m\}.$$

From this (c) follows. $\qquad\square$

Before giving examples of classes of modules with semilocal endomorphism rings, we study some properties of these modules. In this book we shall not need the definition of rings of arbitrary stable range $n \geq 1$, but only the definition of rings of stable range 1 (for the definition of stable range $n > 1$ see the notes at the end of this chapter). A ring E is said to have *stable range* 1 if, whenever $a, b \in E$ and $Ea + Eb = E$, there exists $t \in E$ with $a + tb \in U(E)$. It would be possible to prove that this condition is left-right symmetric, that is, a ring E has stable range 1 if and only if, whenever $a, b \in E$ and $aE + bE = E$, there exists $s \in E$ with $a + bs \in U(E)$. Since we shall not need this result, the proof is omitted.

Theorem 4.4 (Bass) *A semilocal ring has stable range* 1.

Proof. An element $x \in E$ is a unit in E if and only if $\overline{x} = x + J(E)$ is a unit in $\overline{E} = E/J(E)$. Therefore we may substitute the semilocal ring E with the semisimple artinian ring $\overline{E}$, that is, we may assume that E is semisimple artinian. Let a and b be elements of a semisimple artinian ring E and suppose $Ea + Eb = E$. The left ideal $Ea \cap Eb$ is a direct summand of Eb, so that $Eb = (Ea \cap Eb) \oplus I$ for a left ideal I of E. Then $_EE = Ea \oplus I$, as is easily seen. Let $\mu \colon {}_EE \to Ea$ be the epimorphism given by right multiplication by a. Then $\ker \mu$ is a direct summand of $_EE$, so that $_EE = C \oplus \ker \mu$ for a left ideal C of E, and the restriction $\mu|_C \colon C \to Ea$ is an isomorphism. Then $_EE \cong Ea \oplus \ker \mu$ and $_EE = Ea \oplus I$ imply $\ker \mu \cong I$ because $_EE$ is semisimple artinian. Let $f \colon \ker \mu \to I$ be an isomorphism. Then $(\mu|_C, f) \colon C \oplus \ker \mu \to Ea \oplus I$ is an automorphism of $_EE$, hence it coincides with right multiplication by an element $u \in U(E)$. From $_EE = C \oplus \ker \mu$ we have that $1 = c + k$, where $c \in C$ and $ka = 0$. Hence $a = ca$, so that

$$u = (\mu|_C, f)(1) = (\mu|_C, f)(c + k) = ca + f(k) = a + f(k).$$

Since $f(k) \in I \subseteq Eb$, it follows that $f(k) = tb$ for some $t \in E$. Thus $a + tb = u \in U(E)$. $\qquad\square$

The reason why we have introduced rings with stable range 1 is given in the following theorem: if a module A has the property that its endomorphism ring has stable range 1, then A cancels from direct sums.

Theorem 4.5 (Evans) *Let R be a ring and let A_R be an R-module. Suppose that $E = \mathrm{End}(A_R)$ has stable range 1. If B_R and C_R are arbitrary R-modules and $A \oplus B \cong A \oplus C$, then $B \cong C$.*

Proof. Let $f_{M,N}$ denote an R-homomorphism from N to M. Since $A \oplus B \cong A \oplus C$, there are two inverse isomorphisms

$$F = \begin{pmatrix} f_{A,A} & f_{A,B} \\ f_{C,A} & f_{C,B} \end{pmatrix} \colon A \oplus B \to A \oplus C$$

and

$$G = \begin{pmatrix} g_{A,A} & g_{A,C} \\ g_{B,A} & g_{B,C} \end{pmatrix} : A \oplus C \to A \oplus B.$$

Since GF is the identity on $A \oplus B$, we have

$$\begin{pmatrix} g_{A,A}f_{A,A} + g_{A,C}f_{C,A} & g_{A,A}f_{A,B} + g_{A,C}f_{C,B} \\ g_{B,A}f_{A,A} + g_{B,C}f_{C,A} & g_{B,A}f_{A,B} + g_{B,C}f_{C,B} \end{pmatrix} = \begin{pmatrix} 1_A & 0 \\ 0 & 1_B \end{pmatrix},$$

where 1_M is the identity of M. From $g_{A,A}f_{A,A} + g_{A,C}f_{C,A} = 1_A$ it follows that $Ef_{A,A} + Eg_{A,C}f_{C,A} = E$. Hence there exists $t \in E$ such that

$$u = f_{A,A} + tg_{A,C}f_{C,A}$$

is an automorphism of A. Consider the mapping

$$G' = \begin{pmatrix} 1_A & tg_{A,C} \\ g_{B,A} & g_{B,C} \end{pmatrix} : A \oplus C \to A \oplus B.$$

Then

$$G'F = \begin{pmatrix} u & v_{A,B} \\ 0 & 1_B \end{pmatrix}$$

is clearly an automorphism of $A \oplus B$. Since $F : A \oplus B \to A \oplus C$ is an isomorphism, it follows that $G' : A \oplus C \to A \oplus B$ is an isomorphism as well. But then

$$G'' = \begin{pmatrix} 1_A & 0 \\ -g_{B,A} & 1_B \end{pmatrix} G' \begin{pmatrix} 1_A & -tg_{A,C} \\ 0 & 1_C \end{pmatrix}$$

is an isomorphism. Since

$$G'' = \begin{pmatrix} 1_A & 0 \\ 0 & g_{B,C} - g_{B,A}tg_{A,C} \end{pmatrix},$$

the homomorphism $g_{B,C} - g_{B,A}tg_{A,C}$ is an isomorphism of C onto B. $\quad\square$

From Theorems 4.4 and 4.5 we obtain:

Corollary 4.6 *Let A_R be a module over a ring R and suppose $\mathrm{End}(A_R)$ semilocal. If B_R and C_R are arbitrary R-modules and $A \oplus B \cong A \oplus C$, then $B \cong C$.*
$\quad\square$

We now recall a well-known technique that goes back to [Dress]. We shall often use it in the sequel. Given a ring S, let Mod-S denote the category of all right S-modules, Proj-S the full subcategory of Mod-S whose objects are all projective right S-modules, and proj-S the full subcategory of Mod-S whose objects are all finitely generated projective right S-modules.

If M_S is a right S-module, let $\mathrm{Add}(M_S)$ denote the full subcategory of Mod-S whose objects are all the modules isomorphic to direct summands of direct sums of copies of M, and let add(M_S) denote the full subcategory of Mod-S whose objects are all the modules isomorphic to direct summands of direct sums M^n of a finite number of copies of M. For example, Proj-S = Add(S_S) and proj-S = add(S_S).

Let M_S be a right S-module and let $E = \text{End}(M_S)$ be its endomorphism ring, so that $_E M_S$ is a bimodule. Clearly

Theorem 4.7 *The functors*

$$\text{Hom}_S(M, -): \text{Mod-}S \to \text{Mod-}E \quad and \quad - \otimes_E M: \text{Mod-}E \to \text{Mod-}S$$

induce an equivalence between the full subcategory add(M_S) *of* Mod-S *and the full subcategory* proj-E *of* Mod-E. *Moreover, if* M_S *is finitely generated, they induce an equivalence between the full subcategory* Add(M_S) *of* Mod-S *and the full subcategory* Proj-E *of* Mod-E.   □

If $e \in E$ is idempotent, the object eE of proj-E corresponds to the direct summand eM of M_S, which is an object of add(M_S).

Let us go back to modules with semilocal endomorphism ring.

Proposition 4.8 (*n*-th root property) *Let R be a ring and let A_R, B_R be R-modules. If A_R has a semilocal endomorphism ring, n is a positive integer and $A^n \cong B^n$, then $A \cong B$.*

Proof. We shall suppose that we have a module $M = \oplus_{i=1}^n A_i = \oplus_{i=1}^n B_i$ where each $A_i \cong A$ and $B_i \cong B$, A has a semilocal endomorphism ring, and prove that $A_1 \cong B_1$.

Note that M has semilocal endomorphism ring, because

$$\text{End}(M_R) \cong \mathbf{M}_n(\text{End}(A_R))$$

(Example (3) on page 6). Let $\{d_1, \ldots, d_n\}$ and $\{e_1, \ldots, e_n\}$ be the two complete sets of orthogonal idempotents in $E = \text{End}(M_R)$ with $d_i M = A_i$ and $e_i M = B_i$ for every i. By Theorem 4.7 $d_i E \cong d_1 E$ and $e_i E \cong e_1 E$ for every $i = 1, \ldots, n$, so that $d_i E/d_i J(E) \cong d_1 E/d_1 J(E)$ and $e_i E/e_i J(E) \cong e_1 E/e_1 J(E)$. Hence

$$E/J(E) \cong \oplus_{i=1}^n d_i E/d_i J(E) \cong \oplus_{i=1}^n e_i E/e_i J(E).$$

Since E is semilocal, $E/J(E)$ is a semisimple right E-module of finite length, and therefore $d_1 E/d_1 J(E) \cong e_1 E/e_1 J(E)$. By Proposition 3.3(b) $d_1 E \cong e_1 E$, so that $A_1 \cong B_1$ by Theorem 4.7, as desired.   □

For the next proposition recall that a ring is semilocal if and only if it has finite dual Goldie dimension (Proposition 2.43).

Proposition 4.9 *If A_R is a module with a semilocal endomorphism ring and* $\mathrm{codim}(\mathrm{End}(A_R)) = n$, *then A_R has at most 2^n isomorphism classes of direct summands.*

Proof. Set $E = \mathrm{End}(A_R)$. Every direct summand of A_R has the form eA for some idempotent $e \in E$, and in the equivalence of Theorem 4.7 the module eA corresponds to the module eE. Since equivalences preserve isomorphism, the number of non-isomorphic direct summands of A_R is the same as the number of non-isomorphic direct summands of E_E. Hence it is sufficient to show that E_E has at most 2^n non-isomorphic direct summands eE. By Proposition 3.3(b) it suffices to show that $E/J(E)$ has at most 2^n non-isomorphic direct summands. This is obvious, because $E/J(E)$ is semisimple artinian and

$$\dim(E/J(E)) = \mathrm{codim}(E) = n\,. \qquad \square$$

In a semilocal ring every set of orthogonal idempotents is finite. Therefore every direct sum decomposition of a module with a semilocal endomorphism ring has only finitely many direct summands. By Proposition 4.9 a module with a semilocal endomorphism ring has only finitely many direct sum decompositions up to isomorphism.

Proposition 4.9 will be generalized in Corollary 4.11. First we shall prove that a semilocal ring has only finitely many isomorphism classes of finitely generated indecomposable projective modules.

Theorem 4.10 *Let R be a semilocal ring. Then there are, up to isomorphism, only finitely many finitely generated indecomposable projective R-modules.*

Proof. Since R is semilocal, there are only finitely many simple R-modules up to isomorphism, say, $T_1, \ldots, T_n$. If P_R is a finitely generated projective R-module, then $P/PJ(R) \cong T_1^{r_1} \oplus \cdots \oplus T_n^{r_n}$ with $(r_1, \ldots, r_n) \in \mathbf{N}^n$. Set $\varphi(P_R) = (r_1, \ldots, r_n)$. Partially order $\mathbf{N}^n$ by setting $(r_1, \ldots, r_n) \le (s_1, \ldots, s_n)$ if $r_i \le s_i$ for every $i = 1, 2, \ldots, n$.

Note that if P_R, Q_R are finitely generated projective R-modules, then P is isomorphic to a direct summand of Q_R if and only if $\varphi(P_R) < \varphi(Q_R)$, because if $\varphi(P_R) \le \varphi(Q_R)$, there is an epimorphism $Q/QJ(R) \to P/PJ(R)$, which can be lifted to a homomorphism $f \colon Q \to P$ because Q_R is projective. Since $f(Q) + PJ(R) = P$, it follows that f is an epimorphism by Nakayama's Lemma. Hence f splits.

In particular, P_R is isomorphic to Q_R if and only if $\varphi(P_R) = \varphi(Q_R)$, because if $\varphi(P_R) = \varphi(Q_R)$, then P_R is isomorphic to a direct summand of Q_R and Q_R is isomorphic to a direct summand of P_R. But a finitely generated projective module over a semilocal ring cannot be isomorphic to a proper direct summand of itself, because every non-zero finitely generated projective module has non-zero finite dual Goldie dimension (Proposition 2.43).

Thus if $\{\, P_\lambda \mid \lambda \in \Lambda \,\}$ is a family of pairwise non-isomorphic finitely generated indecomposable projective R-modules, then $\{\, \varphi(P_\lambda) \mid \lambda \in \Lambda \,\}$ is a family of pairwise incomparable elements of $(\mathbf{N}^n, \leq)$. But it is easily seen that in $(\mathbf{N}^n, \leq)$ there are no infinite families of pairwise incomparable elements. $\square$

From Theorems 4.7 and 4.10 we obtain

Corollary 4.11 *Let A_R be a module over an arbitrary ring R with $\mathrm{End}(A_R)$ semilocal. Then there are, up to isomorphism, only finitely many indecomposable R-modules in $\mathrm{add}(A_R)$.* $\square$

4.3   Examples

In this section we shall give examples of classes of modules with semilocal endomorphism rings. The first example is the class of artinian modules.

Theorem 4.12 *Every artinian module has semilocal endomorphism ring.*

Proof.   Every artinian module A has finite Goldie dimension, and injective endomorphisms of A are bijective (Lemma 2.16(b)). Now apply Theorem 4.3(a).
$\square$

Therefore artinian modules have the n-th root property, have only finitely many isomorphism classes of direct summands, and cancel from direct sums. The next result provides an upper bound to the number of isomorphism classes of direct summands of an artinian module.

Proposition 4.13 *Let A_R be an artinian module and let n be the composition length of the socle of A_R, that is, $n = \dim(\mathrm{soc}(A_R))$. Then there are at most 2^n isomorphism classes of direct summands of A_R.*

Proof.   If A_R is an artinian module and $n = \dim(\mathrm{soc}(A_R))$, then

$$n = \dim(A_R).$$

Since injective endomorphisms of A_R are bijective (Lemma 2.16(b)), Theorem 4.3(a) implies that $\mathrm{codim}(\mathrm{End}(A_R)) \leq n$. Therefore A_R has at most 2^n isomorphism classes of direct summands by Proposition 4.9. $\square$

From Lemma 2.17(b) and Theorem 4.3(b) we obtain a second class of examples of modules with semilocal endomorphism rings:

Proposition 4.14 *Every noetherian module of finite dual Goldie dimension has a semilocal endomorphism ring.* □

Corollary 4.15 *Let R be a semilocal ring and let A_R be a noetherian R-module. Then $\text{End}(A_R)$ is semilocal.*

Proof. Since A_R is finitely generated and R_R has finite dual Goldie dimension, A_R also has finite dual Goldie dimension (Propositions 2.43 and 2.42). Now apply Proposition 4.14. □

We say that a module M_R is *biuniform* if it is uniform and couniform, that is, if it has Goldie dimension 1 and dual Goldie dimension 1. For instance, every non-zero uniserial module is biuniform. The next corollary follows immediately from Theorem 4.3(c).

Corollary 4.16 *Let A be a module that is the direct sum of d biuniform modules. Then $\text{End}(A)$ is a semilocal ring and $\text{codim}(\text{End}(A)) \leq 2d$.* □

In particular,

Corollary 4.17 *Let A be a serial module of finite Goldie dimension d. Then $\text{End}(A)$ is a semilocal ring and $\text{codim}(\text{End}(A)) \leq 2d$.* □

Therefore serial modules of finite Goldie dimension, and more generally the modules that are finite direct sums of biuniform modules, have the n-th root property, have only finitely many isomorphism classes of direct summands, and cancel from direct sums.

[Herbera and Shamsuddin] have determined other interesting examples of modules with semilocal endomorphism rings. We present their results without proofs because they will not be used in the sequel of this book. For the proofs see [Herbera and Shamsuddin].

(1) *If P_R is a finitely generated quasi-projective module, then $\text{End}(P_R)$ is semilocal if and only if P_R has finite dual Goldie dimension.*

Recall that a module P_R is *quasi-projective* in case for each epimorphism $f\colon P_R \to Q_R$ and each homomorphism $g\colon P_R \to Q_R$ there is an endomorphism h of P_R such that $fh = g$.

(2) *The endomorphism ring of any linearly compact module is semilocal.*

A module A_R is *linearly compact* if whenever $\{\, x_i + A_i \mid i \in I \,\}$ is a family of cosets of A_R ($x_i \in A$ and A_i a submodule of A_R) with the finite intersection property, then $\bigcap_{i \in I} x_i + A_i \neq \emptyset$. In terms of congruences, note that $x \in x_i + A_i$ if and only if $x \equiv x_i \pmod{A_i}$. Thus a module A_R is linearly compact if and only if any finitely soluble system of congruences

$$x \equiv x_i \pmod{A_i}, \qquad i \in I,$$

is soluble.

Therefore linearly compact modules have the n-th root property, have only finitely many isomorphism classes of direct summands, and cancel from direct sums. Every artinian module is linearly compact, and therefore this result of Herbera and Shamsuddin's generalizes Theorem 4.12.

(3) *If A is a finitely generated right module of finite Goldie dimension over a strongly π-regular ring R, then $\mathrm{End}(A_R)$ is semilocal.*

(4) *If A_R is a finitely generated module of finite dual Goldie dimension over a ring R and $\mathbf{M}_n(R)$ is a right repetitive ring for every $n \geq 1$, then $\mathrm{End}_R(A)$ is semilocal.*

A ring S is *right repetitive* [Goodearl 87] if for any elements $a, b \in S$ the right ideal $I = \sum_{i \geq 0} a^i bS$ is finitely generated. Goodearl has shown that $\mathbf{M}_n(S)$ is a right repetitive ring for every $n \geq 1$ if and only if any surjective endomorphism of a finitely generated right S-module is an automorphism.

(5) *If R is a commutative ring and A_R is a module satisfying $AB5^*$, then $\mathrm{End}(A_R)$ is direct product of semilocal rings, 1 is in the stable range of $\mathrm{End}(A_R)$, A_R cancels from direct sums and satisfies the n-th root property* [Herbera and Shamsuddin, Corollary 9].

A module A_R satisfies $AB5^*$ if $\bigcap_{i \in I}(B + A_i) = B + \bigcap_{i \in I} A_i$ for every submodule B of A_R and every inverse system $\{A_i \mid i \in I\}$ of submodules of A_R.

4.4   Notes on Chapter 4

Theorem 4.2 is due to [Camps and Dicks, Theorem 1], who gave various applications of their result. One of these applications was the proof that the endomorphism ring of an artinian module is semilocal (Theorem 4.12), a result conjectured by Pere Menal. Theorem 4.3 is due to [Herbera and Shamsuddin]. They extended some of the ideas of Camps and Dicks, studying the modules with semilocal endomorphism rings. For the decompositions of projective modules over semilocal rings, see [Facchini and Herbera].

Theorem 4.4 is due to [Bass 64] and Theorem 4.5 is due to [Evans, Theorem 2]. Corollary 4.6 generalizes a result of [Warfield 69a, Proposition 2], who proved that a module with a *local* endomorphism ring cancels from direct sums. He also proved that every finitely generated projective module over a semilocal ring cancels from direct sums [Warfield 79, Lemma 2]. This immediately follows also from Example (3) on page 6, Proposition 1.13 and Corollary 4.6. Propositions 4.8 and 4.9 appear in [Facchini, Herbera, Levy and Vámos, Proposition 2.1], but it was W. Dicks who realized that they hold for arbitrary modules with semilocal endomorphism rings and not only for artinian modules. Theorem 4.10 is due to [Fuller and Shutters, Theorem 9] and Corollary 4.11 to [Wiegand]. The examples in Section 4.3 are taken from [Herbera and Shamsuddin], apart from Theorem 4.12 (artinian modules have semilocal endomorphism rings), which was proved by [Camps and Dicks, Corollary 6] as we have already said in the first paragraph of this section.

We shall now consider the *stable range conditions*. The definition of left stable range was introduced by [Bass 64] in algebraic K-theory. Let R be an arbitrary ring. An n-tuple $(a_1, \ldots, a_n) \in R^n$ is called *left unimodular* if $a_1, \ldots, a_n$ generate $_R R$, that is, $\sum_{i=1}^n Ra_i = R$. Similarly for *right unimodular* n-tuples. A ring R is said to have *left stable range* $\leq n$ if for each $m > n$ and left unimodular m-tuple $(a_1, \ldots, a_m) \in R^m$ there exist $b_1, \ldots, b_{m-1} \in R$ such that $(a_1 + b_1 a_m, a_2 + b_2 a_m, \ldots, a_{m-1} + b_{m-1} a_m)$ is a left unimodular $(m-1)$-tuple. A similar definition is given for *right stable range* $\leq n$. Since a ring has left stable range $\leq n$ if and only if it has right stable range $\leq n$ [Vasershtein, Theorem 2], we shall just say that R has stable range $\leq n$, and write $\operatorname{sr}(R) \leq n$.

With these definitions, if a ring R has stable range $\leq n$, then R has stable range $\leq n+1$, so that we are interested in the least positive integer n such that R has stable range $\leq n$. This least integer n is called the *stable range* $\operatorname{sr}(R)$ of R. If no such integer exists, the stable range of R is said to be infinite, $\operatorname{sr}(R) = \infty$. For instance, if R is a commutative noetherian ring of Krull dimension d, then $\operatorname{sr}(R) \leq d+1$ [Bass 68].

Now let X be a topological space, $n \geq 1$ an integer,

$$\mathbf{S}^n = \{\, x \in \mathbf{R}^{n+1} \mid \|x\| = 1 \,\}$$

the n-dimensional sphere and

$$\mathbf{S}^{n-1} = \{\, x \in \mathbf{S}^n \mid x_{n+1} = 0 \,\}.$$

A continuous map $f \colon X \to \mathbf{S}^n$ is said to be *non-essential* if there is a continuous map $g \colon X \to \mathbf{S}^{n-1}$ such that f and g agree on $f^{-1}(\mathbf{S}^{n-1})$. Otherwise f is said to be *essential*. The *dimension* $\dim(X)$ of the topological space X is the greatest integer n for which there exists an essential mapping $f \colon X \to \mathbf{S}^n$. If no such integer exists the dimension of X is defined to be infinite, $\dim(X) = \infty$. For instance, one can prove that $\dim(\mathbf{R}^n) = n$. For an arbitrary topological space X let $\mathcal{C}(X)$ and $\mathcal{C}^*(X)$ denote the ring of all continuous functions $f \colon X \to \mathbf{R}$ and the ring of all bounded continuous functions $f \colon X \to \mathbf{R}$ respectively. Then $\operatorname{sr}(\mathcal{C}(X)) = \operatorname{sr}(\mathcal{C}^*(X)) = \dim(X) + 1$ [Vasershtein]. Hence the stable range of the rings $\mathcal{C}(X)$ and $\mathcal{C}^*(X)$ determines the dimension of the topological space X, and conversely.

We have seen that semilocal rings have stable range 1 (Theorem 4.4). Another important class of rings with stable range 1 is given by strongly π-regular rings: every strongly π-regular ring has stable range 1 [Ara 96, Theorem 4]. In particular, the endomorphism ring of a Fitting module has stable range 1 (Proposition 3.18). From this and Theorem 4.5 it follows that Fitting modules cancel from direct sums [Ara 96, Corollary 6].

Chapter 5

Serial Rings

In this chapter we shall give some results about the structure of serial rings, in particular noetherian serial rings (Section 5.4). We begin by considering the case of chain rings, since this is considerably simpler than that of arbitrary serial rings.

5.1 Chain rings and right chain rings

Let R be a right chain ring and J its Jacobson radical, so that J is the unique maximal right ideal of R. The R-module J^n/J^{n+1} is a uniserial right module and is a right vector space over the division ring R/J, that is, it is a semisimple R-module. Thus J^n/J^{n+1} must be either zero or a simple right R-module, that is, there are no right ideals properly between J^n and J^{n+1}. Therefore the unique right ideals of R containing J^n are the powers J^i, $i = 0, 1, 2, \ldots, n$.

Lemma 5.1 *Let R be a right chain ring and let A be a right ideal of R. Then there is no prime ideal P with $\bigcap_{n \geq 1} A^n \subset P \subset A$.*

Proof. If P is a prime ideal of R and $P \subset A$, then $P \subset A^n$ for every n, so that $P \subseteq \bigcap_{n \geq 1} A^n$. $\qquad \square$

Recall that an ideal P of an arbitrary ring R is said to be *completely prime* if it is proper and $xy \in P$ implies $x \in P$ or $y \in P$ for every $x, y \in R$. Therefore an ideal P of R is completely prime if and only if R/P is a domain. Every completely prime ideal is prime. We shall see that it is possible to localize a chain ring at a completely prime ideal in the same way that a commutative ring can be localized at an arbitrary prime ideal (Section 6.4).

We collect some properties of completely prime ideals of right chain rings in the following proposition.

Proposition 5.2 *Let R be a right chain ring. Then:*

(a) *A proper ideal P is completely prime if and only if $x^2 \in P$ implies $x \in P$ for every $x \in R$.*

(b) *If $x \in R$, then there is no completely prime ideal P with*

$$\bigcap_{n \geq 1} x^n R \subset P \subset xR.$$

(c) *If A is a non-zero proper ideal of R and $A^2 = A$, then A is completely prime.*

(d) *If A is a proper ideal of R which is not nilpotent, then $\bigcap_{n \geq 1} A^n$ is a completely prime ideal of R.*

(e) *If P is a completely prime ideal of R and $x \in R \setminus P$, then $P = Px$.*

(f) *If P is a completely prime ideal which is right principal, then either P is the Jacobson radical $J(R)$ of R or $P = 0$.*

Proof. (a) It is sufficient to prove that if P is a proper ideal with the property that $x^2 \in P$ implies $x \in P$ for every $x \in R$, then P is completely prime. Let P be a proper ideal with this property, and suppose that $xy \in P$ for some $x, y \in R$. Since R is a right chain ring, either $x = ys$ or $y = xs$ for some $s \in R$. If $x = ys$, then $x^2 = xys \in P$, so that $x \in P$. Suppose $y = xs$. Then $(yx)^2 = y(xy)x \in P$, so that $yx \in P$. It follows that $y^2 = yxs \in P$, and thus $y \in P$. This shows that P is completely prime.

(b) Let P be a completely prime ideal of R properly contained in xR. Then $x \notin P$, so that $x^n \notin P$ for every $n \geq 1$. Since R is a right chain ring, it follows that $P \subset x^n R$ for every $n \geq 1$. Therefore $P \subseteq \bigcap_{n \geq 1} x^n R$.

(c) Let $A = A^2$ be a non-zero proper ideal of R. By (a) it is sufficient to prove that $x \in R$ and $x^2 \in A$ imply $x \in A$. Suppose $x \in R$, $x^2 \in A$ and $x \notin A$. Then $A \subset xR$, so that $A \subseteq xJ(R)$. Therefore $A = A^2 \subseteq xRA = xA \subseteq x^2 J(R)$, hence $x^2 \neq 0$. Thus $A \subseteq x^2 J(R) \subset x^2 R \subseteq A$, contradiction.

(d) Let A be a non-nilpotent proper ideal of R and set $P = \bigcap_{n \geq 1} A^n$. We must show that the ideal P is completely prime in R. If $P = A^n$ for some $n \geq 1$, then $P = A^{2n}$, so that $P = P^2$ and the conclusion follows from (c). Hence we can suppose $P \subset A^n$ for every $n \geq 1$. If P is not completely prime, there exists an element $x \in R$ with $x^2 \in P$ and $x \notin P$. Then $x \notin A^m$ for some positive integer m, and $x^2 R \subseteq P \subset A^{2m} \subseteq xRA^m = xA^m \subseteq x^2 R$, contradiction.

(e) This is an elementary proof.

(f) Let P be a completely prime right principal ideal of R. Then $P \subseteq J(R)$ because P is proper. If $P \subset J(R)$, then let x be an element of $J(R)$ not in P, so that $P = Px$ by (e). Apply Nakayama's Lemma to the finitely generated module P_R. Then $P = Px \subseteq PJ(R)$ implies $P = 0$. $\square$

The final part of this section is devoted to chain rings with chain conditions.

Proposition 5.3 *Let R be a chain ring and $J = J(R)$ be its Jacobson radical.*

(a) *The ring R is right artinian if and only if it is left artinian, if and only if J is nilpotent.*

(b) *The ring R is right noetherian if and only if it is left noetherian, if and only if $\bigcap_{n \geq 1} J^n = 0$.*

Proof. (a) Follows immediately from the fact that in a right artinian ring the Jacobson radical is a nilpotent ideal.

(b) If R is a right noetherian chain ring, the completely prime ideals of R are just J and (possibly) 0 (Proposition 5.2(f)). If J is nilpotent, then $\bigcap_{n \geq 1} J^n = 0$. If J is not nilpotent, then $\bigcap_{n \geq 1} J^n = 0$ by Proposition 5.2(d).

Conversely, suppose that $\bigcap_{n \geq 1} J^n = 0$. If A is a non-zero right ideal of A, then it cannot be that $A \subseteq J^n$ for every n. Hence $A \supseteq J^m$ for some m. In the first paragraph of this section we have already remarked that in a right chain ring the unique right ideals containing J^m are the powers J^i, $i = 0, 1, 2, \ldots, m$. This shows that $A = J^i$ for some $i \geq 0$. It follows immediately that R is right noetherian. □

From the previous proof it follows that in a right noetherian chain ring R, which is necessarily left noetherian also, every one-sided ideal is two-sided. The ideals of R are just zero and the powers J^n of J, so that the lattice of its right ideals coincides with the lattice of its left ideals and is either finite (if R is artinian) or order antiisomorphic to $\omega + 1$ (if R is not artinian). Moreover, R has one completely prime ideal (the ideal J) if R is artinian, and two completely prime ideals (the ideals J and $0 = \bigcap_{n \geq 1} J^n$) if R is not artinian. In particular, if p is any element of J not in J^2, then $Rp = J$ and $pR = J$. Hence $Rp = pR$, so that $J^n = Rp^n = p^n R$ for every n.

5.2 Modules over artinian serial rings

In the last paragraph of Section 5.1 we saw that if R is an artinian chain ring and p is an element of $J = J(R)$ not in J^2, then the right (or left) ideals of R are only the ideals $J^n = Rp^n = p^n R$ for $n \geq 0$. As an exercise, the reader could prove that from this remark the next lemma follows easily. (We shall prove Lemma 5.5, which is an improvement of Lemma 5.4.)

Lemma 5.4 *If R is an artinian chain ring, then R is self-injective, that is, R_R is an injective module.* □

Lemma 5.4 cannot be extended to artinian serial rings by simply saying that every artinian serial ring is injective as a module over itself. For instance, let $R = \mathbf{UT}_2(k)$ be the ring of 2×2 upper triangular matrices over a field k, so that R is an artinian serial ring (Example 1.21). The Jacobson radical of R is

$$J = \begin{pmatrix} 0 & k \\ 0 & 0 \end{pmatrix}.$$

Let $\varphi \colon J_R \to R_R$ be the homomorphism defined by

$$\varphi \begin{pmatrix} 0 & a \\ 0 & 0 \end{pmatrix} = \begin{pmatrix} 0 & 0 \\ 0 & a \end{pmatrix}$$

for every $a \in k$. Then φ does not extend to an endomorphism of R_R, otherwise it would coincide with left multiplication by some element of R, which it does not. Therefore R_R is not injective.

Let R be a serial ring and let $\{e_1, \dots, e_n\}$ be a complete set of orthogonal primitive idempotents of R. Since $R_R = \oplus_{i=1}^{n} e_i R$, the module R_R is injective if and only if $e_i R$ is injective for every i. The next lemma describes which indecomposable direct summands $e_i R$ of an artinian serial ring R are injective.

Lemma 5.5 *Let R be an artinian serial ring with Jacobson radical J and complete set of orthogonal primitive idempotents $\{e_1, \dots, e_n\}$. Then $e_i R$ is injective if and only if $e_i R$ is not isomorphic to $e_k J$ for every $k = 1, 2, \dots, n$.*

Proof. If $e_i R$ is isomorphic to $e_k J$ for some $k = 1, 2, \dots, n$, then $e_i R$ is isomorphic to a submodule of $e_k R$ that is not a direct summand of $e_k R$. In particular, $e_i R$ is not injective.

Conversely, suppose $e_i R \not\cong e_k J$ for every $k = 1, 2, \dots, n$. In order to show that $e_i R$ is injective, we must prove that every homomorphism $I \to e_i R$ from a right ideal I of R extends to a homomorphism $R_R \to e_i R$. Equivalently, we must show that $\operatorname{Ext}_R^1(R/I, e_i R) = 0$. Since R is artinian, it suffices to show that $\operatorname{Ext}_R^1(S_R, e_i R) = 0$ for every simple right R-module S_R by induction on the composition length of R/I. Hence we only have to show that for every $k = 1, 2, \dots, n$ every homomorphism $\varphi \colon e_k J \to e_i R$ extends to a homomorphism $\psi \colon e_k R \to e_i R$, that is, $\psi \varepsilon = \varphi$, where $\varepsilon \colon e_k J \to e_k R$ is the embedding of $e_k J$ into $e_k R$.

Let $\varepsilon_j \colon e_j R \to R$ be the embedding of $e_j R$ into R, and let $\pi_j \colon R \to e_j R$ be the canonical projection for every $j = 1, 2, \dots, n$. Since $e_k J$ is cyclic and uniserial, there exists an index $\ell = 1, 2, \dots, n$ such that $e_k J$ is a homomorphic image of $e_\ell R$. Applying the functor $\operatorname{Hom}_R(-, R)$ to the exact sequence $e_\ell R \to e_k J \to 0$ we see that $\operatorname{Hom}_R(e_k J, R)$ is isomorphic to a submodule of $\operatorname{Hom}_R(e_\ell R, R) \cong R e_\ell$. Hence $\operatorname{Hom}_R(e_k J, R)$ is a uniserial left R-module, so that one of $\varepsilon_i \varphi$ and $\varepsilon_k \varepsilon$ is a multiple of the other.

If $\varepsilon_i \varphi = r \varepsilon_k \varepsilon$ for some $r \in R$, then $\psi = \pi_i r \varepsilon_k$ is the required extension of φ, because $\psi \varepsilon = \pi_i r \varepsilon_k \varepsilon = \pi_i \varepsilon_i \varphi = \varphi$.

Suppose $\varepsilon_k \varepsilon = r\varepsilon_i \varphi$ for some $r \in R$ and set $f = \pi_k r\varepsilon_i : e_i R \to e_k R$. Then

$$f\varphi = \pi_k r\varepsilon_i \varphi = \pi_k \varepsilon_k \varepsilon = \varepsilon, \tag{5.1}$$

so that $e_k J = \varepsilon(e_k J) = f\varphi(e_k J) \subseteq f(e_i R)$. If $e_k J = f(e_i R)$, then $f\varphi = \varepsilon$ forces $f\varphi = 1_{e_k J}$, so that $\varphi : e_k J \to e_i R$ is a splitting monomorphism. Since $e_i R$ is uniserial, it follows that either $e_k J = 0$ or $e_k J \cong e_i R$. As $e_k J \not\cong e_i R$, we have $e_k J = 0$, hence $\varphi : e_k J \to e_i R$ trivially extends to a homomorphism $\psi : e_k R \to e_i R$. Thus we may suppose $e_k J \subset f(e_i R)$, that is, $f : e_i R \to e_k R$ is an epimorphism. Since $e_k R$ is a non-zero projective module and $e_i R$ is indecomposable, it follows that f is an isomorphism. Thus from (5.1) we obtain that $\varphi = f^{-1}\varepsilon$, i.e., f^{-1} is the required extension of φ. $\qquad\square$

As a consequence of Lemma 5.5 and Proposition 2.28 we find that the Krull-Schmidt-Remak-Azumaya Theorem holds for arbitrary modules over an artinian serial ring:

Theorem 5.6 *Let R be an artinian serial ring. Then every module M_R is a direct sum of cyclic uniserial modules, and any two direct sum decompositions of M_R into direct sums of non-zero uniserial modules are isomorphic.*

Proof. We shall apply Proposition 2.28 to the class $\mathcal{C}$ of all non-zero right R-modules and the class $\mathcal{B}$ of all non-zero cyclic uniserial right R-modules. Firstly, we must show that every non-zero module M_R over the artinian serial ring R has a non-zero cyclic uniserial direct summand. To see this, we can suppose M faithful without loss of generality. Since the Jacobson radical J of R is nilpotent, there exists an $m \geq 1$ such that $J^m = 0$ and $J^{m-1} \neq 0$. Then $e_i J^{m-1} \neq 0$ for some $i = 1, 2, \ldots, n$, so that $e_i R$ is a uniserial module of composition length m and $e_j R$ has composition length $\leq m$ for every $j = 1, 2, \ldots, n$. By Lemma 5.5 the right ideal $e_i R$ is injective. As M is faithful, there exists $x \in M$ with $xe_i J^{m-1} \neq 0$, hence $xe_i R$ is a uniserial module of composition length m that is a homomorphic image of $e_i R$. Thus $xe_i R \cong e_i R$ is injective and is the desired non-zero cyclic uniserial direct summand of M.

Secondly, we must prove that for every proper pure submodule P of a module M_R there exists a non-zero cyclic uniserial submodule D of M such that $P \cap D = 0$ and $P + D = P \oplus D$ is pure in M. If P is such a proper submodule, then M/P has a non-zero cyclic uniserial direct summand M'/P by the first part of the proof. From Proposition 1.31(b) it follows that M' is pure in M. By Proposition 1.31(c) P is a pure submodule of M'. Since M'/P is finitely presented, hence pure-projective, the pure submodule P is a direct summand of M', that is, $M' = P \oplus D$ for a non-zero cyclic uniserial submodule D of M. Hence D has the required properties, so that from Proposition 2.28 we obtain the first part of the statement.

Since every uniserial R-module is artinian, the uniqueness in the second part of the statement follows from Theorem 2.18. $\qquad\square$

We leave the proof of the following result to the reader as an exercise. It follows easily from Theorem 5.6 and Example 3.11. It gives a complete set of invariants for modules over a fixed artinian serial ring R.

Corollary 5.7 *Let R be an artinian serial ring and let $\mathcal{P} = \{P_1, \dots, P_m\}$ be a set of representatives of the indecomposable projective right R-modules up to isomorphism. Then for every right R-module M_R there is a unique set $\{(P_\lambda, n_\lambda) \mid \lambda \in \Lambda\}$ such that $M_R \cong \oplus_{\lambda \in \Lambda} P_\lambda / P_\lambda J(R)^{n_\lambda}$, $P_\lambda \in \mathcal{P}$ and n_λ is a positive integer $\leq \text{length}(P_\lambda)$ for every $\lambda \in \Lambda$.* $\qquad\square$

5.3 Nonsingular and semihereditary serial rings

The aim of this section is to prove that a right serial ring is right nonsingular if and only if it is right semihereditary. First of all we introduce the singular submodule of a module.

Lemma 5.8 *Let M and N be right R-modules and let $f\colon M \to N$ be a homomorphism.*

(a) *If N' is an essential submodule of N, then $f^{-1}(N')$ is an essential submodule of M.*

(b) *If M' is a superfluous submodule of M, then $f(M')$ is a superfluous submodule of N.*

Proof. (a) Let P be a non-zero submodule of M. If $f(P) = 0$, then $P \subseteq f^{-1}(N')$, so that $f^{-1}(N') \cap P \neq 0$. If $f(P) \neq 0$, then $N' \cap f(P) \neq 0$, hence there exists $x \in P$ with $f(x) \neq 0$ and $f(x) \in N'$. Then $x \neq 0$ and $x \in f^{-1}(N')$, i.e., $f^{-1}(N') \cap P \neq 0$.

(b) Let Q be a submodule of N such that $f(M') + Q = N$. Then

$$M' + f^{-1}(Q) = M,$$

because if $x \in M$, then $f(x) \in N = f(M') + Q$, so that $f(x) = f(y) + q$ for some $y \in M'$ and $q \in Q$, and this forces $x - y \in f^{-1}(Q)$, that is, $x \in M' + f^{-1}(Q)$. Now M' is superfluous in M, so that $f^{-1}(Q) = M$. In particular $M' \subseteq f^{-1}(Q)$, hence $f(M') \subseteq Q$. Thus $f(M') + Q = N$ implies $Q = N$. $\qquad\square$

Lemma 5.9 *If M is a right module over an arbitrary ring R, the set*

$$Z(M) = \{\, x \in M \mid \text{ann}_R(x) \text{ is essential in } R_R \,\}$$

is a submodule of M.

Proof. Since R is essential in R_R, we have $0 \in Z(M)$. If $x, y \in Z(M)$, then $\mathrm{ann}_R(x - y)$ is essential in R_R because it contains $\mathrm{ann}_R(x) \cap \mathrm{ann}_R(y)$. If $x \in Z(M)$ and $r \in R$, apply Lemma 5.8(a) to the homomorphism $f: R_R \to R_R$ given by left multiplication by r. Then $f^{-1}(\mathrm{ann}_R(x))$ is essential in R_R, and $\mathrm{ann}_R(xr) \supseteq f^{-1}(\mathrm{ann}_R(x))$. Hence $\mathrm{ann}_R(xr)$ is essential and $xr \in Z(M)$. □

The submodule $Z(M) = \{ x \in M \mid \mathrm{ann}_R(x) \text{ is essential in } R_R \}$ is called the *singular submodule* of M. The module M is a *nonsingular module* if $Z(M) = 0$.

Proposition 5.10 *A module M is singular if and only if $M \cong A/B$ for some module A and some essential submodule B of A.*

Proof. Suppose M is singular. Then $M \cong A/B$ for some free module A and some submodule B of A. Suppose $A = \oplus_{\lambda \in \Lambda} x_\lambda R$ with $x_\lambda R \cong R_R$ for every λ. Since M is singular, for every $\lambda \in \Lambda$ there is an essential right ideal I_λ of R such that $x_\lambda I_\lambda \subseteq B$. Then $\oplus_{\lambda \in \Lambda} x_\lambda I_\lambda$ is essential in $\oplus_{\lambda \in \Lambda} x_\lambda R = A$, hence B is essential in A.

Conversely, suppose $M \cong A/B$ with B essential in A. If $x \in A$, apply Lemma 5.8(a) to the homomorphism $f: R_R \to A$ given by left multiplication by x. Then $f^{-1}(B)$ is essential in R_R. But $f^{-1}(B) = \mathrm{ann}_R(x + B)$, so that $x + B \in Z(A/B)$. This shows that A/B, hence M, is singular. □

It is easily seen that if R is a ring, $r \in R$ and $x \in Z(R_R)$, then $rx \in Z(R_R)$. Thus $Z(R_R)$ is a two-sided ideal of R, called the *right singular ideal* of R. A ring R is *right nonsingular* if $Z(R_R) = 0$. Similarly, R is left nonsingular if $Z(_R R) = 0$.

Proposition 5.11 *Let R be a right serial ring and let $\{e_1, \ldots, e_n\}$ be a complete set of orthogonal primitive idempotents of R. The following conditions are equivalent:*
(a) *The ring R is right nonsingular.*
(b) *For every $i, j, k = 1, 2, \ldots, n$, if $r \in e_i R e_j$ and $s \in e_j R e_k$ are non-zero, then rs is non-zero.*

Proof. (a) $\Rightarrow$ (b). Suppose R is right nonsingular and (b) false, so there exist indices $i, j, k = 1, \ldots, n$ and non-zero elements $r \in e_i R e_j$, $s \in e_j R e_k$ with $rs = 0$. Then sR is an essential submodule of $e_j R$, so that $I = sR \oplus (1 - e_j)R$ is an essential right ideal of R. But $rI = 0$. This is a contradiction, because R is right nonsingular.

(b) $\Rightarrow$ (a). Suppose that R is not right nonsingular, so that $Z(R_R) \neq 0$. Then there exist $i, j = 1, 2, \ldots, n$ such that $e_i Z(R_R) e_j \neq 0$. Let r be a non-zero element of $e_i Z(R_R) e_j$. As $r \in Z(R_R)$, the right ideal $\mathrm{r.ann}_R(r)$ is essential in R_R. Hence there exists a non-zero element $t \in \mathrm{r.ann}_R(r) \cap e_j R$. Since t is non-zero, there exists an index $k = 1, 2, \ldots, n$ such that $te_k \neq 0$. Then $s = te_k$ is a non-zero element of $e_j R e_k$ and $rs = rte_k = 0$. □

In particular, a right chain ring is right nonsingular if and only if it is a domain. From Proposition 5.11 and its left-handed version we obtain

Corollary 5.12 *A serial ring is right nonsingular if and only if it is left non-singular.* □

We briefly review some notions about the decomposition theory of a ring R [Anderson and Fuller, pp. 99–100]. Let $E = \{e_1, \ldots, e_n\}$ be a complete set of orthogonal primitive idempotents of an arbitrary ring R. On the set E define a relation $\sim$ by setting $e_i \sim e_j$ if there is an index $k = 1, 2, \ldots, n$ such that $e_k R e_i \neq 0$ and $e_k R e_j \neq 0$. Then $\sim$ is a reflexive and symmetric relation on E. The transitive closure $\approx$ of $\sim$, i.e., the smallest transitive relation that contains $\sim$, is an equivalence relation on E. Let $E_1, \ldots, E_m$ be the $\approx$ equivalence classes of E, and for each $t = 1, \ldots, m$ let u_t be the sum of the elements in E_t. The idempotents $u_1, \ldots, u_m$ are called the *block idempotents* of R determined by E. For the proof of the following theorem see [Anderson and Fuller, Theorem 7.9].

Theorem 5.13 *Let R be a ring with a complete set of pairwise orthogonal primitive idempotents $E = \{e_1, \ldots, e_n\}$, and let $u_1, \ldots, u_m$ be the block idempotents of R determined by E. Then $\{u_1, \ldots, u_m\}$ is a complete set of pairwise orthogonal central idempotents, and $R = u_1 R u_1 \times \cdots \times u_m R u_m$ is the unique ring decomposition of R into indecomposable rings.* □

For right serial rings the situation is considerably simpler.

Lemma 5.14 *Let R be a right serial ring and let $E = \{e_1, \ldots, e_n\}$ be a complete set of orthogonal primitive idempotents of R. Then for every $e_i, e_j \in E$, $e_i \sim e_j$ if and only if either $e_i R e_j \neq 0$ or $e_j R e_i \neq 0$. Moreover, if R is a nonsingular serial ring, the relations $\sim$ and $\approx$ coincide.*

Proof. It is clear that if either $e_i R e_j \neq 0$ or $e_j R e_i \neq 0$, then $e_i \sim e_j$. Conversely, suppose R is right serial and $e_i \sim e_j$, so that there exists an index $k = 1, 2, \ldots, n$ with $e_k R e_i \neq 0$ and $e_k R e_j \neq 0$. Let $e_k r e_i \in e_k R e_i$ and $e_k s e_j \in e_k R e_j$ be two non-zero elements. Since $e_k R$ is uniserial, there exists $t \in R$ such that either $e_k r e_i t = e_k s e_j$ or $e_k r e_i = e_k s e_j t$. If $e_k r e_i t = e_k s e_j$, then $e_k r e_i \cdot e_i t e_j = e_k s e_j$, so that $e_i t e_j \neq 0$. In particular, $e_i R e_j \neq 0$. Similarly, $e_k r e_i = e_k s e_j t$ implies $e_j R e_i \neq 0$.

Now assume that R is a nonsingular serial ring. In order to prove that the relations $\sim$ and $\approx$ coincide, we must show that $\sim$ is transitive. Suppose $e_i \sim e_j$ and $e_j \sim e_k$. If $e_i R e_j \neq 0$ and $e_j R e_k \neq 0$, then $e_i R e_k \neq 0$ by Proposition 5.11, so that $e_i \sim e_k$. Similarly, if $e_j R e_i \neq 0$ and $e_k R e_j \neq 0$, then $e_k R e_i \neq 0$, so that $e_i \sim e_k$. If $e_j R e_i \neq 0$ and $e_j R e_k \neq 0$, then $e_i \sim e_k$ by the definition of $\sim$. Finally, if $e_i R e_j \neq 0$ and $e_k R e_j \neq 0$, there exist $u, v \in R$ such that $e_i u e_j \neq 0$ and $e_k v e_j \neq 0$. Then $R e_j$ uniserial implies that there exists $w \in R$ such that either $w e_i u e_j = e_k v e_j$ or $e_i u e_j = w e_k v e_j$. In the first case $e_k w e_i \neq 0$, and in the second case $e_i w e_k \neq 0$. Therefore $e_i \sim e_k$. □

From Theorem 5.13 and Lemma 5.14 we obtain

Corollary 5.15 *Let R be a nonsingular serial ring and $E = \{e_1, \ldots, e_n\}$ a complete set of orthogonal primitive idempotents of R. The following conditions are equivalent:*
(a) *The ring R is indecomposable.*
(b) *For every $i, j = 1, 2, \ldots, n$ either $e_i R e_j \neq 0$ or $e_j R e_i \neq 0$.* □

A ring R is *right hereditary* if every right ideal of R is a projective module, and is *right semihereditary* if every finitely generated right ideal is projective.

Theorem 5.16 *Let R be a right serial ring. Then R is right nonsingular if and only if R is right semihereditary.*

Proof. If R is a right semihereditary ring and $r \in Z(R_R)$, then $\mathrm{r.ann}_R(r)$ is essential and $R/\mathrm{r.ann}_R(r) \cong rR$ is projective. Therefore $\mathrm{r.ann}_R(r)$ is an essential direct summand of R_R. Thus $\mathrm{r.ann}_R(r) = R$, $r = 0$, and R_R is nonsingular.

Conversely, let R be a right nonsingular, right serial ring. Then

$$R = e_1 R \oplus \cdots \oplus e_n R$$

with every $e_i R$ uniserial. We shall show by induction on $m = 0, 1, \ldots, n$ that every finitely generated submodule of $e_1 R \oplus \cdots \oplus e_m R$ is projective. The case $m = 0$ is trivial. Let M be a finitely generated submodule of $e_1 R \oplus \cdots \oplus e_m R$, and let $\pi_m : e_1 R \oplus \cdots \oplus e_m R \to e_m R$ be the canonical projection. Then $\pi_m(M)$ is a finitely generated submodule of the uniserial module $e_m R$, hence it is a cyclic uniserial module, from which $\pi_m(M) \cong e_i R/N$ for some $i = 1, \ldots, n$ and some submodule N of $e_i R$. If $N \neq 0$, then N is essential in the uniserial module $e_i R$, so that $e_i R/N$ is singular by Proposition 5.10. Then $\pi_m(M)$ is a singular submodule of the nonsingular module $e_m R$, so that $\pi_m(M) = 0$. If $N = 0$, then $\pi_m(M) \cong e_i R$. In both cases $\pi_m(M)$ is projective, so that $M \cong \pi_m(M) \oplus (M \cap \ker \pi_m)$. As $M \cap \ker \pi_m$ is a finitely generated submodule of $\ker \pi_m = e_1 R \oplus \cdots \oplus e_{m-1} R$, by the induction hypothesis we conclude that $M \cap \ker \pi_m$, hence M, is projective. □

5.4 Noetherian serial rings

In Proposition 5.3(b) we saw that a chain ring is right noetherian if and only if it is left noetherian. This proposition cannot be extended to serial rings. For instance, if $\mathbf{Z}$ is the ring of integers, p is a prime number, $\mathbf{Z}_p$ is the localization of $\mathbf{Z}$ at its maximal ideal (p) and $\mathbf{Q}$ is the field of rational numbers, then

$$R = \begin{pmatrix} \mathbf{Z}_p & \mathbf{Q} \\ 0 & \mathbf{Q} \end{pmatrix}$$

is a serial right noetherian ring that is not left noetherian. In this section we shall study the structure of (two-sided) noetherian serial rings. The symbol R will denote a serial ring with Jacobson radical J in the whole section.

Lemma 5.17 *Let R be a serial, right noetherian ring. Then every uniform right R-module is uniserial. In particular, every indecomposable injective right R-module is uniserial and every injective right R-module is serial.*

Proof. In order to show that every uniform right R-module is uniserial, it suffices to prove that every finitely generated submodule N_R of a uniform module M_R is uniserial. Since N_R is finitely generated, it is finitely presented, hence serial (Corollary 3.30). But M_R uniform implies N_R indecomposable (Lemma 2.24(d)). Thus N_R is uniserial. In particular, every indecomposable injective module is uniform (Lemma 2.24(b)). Finally, every injective right module over a right noetherian ring is a direct sum of indecomposable modules [Anderson and Fuller, Theorem 25.6].   □

Let e be a primitive idempotent of a serial ring R, so that eR is a uniserial right module. Then for each integer $n \geq 0$ the module $eJ(R)^n/eJ(R)^{n+1}$ is a uniserial module annihilated by $J(R)$, that is, it is a uniserial module over the semisimple artinian ring $R/J(R)$. Hence $eJ(R)^n/eJ(R)^{n+1}$ is either zero or a simple module, so that $eR/eJ(R)^n$ is a uniserial module of finite length $\leq n$. It follows that $eR/\bigcap_{n\geq 0}(eJ(R)^n)$ is either a module of finite length or a module whose lattice of submodules is order antiisomorphic to $\omega + 1$. In both cases $eR/\bigcap_{n\geq 0}(eJ(R)^n)$ is a noetherian uniserial module. Therefore:

Lemma 5.18 *If R is a serial ring and $\bigcap_{n\geq 0} J(R)^n = 0$, then R is a noetherian ring. In particular, $R/\bigcap_{n\geq 0} J(R)^n$ is a noetherian ring for every serial ring R.*

Proof. As a right module $R_R = e_1 R \oplus \cdots \oplus e_n R$ for some orthogonal primitive idempotents $e_1, \ldots, e_n$. From $\bigcap_{n\geq 0} J(R)^n = 0$, it follows that each $e_i R$ is noetherian. Therefore R_R is noetherian. Similarly for $_R R$. In particular, if R is an arbitrary serial ring, then $J\left(R/\bigcap_{n\geq 0} J(R)^n\right) = J(R)/\bigcap_{n\geq 0} J(R)^n$.   □

We shall now define an "order" on the set of simple right modules over a serial ring R (up to isomorphism). It is essentially the order in which they appear as composition factors of an indecomposable projective right R-module. The next lemma gives some equivalent conditions and is much more precise.

Lemma 5.19 *Let R be a serial ring and let S, T be simple right R-modules. The following conditions are equivalent:*

(a) $\operatorname{Ext}_R^1(S, T) \neq 0$;

(b) *there exists a uniserial module M of composition length 2 with $\operatorname{soc}(M) \cong T$ and $M/MJ \cong S$;*

(c) *if P is the projective cover of S, then $T \cong PJ/PJ^2$.*

Proof. (a) $\Rightarrow$ (b). If $\text{Ext}_R^1(S, T) \neq 0$, there is an exact sequence

$$0 \longrightarrow T \overset{\varphi}{\longrightarrow} M \longrightarrow S \longrightarrow 0$$

that does not split. In particular M is not a semisimple module (Proposition 1.1), so that $MJ \neq 0$. Since $SJ = 0$, we have that $MJ \subseteq \varphi(T)$. But $\varphi(T)$ is simple, hence $MJ = \varphi(T)$. In particular, M is local (Lemma 1.14(b)) and $\varphi(T)$ is its unique maximal submodule. As T is simple, the unique proper submodules of M are just 0 and $\varphi(T)$. Thus M is a uniserial module of composition length 2.

(b) $\Rightarrow$ (c). Let M be a uniserial module of composition length 2 with $\text{soc}(M) \cong T$ and $M/MJ \cong S$, and let P be the projective cover of S. Then M is a homomorphic image of P, so that $M \cong P/PJ^2$ and $T \cong PJ/PJ^2$.

(c) $\Rightarrow$ (a). If P is a projective cover of S and $T \cong PJ/PJ^2$, there is an exact sequence $0 \rightarrow T \rightarrow P/PJ^2 \rightarrow S \rightarrow 0$, which is not splitting. $\qquad\square$

If S and T are simple modules, we say that T is a *successor* of S (or that S is a *predecessor* of T) if S and T satisfy the equivalent conditions of Lemma 5.19.

Lemma 5.20 *Let R be a serial, right noetherian ring and let S be a simple right R-module. Then S has at most one successor and one predecessor up to isomorphism. Moreover, if S is not projective, then S has one successor.*

Proof. If P is a projective cover of S, then every simple module T that is a successor of S must be isomorphic to PJ/PJ^2 (Lemma 5.19). This proves the uniqueness of T. Also if S is not projective, then $PJ \neq 0$. Since R is right noetherian, $PJ \neq PJ^2$. But P is uniserial, so that PJ/PJ^2 is a simple module and is a successor of S.

In order to show that the predecessor of S, if it exists, is unique up to isomorphism, let M be a uniserial module of composition length 2 with $\text{soc}(M) \cong S$. Then M is an essential extension of S, so that M is isomorphic to a submodule of the injective envelope $E(S)$ of S. Thus M is isomorphic to a submodule of $\text{soc}_2(E(S))$. By Lemma 5.17 the module $E(S)$ is uniserial, so that both $\text{soc}_1(E(S))$ and $\text{soc}_2(E(S))/\text{soc}_1(E(S))$ are semisimple uniserial modules, that is, they are either simple or zero. Thus $\text{soc}_2(E(S))$ is a uniserial module of composition length at most 2. It follows that $M \cong \text{soc}_2(E(S))$. Thus a uniserial module M of composition length 2 with $\text{soc}(M) \cong S$, if it exists, is isomorphic to $\text{soc}_2(E(S))$. Therefore the predecessor of S, if it exists, is necessarily isomorphic to $\text{soc}_2(E(S))/\text{soc}_1(E(S))$. $\qquad\square$

Denote the successor of a simple module S by $\sigma(S)$. More generally, if S is a simple right R-module and P is the projective cover of S, a simple module T is said to be a *successor of degree k* of S if $T \cong PJ^k/PJ^{k+1}$. Hence the successor of degree k of S exists if and only if $PJ^k \neq PJ^{k+1}$, and, if it exists, it is uniquely determined up to isomorphism and we shall write $PJ^k/PJ^{k+1} = \sigma_k(S)$. Thus

$\sigma_k(S)$ is defined only for the integers k with $PJ^k \neq PJ^{k+1}$. If R is serial and right noetherian, then two cases may occur: the submodules of the uniserial noetherian module P can be either $P \supset PJ \supset PJ^2 \supset \cdots \supset PJ^n = 0$ or $P \supset PJ \supset PJ^2 \supset \cdots \supset PJ^n \supset PJ^{n+1} \supset \cdots \supset \bigcap_{n \geq 0} PJ^n \supseteq \cdots$ (and in Theorem 5.28 we shall see that if R is noetherian then $\bigcap_{n \geq 0} PJ^n = 0$.) In the first case P is a uniserial module of finite composition length n and $\sigma_k(S)$ is defined only for $k < n$. In the second case P is a noetherian non-artinian uniserial module and $\sigma_k(S)$ is defined for every integer $k \geq 1$.

Also note that if $\sigma(S) = T$ and $\sigma(T) = U$, it might not follow that $\sigma_2(S) \cong U$, because if P is the projective cover of S it can happen that $PJ^2 = 0$.

Lemma 5.21 *Let e and f be primitive idempotents of a serial ring R and let k be a positive integer. Then $\sigma_k(eR/eJ) = fR/fJ$ if and only if $eRf \cap J^k \not\subseteq J^{k+1}$. In this case, if $t \in (eRf \cap J^k) \setminus J^{k+1}$, then left multiplication by t is a non-zero homomorphism $fR \to eR$ whose image is eJ^k.*

Proof. $\sigma_k(eR/eJ) = fR/fJ$ if and only if there is a homomorphism $\varphi \colon fR \to eR$ such that $\varphi(fR) \subseteq eJ^k$ and $\varphi(fR) \not\subseteq eJ^{k+1}$. Since homomorphisms $fR \to eR$ are given by left multiplication by elements of eRf, this happens if and only if there is an element $t \in eRf$ such that $t \in eJ^k$ and $t \notin eJ^{k+1}$. This is equivalent to $eRf \cap J^k \not\subseteq J^{k+1}$. $\qquad\square$

Our aim is now to prove Theorems 5.28 and 5.29, which describe the structure of noetherian serial rings. The proof of these theorems is subdivided into some technical lemmas.

Lemma 5.22 *If e and f are primitive idempotents of a serial ring R and k is a positive integer, then $eR/eJ \cong fJ^k/fJ^{k+1}$ if and only if $Rf/Jf \cong J^k e/J^{k+1} e$.*

Proof. By Lemma 5.21, for any primitive idempotents e, f of a serial ring R we have $eJ^k/eJ^{k+1} \cong fR/fJ$ if and only if $eRf \cap J^k \not\subseteq J^{k+1}$. If we apply this result to the opposite ring R^0 of R we obtain that $J^k e/J^{k+1} e \cong Rf/Jf$ if and only if $fRe \cap J^k \not\subseteq J^{k+1}$. Thus $eR/eJ \cong fJ^k/fJ^{k+1}$ if and only if $eR/eJ = \sigma_k(fR/fJ)$, if and only if (by Lemma 5.21) $fRe \cap J^k \not\subseteq J^{k+1}$, if and only if $J^k e/J^{k+1} e \cong Rf/Jf$. $\qquad\square$

Lemma 5.23 *If e is a primitive idempotent of a noetherian serial ring R and the simple module eR/eJ has no predecessor, then eR is artinian.*

Proof. If eR is not artinian, then $\sigma_k(eR/eJ)$ is defined for all $k \geq 1$. Since there are only a finite number of isomorphism classes of simple modules, there must be positive integers k, n such that $\sigma_n(eR/eJ) \cong \sigma_{n+k}(eR/eJ)$. By the uniqueness of predecessors (Lemma 5.20), it follows that

$$eR/eJ \cong \sigma_k(eR/eJ),$$

which implies that eR/eJ itself has a predecessor. $\qquad\square$

Lemma 5.24 *If R is a serial, right noetherian ring with Jacobson radical J and e is a primitive idempotent such that $\bigcap_{n>0} eJ^n \neq 0$, then there exists a primitive idempotent f such that fR/fJ is a simple module with no predecessor and $eRf \neq 0$.*

Proof. Since R is a serial, right noetherian ring, the indecomposable projective module eR is a noetherian cyclic uniserial module whose lattice of submodules is well-ordered under inverse inclusion. Let

$$eR = P_0 \supset P_1 \supset P_2 \supset \cdots \supset P_\omega \supset P_{\omega+1} \supseteq \cdots$$

be the submodules of eR, so that all the submodules P_α are cyclic, $P_{\alpha+1} = P_\alpha J$ for every ordinal $\alpha \leq \omega$, and $P_\omega = \bigcap_{n>0} eJ^n$. Then $P/P_{\omega+1}$ is an essential extension of the simple module $S = P_\omega/P_{\omega+1}$. If S has a predecessor, there exists a uniserial module M of composition length 2 with $\text{soc}(M) \cong S$. The injective envelope E of S is a uniserial module by Lemma 5.17. Now $P/P_{\omega+1}$ and M are isomorphic to submodules of E, the module M has finite composition length and $P/P_{\omega+1}$ does not have finite composition length. It follows that M is isomorphic to a submodule of $P/P_{\omega+1}$. This is a contradiction, because $P/P_{\omega+1}$ does not have any submodules of composition length 2. Hence S has no predecessor. Let f be a primitive idempotent of R such that $S \cong fR/fJ$. Then there is a non-zero homomorphism $fR \to P_\omega/P_{\omega+1}$, so that there is a non-zero homomorphism $fR \to eR$. Thus $eRf \neq 0$. $\qquad\square$

Recall that in Section 2.11 we defined $\delta(M)$ as the largest Loewy submodule of a module M. If R is a ring, then we can regard R as a right or a left module over itself, obtaining in this way a Loewy right ideal $\delta(R_R)$ and a Loewy left ideal $\delta(_RR)$. Since $\delta(M)$ is a fully invariant submodule of M, both $\delta(R_R)$ and $\delta(_RR)$ are two-sided ideals of R.

Lemma 5.25 *If R is a noetherian serial ring, then $\delta(_RR) = \delta(R_R)$.*

Proof. By symmetry it is sufficient to show that $\delta(_RR) \subseteq \delta(R_R)$.

The ring $R/\delta(R_R)$ is a noetherian serial ring whose socle as a right module over itself is zero. If we can show that $\text{soc}(R_R) = 0$ implies $\text{soc}(_RR) = 0$ for every noetherian serial ring R, it will follow that the socle of the left $R/\delta(R_R)$-module $R/\delta(R_R)$ is zero. Then the socle of the left R-module $R/\delta(R_R)$ is zero. Thus the δ-torsion submodule of the left R-module $R/\delta(R_R)$ is zero, hence $\delta(_RR) \subseteq \delta(R_R)$.

This shows that we may suppose that R is a noetherian serial ring with $\text{soc}(R_R) = 0$ and prove that $\text{soc}(_RR) = 0$. If $\bigcap_{n\geq 1} J^n \neq 0$, then there is a primitive idempotent e such that $e\left(\bigcap_{n\geq 1} J^n\right) \neq 0$. Then $\bigcap_{n\geq 1} eJ^n \neq 0$, so that by Lemma 5.24 there exists a primitive idempotent f with $eRf \neq 0$ and fR/fJ with no predecessor. By Lemma 5.23 the module fR is artinian.

But this contradicts the fact that $\text{soc}(R_R) = 0$. The contradiction proves that $\bigcap_{n \geq 1} J^n = 0$.

Let $\{e_1, \ldots, e_n\}$ be a complete set of primitive orthogonal idempotents of R. As $\bigcap_{n \geq 1} J^n = 0$, the submodules of the indecomposable projective left module Re_i are only 0 and the modules $J^n e_i$. Suppose $\text{soc}(_R R) \neq 0$. Then $\text{soc}(Re_i) \neq 0$ for some index i, so that Re_i has a simple submodule $J^m e_i$. Then Re_i is artinian. Now consider the simple right module $e_i R / e_i J$. Since R is a noetherian serial ring and $\text{soc}(R_R) = 0$, we know that $e_i J^n \neq e_i J^{n+1}$ for every n, so that $\sigma_n(e_i R / e_i J)$ is defined for every $n > 0$. As there are only a finite number of isomorphism classes of simple modules, there must be positive integers ℓ and k such that $\sigma_\ell(e_i R / e_i J) = \sigma_{\ell+k}(e_i R / e_i J)$. By the uniqueness of the predecessor we obtain that $e_i R / e_i J = \sigma_k(e_i R / e_i J)$. Thus $e_i R / e_i J = \sigma_{tk}(e_i R / e_i J)$ for every $t \geq 1$, that is, $e_i R / e_i J \cong e_i J^{tk} / e_i J^{tk+1}$ for every $t \geq 1$. By Lemma 5.22 we have that $Re_i / Je_i \cong J^{tk} e_i / J^{tk+1} e_i$ for every $t \geq 1$. Thus Re_i is not artinian. This contradiction shows that $\text{soc}(_R R) = 0$. $\square$

Corollary 5.26 *If R is a noetherian serial ring and $r \in R$, then Rr is an artinian left module if and only if rR is an artinian right module.*

Proof. By Corollary 2.60 the module rR is artinian if and only if $r \in \delta(R_R)$. Now apply Lemma 5.25. $\square$

Lemma 5.27 *Let e and f be two primitive idempotents of a noetherian serial ring R. Suppose that eR is not artinian and fR is artinian. Then*

$$eRf = fRe = 0.$$

Proof. Assume there is a non-zero element $t \in eRf$. Since Rf is artinian by Corollary 5.26, Rf must have finite length, so that $J^n f = 0$ for some positive n. Hence $t \notin J^n$. Let k be the non-negative integer such that $t \in J^k$ and $t \notin J^{k+1}$. By Lemma 5.21 left multiplication by t is a non-zero homomorphism $fR \to eR$ whose image is eJ^k. Since both fR and the cokernel eR/eJ^k have finite composition length, eR also has finite composition length, which is a contradiction. This proves that $eRf = 0$.

Since eR is not artinian and fR is artinian, Re is not artinian and Rf is artinian by Corollary 5.26. If R^0 is the opposite ring of R, then fR^0 is an artinian right ideal in R^0 and eR^0 is not artinian. Apply what we have proved in the first part of the proof to R^0. Then $eR^0 f = 0$, i.e., $fRe = 0$. $\square$

We are ready for the two main results of this section.

Theorem 5.28 *A serial ring R is noetherian if and only if $\bigcap_{n > 0} J^n = 0$.*

Proof. One of the two implications was already proved in Lemma 5.18. Conversely, let R be a noetherian serial ring and suppose $\bigcap_{n>0} J^n \neq 0$. Then there is a primitive idempotent e such that

$$e \left(\bigcap_{n>0} J^n \right) \neq 0,$$

so that

$$\bigcap_{n>0} eJ^n \neq 0.$$

In particular, eR is not artinian. By Lemma 5.24 the ring R has a primitive idempotent f such that fR/fJ is a simple module with no predecessor and $eRf \neq 0$. By Lemma 5.23 fR is artinian. This contradicts Lemma 5.27. $\square$

Theorem 5.29 *A noetherian serial ring is the direct product of an artinian serial ring and a hereditary serial ring with no simple left or right ideals.*

Proof. Let $e_1, \ldots, e_n$ be a complete set of primitive orthogonal idempotents of R. Without loss of generality we may suppose that $e_1 R, \ldots, e_m R$ are not artinian modules and $e_{m+1} R, \ldots, e_n R$ are artinian modules. Set

$$e = e_1 + \ldots + e_m.$$

By Lemma 5.27 $eR(1-e) = (1-e)Re = 0$, whence there is a ring decomposition $R = eRe \times (1-e)R(1-e)$, where eRe is a serial ring with no projective indecomposable artinian modules and $(1-e)R(1-e)$ is an artinian serial ring. Hence we only have to show that if R is a noetherian serial ring, $e_1, \ldots, e_m$ is a complete set of primitive orthogonal idempotents of R and $e_i R$ is not artinian for every $i = 1, \ldots, m$, then R is a hereditary ring with no simple left or right ideals. Since the lattice of submodules of each $e_i R$ is order antiisomorphic to $\omega + 1$, it is obvious that $\delta(e_i R_R) = 0$, so that $\delta(R_R) = 0$. From Lemma 5.25 we get that $\delta(_R R) = \delta(R_R) = 0$, hence R has no simple left or right ideals. Let I be any right ideal of R. Then I is a finitely presented R-module, hence it is a finite direct sum of cyclic uniserial right ideals of R (Corollary 3.30). A cyclic uniserial right ideal of R is a homomorphic image of some $e_i R$. But the lattice of submodules of $e_i R$ is order antiisomorphic to $\omega + 1$, so that every proper homomorphic image of $e_i R$ is artinian. As $\delta(R_R) = 0$, the ring R has no artinian left ideals. Therefore every cyclic uniserial right ideal of R is isomorphic to some $e_i R$, hence it is projective. Therefore I itself must be projective, and R is hereditary. $\square$

5.5   Notes on Chapter 5

Section 5.1 contains some elementary properties of chain rings and their prime ideals. For further results about chain rings see [Bessenrodt, Brungs and Törner]. Lemma 5.5 and Theorem 5.6 are due to [Eisenbud and Griffith, Theorems 1.2 and 1.4], who strengthened a result of Nakayama. [Nakayama, Theorem 17] proved that every finitely generated module over an artinian serial ring is serial, and [Eisenbud and Griffith, Theorem 1.2] extended this result to arbitrary modules.

The singular submodule of a module was introduced by [Johnson 51, 57]. Proposition 5.11 and Corollary 5.15 are due to [Facchini and Puninski 96, Lemma 2.2 and Corollary 2.3]. In Corollary 5.15 a characterization of indecomposable nonsingular serial rings is given. For further characterizations see [Upham, Proposition 2.2]. Corollary 5.12 and Theorem 5.16 were proved by [Warfield 75, Theorems 4.1 and 4.6].

All the results in Section 5.4 are due to [Warfield 75], except for Lemma 5.18, taken from a paper of [Singh, Lemma 2.1].

Chapter 6

Quotient Rings

6.1 Quotient rings of arbitrary rings

In this section we briefly review some well-known results about quotient rings and Goldie's Theorem. The proofs can be found in almost every book of ring theory. We refer to [Goodearl and Warfield 89, Chapter 9].

If R is a ring, a *multiplicatively closed subset* S of R is a subset $S \subseteq R$ such that $1_R \in S$ and $xy \in S$ whenever $x \in S$ and $y \in S$. A *saturated multiplicatively closed subset* is a multiplicatively closed subset S of R such that for every $x, y \in R$, if $xy \in S$, then both x and y belong to S. Given a ring R and a multiplicatively closed subset S of R, a *right quotient ring of R with respect to S* is a ring $R[S^{-1}]$ together with a ring homomorphism $\varphi \colon R \to R[S^{-1}]$ satisfying the following conditions:

(a) for all $s \in S$, $\varphi(s)$ is an invertible element in $R[S^{-1}]$;

(b) every element of $R[S^{-1}]$ can be written in the form $\varphi(r)\varphi(s)^{-1}$ for some $r \in R$, $s \in S$;

(c) an element $r \in R$ belongs to $\ker \varphi$ if and only if $rs = 0$ for some $s \in S$.

The definition of a *left quotient ring of R with respect to S* is similar.

Proposition 6.1 *Let R be a ring and let S be a multiplicatively closed subset of R. Suppose that there exists a right quotient ring $\varphi \colon R \to R[S^{-1}]$ of R with respect to S. If R' is any ring and $f \colon R \to R'$ is a ring homomorphism such that $f(s)$ is invertible in R' for every $s \in S$, then there exists a unique ring homomorphism $g \colon R[S^{-1}] \to R'$ such that $f = g\varphi$.* □

A multiplicatively closed subset S of a ring R is a *right Ore set* if S is a proper subset of R and for each $r \in R$ and $s \in S$, there exist $r' \in R$, $s' \in S$ such that $rs' = sr'$. A *right denominator set* is a right Ore set with the property that for every $r \in R$, if there exists an element $s \in S$ such that $sr = 0$, there exists an element $s' \in S$ such that $rs' = 0$. Similarly for *left Ore sets* and *left denominator sets.*

Theorem 6.2 *If R is a ring and S is a multiplicatively closed subset of R, then there exists a right quotient ring of R with respect to S if and only if S is a right denominator set. In this case if $\varphi\colon R \to T$ and $\varphi'\colon R \to T'$ are two right quotient rings of R with respect to S, then there exists a unique isomorphism $\psi\colon T \to T'$ such that $\psi\varphi = \varphi'$.* $\qquad\square$

If R is a ring and S is a right denominator set, the right quotient ring $R[S^{-1}]$ can be constructed in the following way. Let $R \times S$ be the cartesian product and let $\sim$ be the equivalence relation on $R \times S$ defined by $(r, s) \sim (r', s')$ if there exist $x, x' \in R$ such that $rx = r'x'$ and $sx = s'x' \in S$. Let $R[S^{-1}]$ be the quotient set $R \times S/\sim$. For every element $(r, s) \in R \times S$ let rs^{-1} denote the equivalence class of (r, s) modulo $\sim$. If $(r, s), (r', s') \in R \times S$, then $sr'' = s's''$ for suitable $r'' \in R$, $s'' \in S$. Set

$$rs^{-1} + r's'^{-1} = (rr'' + r's'')(s's'')^{-1}.$$

Moreover $sr''' = r's'''$ for suitable $r''' \in R$, $s''' \in S$. Set

$$(rs^{-1})(r's'^{-1}) = (rr''')(s's''')^{-1}.$$

Then these operations on $R \times S/\sim$ are well-defined, and $R \times S/\sim$ with these operations is the right quotient ring $R[S^{-1}]$ of R with respect to S. The mapping $\varphi\colon R \to R \times S/\sim$ is defined by $\varphi(r) = r1^{-1}$ for every $r \in R$.

Of course there is also a left-hand version of Theorem 6.2. If S is a multiplicatively closed subset of a ring R and both a right quotient ring and a left quotient ring of R with respect to S exist, then the two quotient rings coincide. This is stated more precisely in the next proposition.

Proposition 6.3 *If S is a right and left denominator set in a ring R, then every right quotient ring of R with respect to S is also a left quotient ring of R with respect to S, and every left quotient ring of R with respect to S is also a right quotient ring of R with respect to S.* $\qquad\square$

Any finite number of elements of $R[S^{-1}]$ can be written in the form $\varphi(r_1)\varphi^{-1}(s), \ldots, \varphi(r_n)\varphi^{-1}(s)$ for suitable $r_1, \ldots, r_n \in R$, $s \in S$, that is, they can be written with a common denominator $\varphi(s)$.

We shall need the following properties:

Proposition 6.4 *Let S be a right denominator set in a ring R and let*

$$\varphi\colon R \to R[S^{-1}]$$

be the canonical homomorphism of R into the right quotient ring $R[S^{-1}]$.

(a) *If I a right ideal of $R[S^{-1}]$, then $\varphi^{-1}(I)$ is a right ideal of R and*

$$I = \varphi(\varphi^{-1}(I))R[S^{-1}].$$

(b) *If I an ideal of $R[S^{-1}]$ and $\varphi^{-1}(I)$ is prime in R, then I is prime.* □

Proposition 6.5 (a) *If R is a right serial ring and $S \subseteq R$ is a right denominator set, then the right quotient ring $R[S^{-1}]$ is a right serial ring.*

(b) *If R is a right chain ring and $S \subseteq R$ is a right denominator set, then the right quotient ring $R[S^{-1}]$ is a right chain ring.*

Proof. If $\{e_1, \ldots, e_n\}$ is a complete set of orthogonal idempotents for R and each $e_i R$ is uniserial, then $\{e_1, \ldots, e_n\}$ is a complete set of orthogonal idempotents for $R[S^{-1}]$, and it is sufficient to show that each $e_i R[S^{-1}]$ is uniserial. This can be proved easily, because $e_i R$ is uniserial and $I = \varphi(\varphi^{-1}(I))R[S^{-1}]$ for every right ideal I of $R[S^{-1}]$ (Proposition 6.4(a)). □

An element $x \in R$ is *regular* if it is not a zero divisor, i.e., $xr = 0$ implies $r = 0$ and $rx = 0$ implies $r = 0$ for every $r \in R$. The set of all regular elements of R is a multiplicatively closed subset of R and we shall denote it $\mathcal{C}_R(0)$. More generally, if R is a ring and I is an ideal of R, let $\mathcal{C}_R(I)$ denote the set of elements of R regular modulo I, that is, the set of elements $x \in R$ such that the map $R/I \to R/I$ given by left multiplication by x and the map $R/I \to R/I$ given by right multiplication by x are both monomorphisms. The right quotient ring of a ring R with respect to $\mathcal{C}_R(0)$ is called the *classical right quotient ring* of R (or simply the right quotient ring of R) and is denoted $Q(R)$. Hence the classical right quotient ring $Q(R)$ of R exists if and only if $\mathcal{C}_R(0)$ is a right Ore set. In this case the classical right quotient ring $Q(R)$ of R is a flat left R-module [McConnell and Robson, Proposition 2.1.16]. A *right annihilator ideal* in R (or, briefly, a *right annihilator*) is a right ideal of the form $\mathrm{r.ann}_R X = \{r \in R \mid Xr = 0\}$ for some subset $X \subseteq R$. A ring R is called a *right Goldie ring* if the right R-module R_R has finite Goldie dimension and R satisfies the a.c.c. (ascending chain condition) on right annihilators. For instance, a right serial ring is a right Goldie ring if and only if it satisfies the a.c.c. on right annihilators.

Theorem 6.6 (Goldie's Theorem) *The following conditions on a ring R are equivalent:*

(a) *R has a classical right quotient ring $Q(R)$ which is semisimple artinian.*

(b) *R is a semiprime right Goldie ring.*

(c) *R is semiprime, $Z(R_R) = 0$ and $\dim(R_R) < \infty$.*

(d) *A right ideal of R is essential if and only if it contains a regular element.*

□

6.2   Nil subrings of right Goldie rings

This section is devoted to proving that nil subrings of right Goldie rings are nilpotent (Corollary 6.9). The proof is subdivided into preliminary results.

A subring N of a ring R is said to be *right T-nilpotent* (respectively *left T-nilpotent*) if for each sequence $x_0, x_1, x_2, \ldots$ of elements of N there exists an index n such that $x_n x_{n-1} \ldots x_1 x_0 = 0$ (respectively $x_0 x_1 \ldots x_{n-1} x_n = 0$).

Proposition 6.7 *Let R be a ring with the a.c.c. on right annihilators and let N be a nil subring of R which is not right T-nilpotent. Then there are two sequences $y_0, y_1, y_2, \ldots$ and $a_1, a_2, a_3, \ldots$ of non-zero elements of N such that:*

(a) $y_n y_{n-1} \ldots y_1 y_0 \neq 0$ *for every $n \geq 0$;*

(b) $y_i y_n y_{n-1} \ldots y_1 y_0 = 0$ *for every $n \geq 0$ and every $i = 0, 1, 2, \ldots, n$;*

(c) *the sum $\sum_{n \geq 1} a_n R$ is direct;*

(d) *if $S_n = \{ a_i \mid i \geq n \}$, then the left annihilators $\mathrm{l.ann}(S_n)$ form a strictly ascending chain.*

Proof. We shall say that an element $x \in N$ has an *infinite chain* if there exists an infinite sequence $x_1, x_2, x_3, \ldots$ of elements of N such that $x_n x_{n-1} \ldots x_2 x_1 x \neq 0$ for each n. In this case we shall say that $\ldots, x_3, x_2, x_1$ is a *chain for x*. Define a sequence $y_0, y_1, y_2, \ldots$ of elements of N such that $y_n y_{n-1} \ldots y_1 y_0$ has an infinite chain for every $n \geq 1$ as follows. Since N is not right T-nilpotent, there exist elements in N with an infinite chain, that is, the set $K_0 = \{ y \in N \mid y$ has an infinite chain $\}$ is non-empty. Let $y_0 \in K_0$ be an element with $\mathrm{r.ann}(y_0)$ maximal in $\{ \mathrm{r.ann}(y) \mid y \in K_0 \}$. Suppose $n \geq 1$ and that we have already defined $y_0, y_1, \ldots, y_{n-1} \in N$ such that $y_{n-1} \ldots y_1 y_0$ has an infinite chain. Then the set

$$K_n = \{ y \in N \mid y y_{n-1} \ldots y_1 y_0 \text{ has an infinite chain} \}$$

is non-empty. Let $y_n \in K_n$ be an element with $\mathrm{r.ann}(y_n)$ maximal in

$$\{ \mathrm{r.ann}(y) \mid y \in K_n \}.$$

Then $y_n y_{n-1} \ldots y_1 y_0 \neq 0$ for every n, because $y_n y_{n-1} \ldots y_1 y_0$ has an infinite chain. Hence (a) holds.

We claim that $\mathrm{r.ann}(y_n) = \mathrm{r.ann}(y_{n+1} y_n)$ for every $n \geq 0$. Since $y_{n+1} \in K_{n+1}$, the element $y_{n+1} y_n \ldots y_1 y_0 \in N$ has an infinite chain. Thus

$$y_{n+1} y_n \in K_n.$$

But $\mathrm{r.ann}(y_n) \subseteq \mathrm{r.ann}(y_{n+1} y_n)$, so that $\mathrm{r.ann}(y_n) = \mathrm{r.ann}(y_{n+1} y_n)$ by the choice of y_n.

Also

$$\mathrm{r.ann}(y_n y_{n-1} \ldots y_i) = \mathrm{r.ann}(y_{n+1} y_n y_{n-1} \ldots y_i)$$

for every $i = 0, 1, 2, \ldots, n$, because if $r \in \text{r.ann}(y_{n+1}y_n y_{n-1} \ldots y_i)$, then

$$y_{n+1} y_n y_{n-1} \ldots y_i r = 0,$$

so that $y_{n-1} \ldots y_i r \in \text{r.ann}(y_{n+1} y_n) = \text{r.ann}(y_n)$, that is, $y_n y_{n-1} \ldots y_i r = 0$. Thus $\text{r.ann}(y_i) = \text{r.ann}(y_n y_{n-1} \ldots y_i)$ for every $i = 0, 1, 2, \ldots, n$.

In order to prove that $y_i y_n y_{n-1} \ldots y_1 y_0 = 0$ for every $i = 0, 1, 2, \ldots, n$, suppose to the contrary that $y_j y_k y_{k-1} \ldots y_1 y_0 \neq 0$ for some k, j with $0 \leq j \leq k$. Then $y_k y_{k-1} \ldots y_1 y_0 \notin \text{r.ann}(y_j) = \text{r.ann}(y_n y_{n-1} \ldots y_j)$ for every $n \geq j$, that is, $y_n y_{n-1} \ldots y_j y_k y_{k-1} \ldots y_1 y_0 \neq 0$. Hence $\ldots, y_{j+3}, y_{j+2}, y_{j+1}$ is a chain for $y_j (y_k y_{k-1} \ldots y_1 y_0) = (y_j y_k y_{k-1} \ldots y_j) y_{j-1} \ldots y_1 y_0$. Thus $y_j y_k y_{k-1} \ldots y_j \in K_j$. But $\text{r.ann}(y_j) \subseteq \text{r.ann}(y_j y_k y_{k-1} \ldots y_j)$, hence from the choice of y_j we obtain that $\text{r.ann}(y_j) = \text{r.ann}(y_j y_k y_{k-1} \ldots y_j)$. As $y_k y_{k-1} \ldots y_j \in N$ is nilpotent, there exists a smallest $m \geq 0$ such that $y_j (y_k y_{k-1} \ldots y_j)^m = 0$. If $m \geq 1$, then $(y_k y_{k-1} \ldots y_j)^{m-1} \in \text{r.ann}(y_j y_k y_{k-1} \ldots y_j) = \text{r.ann}(y_j)$, so that $y_j (y_k y_{k-1} \ldots y_j)^{m-1} = 0$, and this contradicts the choice of m. Hence $m = 0$, that is, $y_j = 0$, which is a contradiction. This proves (b).

Set $a_n = y_n y_{n-1} \ldots y_1 y_0$ for each $n \geq 1$. Then $a_n \neq 0$ for every $n \geq 0$, $y_i a_n = 0$ for every $i \leq n$, and $a_{n+1} = y_{n+1} a_n$ for every n. If the sum $\sum_{n \geq 1} a_n R$ is not a direct sum, there exist $r_i, r_{i+1}, \ldots, r_k \in R$ with

$$a_i r_i + a_{i+1} r_{i+1} + \cdots + a_k r_k = 0$$

and $a_i r_i \neq 0$. Multiplying on the left by y_{i+1} we obtain that $y_{i+1} a_i r_i = 0$. Thus $y_{i+1} y_i \ldots y_1 y_0 r_i = 0$, from which $y_{i-1} \ldots y_1 y_0 r_i \in \text{r.ann}(y_{i+1} y_i) = \text{r.ann}(y_i)$. Hence $a_i r_i = y_i y_{i-1} \ldots y_1 y_0 r_i = 0$, a contradiction that proves (c).

Finally, the ascending chain $\text{l.ann}(S_n)$ of left annihilators is strictly ascending, because $y_{k+1} a_k = a_{k+1} \neq 0$, so that $y_{k+1} \notin \text{l.ann}(S_k)$, but $y_{k+1} a_n = 0$ for every $n \geq k + 1$, i.e., $y_{k+1} \in \text{l.ann}(S_{k+1})$. $\square$

Proposition 6.8 *Let R be a ring with the a.c.c. on right annihilators. Then a subring of R is nilpotent if and only if it is right T-nilpotent.*

Proof. Every nilpotent ring is right T-nilpotent. Conversely, let N be a subring of R that is not nilpotent. There is an $n \geq 0$ such that

$$\text{r.ann}(N^n) = \text{r.ann}(N^{n+i})$$

for every $i \geq 0$. Define a sequence of elements $x_i \in N$ $(i \geq 0)$ such that $N^n x_i x_{i-1} \ldots x_0 \neq 0$ as follows. Since N is not nilpotent, $N^{n+1} \neq 0$, so that there exists $x_0 \in N$ with $N^n x_0 \neq 0$. Suppose $x_0, x_1, \ldots, x_{i-1} \in N$ have been defined and that $N^n x_{i-1} x_{i-2} \ldots x_0 \neq 0$. Then $N^{n+1} x_{i-1} x_{i-2} \ldots x_0 \neq 0$, so that there exists $x_i \in N$ such that $N^n x_i x_{i-1} x_{i-2} \ldots x_0 \neq 0$. The existence of the sequence $x_0, x_1, x_2 \ldots$ shows that N is not right T-nilpotent. $\square$

From the last two propositions we obtain two important results.

Corollary 6.9 (Lanski) *Let R be a right Goldie ring. Then each nil subring of R is nilpotent. In particular, the prime radical $N(R)$ of R is nilpotent.*

Proof. In a right Goldie ring every nil subring is right T-nilpotent by Proposition 6.7. Thus every nil subring is nilpotent by Proposition 6.8. Finally, $N(R)$ is nil by Proposition 1.6.   □

Corollary 6.10 (Herstein and Small) *Let R be a ring with the a.c.c. on right annihilators and the a.c.c. on left annihilators. Then each nil subring of R is nilpotent.*

Proof. In R every nil subring is right T-nilpotent by Proposition 6.7. Now apply Proposition 6.8.   □

6.3   Reduced rank

We give a preliminary definition that will be generalized later. Let R be a semiprime right Goldie ring, so R has a semisimple artinian classical right quotient ring $Q = Q(R)$ by Goldie's Theorem 6.6. If M_R is an R-module, then $M \otimes_R Q$ is a semisimple Q-module. The *reduced rank* of M_R is defined to be $\rho(M) = \dim_Q(M \otimes_R Q)$, the Goldie dimension of the Q-module $M \otimes_R Q$. Therefore $\rho(M) = t$ if and only if $M \otimes_R Q$ is a direct sum of t simple Q-modules, where t may be either a non-negative integer or ∞.

Consider the right R-module homomorphism $\varphi \colon M \to M \otimes_R Q$ defined by $\varphi(x) = x \otimes 1$ for every $x \in M$. Let x be an element of M and consider the monomorphism $R/\mathrm{ann}_R(x) \to M$ defined by $r + \mathrm{ann}_R(x) \mapsto xr$. Since $_R Q$ is flat, the mapping $Q/\mathrm{ann}_R(x)Q \cong R/\mathrm{ann}_R(x) \otimes_R Q \to M \otimes_R Q$ defined by $q + \mathrm{ann}_R(x)Q \mapsto x \otimes_R q$ is a monomorphism. Hence $x \in \ker \varphi$ if and only if $Q/\mathrm{ann}_R(x)Q = 0$, that is, if and only if $\mathrm{ann}_R(x)$ contains a regular element of R. This proves that $\ker \varphi = \{\, x \in M \mid xc = 0 \text{ for some } c \in \mathcal{C}_R(0)\,\}$. From Theorem 6.6(d) it follows that $\ker \varphi = Z(M_R)$. As the Q-module $M \otimes_R Q$ is generated by the image of φ, we obtain that $M \otimes_R Q = 0$ if and only if M is singular.

Proposition 6.11 *Let R be a semiprime right Goldie ring, let M_R be an R-module, let M' be a submodule of M_R and let M_λ, $\lambda \in \Lambda$, be a family of right R-modules. Then:*

(a) $\rho(M_R) = 0$ *if and only if M_R is singular.*

(b) $\rho(M) = \rho(M') + \rho(M/M')$.

(c) $\rho(M_R) = \rho(M_R/Z(M_R))$.

(d) *If M' is essential in M, then $\rho(M') = \rho(M)$.*

(e) $\rho(\oplus_{\lambda \in \Lambda} M_\lambda) = \sum_{\lambda \in \Lambda} \rho(M_\lambda)$.

(f) *If M_R is uniform and nonsingular, then $M \otimes_R Q$ is a simple Q-module, hence $\rho(M_R) = 1$.*

(g) *$\rho(M_R)$ is equal to the Goldie dimension $\dim_R(M_R/Z(M_R))$ of the R-module $M_R/Z(M_R)$.*

Proof. (a) This follows immediately from the remarks in the paragraph before the statement of the proposition.

(b) Since Q is a flat left R-module, the sequence

$$0 \to M' \otimes_R Q \to M \otimes_R Q \to M/M' \otimes_R Q \to 0$$

is an exact sequence of right Q-modules. As Q is semisimple artinian, the exact sequence splits, so that

$$\dim_Q(M \otimes_R Q) = \dim_Q(M' \otimes_R Q) + \dim_Q(M/M' \otimes_R Q).$$

(c) From (b) we obtain that $\rho(M_R) = \rho(Z(M_R)) + \rho(M_R/Z(M_R))$, and $\rho(Z(M_R)) = 0$ by (a).

(d) Consider the exact sequence

$$0 \to M' \otimes_R Q \to M \otimes_R Q \to M/M' \otimes_R Q \to 0.$$

If M' is essential in M, the module M/M' is singular (Proposition 5.10), hence $M/M' \otimes_R Q = 0$. Thus $M' \otimes_R Q \cong M \otimes_R Q$.

(e) This is trivial.

(f) Let M_R be a uniform nonsingular R-module and $x \in M \otimes_R Q$ a non-zero element. Then x can be written in the form $y \otimes s^{-1}$ for some non-zero $y \in M$, $s \in \mathcal{C}_R(0)$. Since yR is essential in M_R, the inclusion $yR \to M_R$ induces an isomorphism $yR \otimes_R Q \to M \otimes_R Q$ as in the proof of (d). Hence $y \otimes 1$ generates the Q-module $M \otimes_R Q$, so that $x = y \otimes s^{-1}$ also generates $M \otimes_R Q$. This proves that $M \otimes_R Q$ is simple, hence $\rho(M_R) = 1$.

(g) By (c) we may assume M_R is nonsingular, and we must prove that $\rho(M_R) = \dim_R(M_R)$. If $\dim_R(M_R) = \infty$, then M_R contains an infinite direct sum $\oplus_{\lambda \in \Lambda} M_\lambda$ of non-zero submodules. Hence

$$\rho(M) \geq \rho(\oplus_{\lambda \subset \Lambda} M_\lambda) = \sum_{\lambda \in \Lambda} \rho(M_\lambda) = \infty$$

by (b), (e) and (a) respectively. If $\dim_R(M_R) = n < \infty$, then M_R contains an essential submodule $\oplus_{i=1}^n M_i$ that is a direct sum of n uniform modules M_i. Therefore

$$\rho(M) = \rho(\oplus_{i=1}^n M_i) = \sum_{i=1}^n \rho(M_i) = n$$

by (d), (e) and (f). $\qquad\square$

For instance, $\rho(U_R) \leq 1$ for any uniserial module U_R.

Now let R be an arbitrary ring. Assume that N is a fixed ideal such that R/N is a semiprime right Goldie ring, and M is a right R-module that is annihilated by some power N^n of N, $n \geq 0$ an integer. Define the *reduced rank* $\rho_N(M)$ of M as follows: let Q be the right quotient ring of R/N, and take any finite chain of submodules

$$M = M_0 \supseteq M_1 \supseteq \cdots \supseteq M_k = 0 \qquad (6.1)$$

of M such that $M_i N \subseteq M_{i+1}$ for every i. For instance one can take $M_{i+1} = MN^i$. Define

$$\rho_N(M) = \sum_{i=0}^{k-1} \rho(M_i/M_{i+1}),$$

where $\rho(M_i/M_{i+1})$ is the reduced rank of M_i/M_{i+1} as an R/N-module. In Proposition 6.13(a) we shall see that $\rho_N(M)$ does not depend on the choice of the chain M_i.

Example 6.12 Let R be a ring, N an ideal such that R/N is a semiprime right Goldie ring, and let U_R be a uniserial R-module that is annihilated by some power N^n of N. Then $\rho_N(U_R) \leq n$. $\square$

The next proposition collects the most important elementary properties of the reduced rank.

Proposition 6.13 *Let R be a ring, let N be an ideal such that R/N is a semiprime right Goldie ring, and let M be a right R-module that is annihilated by some power N^n of N. Then:*

(a) *The reduced rank $\rho_N(M)$ of M is independent of the choice of the chain of submodules (6.1).*

(b) *If M' is a submodule of M, then $\rho_N(M) = \rho_N(M') + \rho_N(M/M')$.*

(c) *$\rho_N(M) = 0$ if and only if for each $x \in M$ there exists $c \in \mathcal{C}_R(N)$ such that $xc = 0$.*

Proof. (a) Let

$$M = M_0' \supseteq M_1' \supseteq \cdots \supseteq M_t' = 0$$

be another chain of submodules of M. By the Schreier Refinement Theorem, the two chains of submodules have a common refinement

$$M = M_0'' \supseteq M_1'' \supseteq \cdots \supseteq M_m'' = 0.$$

Therefore it is sufficient to show that a chain and a refinement of the same chain yield the same reduced rank. This follows immediately from Proposition 6.11(b).

For the proof of (b) argue as in the previous paragraph with a suitable refinement of the chain $M \supseteq M' \supseteq 0$.

(c) Let $M = M_0 \supseteq M_1 \supseteq \cdots \supseteq M_k = 0$ be a finite chain of submodules of M with $M_i N \subseteq M_{i+1}$ for every i. Then $\rho_N(M) = 0$ if and only if $\rho_N(M_i/M_{i+1}) = 0$ for every i. Now the conclusion easily follows by induction on k from the fact that $\mathcal{C}_R(N)$ is multiplicatively closed and that $\rho_N(M_i/M_{i+1}) = 0$ if and only if each element of M_i/M_{i+1} is annihilated by some regular element of R/N. $\qquad\square$

Example 6.14 From Example 6.12 we obtain that if R is a right serial ring and N is a nilpotent ideal such that R/N is a semiprime right Goldie ring, then $\rho_N(R_R)$ is finite. $\qquad\square$

We shall now prove the main result of this section, namely Warfield-Small's Theorem 6.16. For its proof it is convenient to give a preliminary lemma.

Lemma 6.15 *Let R be a ring and let N be its prime radical. Let $c \in R$ be an element such that the map $R_R \to R_R$ given by left multiplication by c is a monomorphism. Suppose that:*

(a) *R/N is right Goldie,*

(b) *N is nilpotent, and*

(c) *$\rho_N(R_R)$ is finite.*

Then

(1) *$c \in \mathcal{C}_R(N)$;*

(2) *for every $a \in R$ there exists $c' \in \mathcal{C}_R(N)$ such that $ac' \in cR$.*

Proof. Let c be an element with the property stated in the hypothesis of the lemma. There is an exact sequence of R-modules

$$0 \to R_R \xrightarrow{c} R_R \to R/cR \to 0,$$

so that $\rho_N(R) = \rho_N(R) + \rho_N(R/cR)$. Observe that all the reduced ranks that appear in this formula are finite by hypothesis (c). Therefore $\rho_N(R/cR) = 0$. From Proposition 6.13(c) it follows that for every $a \in R$ there exists $c' \in \mathcal{C}_R(N)$ such that $ac' \in cR$. This proves (2). In particular, for $a = 1$, there exists an element $c' \in \mathcal{C}_R(N)$ with $c' \in cR$. If $r' \in R$ is such that $c' = cr'$, then $c' + N = (c + N)(r' + N)$ is an invertible element in the classical right quotient ring $Q(R/N)$ of R/N, so that $c + N$ is right invertible in the semisimple artinian ring $Q(R/N)$. Thus c belongs to $\mathcal{C}_R(N)$ by Proposition 2.38. $\qquad\square$

The next theorem is the main result of this section.

Theorem 6.16 *Let R be a ring and let N be its prime radical. Then R has a right artinian classical right quotient ring if and only if*

(a) R/N *is a right Goldie ring,*

(b) N *is nilpotent,*

(c) $\rho_N(R_R)$ *is finite, and*

(d) $\mathcal{C}_R(0) = \mathcal{C}_R(N)$.

Proof. Suppose (a), (b), (c) and (d) hold.

Step 1: The multiplicatively closed subset $\mathcal{C}_R(0)$ of R is a right Ore set.

We must prove that for each $r \in R$ and $c \in \mathcal{C}_R(0)$, there exist $r' \in R$ and $c' \in \mathcal{C}_R(0)$ such that $rc' = cr'$. By Lemma 6.15(2) there exists $c' \in \mathcal{C}_R(N)$ such that $rc' \in cR$. But $\mathcal{C}_R(0) = \mathcal{C}_R(N)$, so that $\mathcal{C}_R(0)$ is a right Ore set.

As $\mathcal{C}_R(0)$ is a right Ore set, R has a classical right quotient ring Q. Let N' be the prime radical of Q.

Step 2: $QNQ = NQ$.

The inclusion $QNQ \supseteq NQ$ is obvious, so that it is sufficient to prove that $QN \subseteq NQ$. Fix $q = rc^{-1} \in Q$ ($r \in R$, $c \in \mathcal{C}_R(0)$) and $x \in N$. Since $\mathcal{C}_R(0)$ is a right Ore set, there exist $r' \in R$, $d \in \mathcal{C}_R(0)$ such that $xd = cr'$. By (d), $c \in \mathcal{C}_R(N)$, so that $c(r' + N) = xd + N = N$ implies $r' \in N$. Therefore $qx = rc^{-1}x = rr'd^{-1} \in NQ$.

Step 3: $NQ \subseteq N'$ and $NQ \cap R = N$.

Let n be a positive integer such that $N^n = 0$. Then $(NQ)^n = N^n Q$ by Step 2, so that the ideal $QNQ = NQ$ of Q is nilpotent. In particular $NQ \subseteq N'$. Since NQ is a nilpotent ideal of Q, $NQ \cap R$ is a nilpotent ideal of R. Therefore $NQ \cap R \subseteq N$. The reverse inclusion $NQ \cap R \supseteq N$ is obvious.

Step 4: Q/NQ is the classical right quotient ring of R/N.

By Step 3 the embedding $R \to Q$ induces an injective ring morphism $R/N \to Q/NQ$. It is easily seen that Q/NQ is a right quotient ring of R/N with respect to the multiplicatively closed subset

$$S = \{\, c + N \mid c \in \mathcal{C}_R(0) \,\} = \{\, c + N \mid c \in \mathcal{C}_R(N) \,\} = \mathcal{C}_{R/N}(0).$$

Therefore Q/NQ is the classical right quotient ring of R/N.

Step 5: Q/NQ is semisimple artinian and $NQ = N'$.

The ring Q/NQ is semisimple artinian because it is the classical right quotient ring of the semiprime right Goldie ring R/N. In particular, NQ is an intersection of prime ideals of Q, so that $NQ \supseteq N'$. In Step 3 we have already seen that $NQ \subseteq N'$.

Step 6: Q is right artinian.

Since the left R-module Q is flat, $I \otimes_R Q$ is isomorphic to the right ideal IQ of Q for every right ideal I of R (in order to see this it is sufficient to apply the exact functor $- \otimes_R Q$ to the exact sequence $0 \to I \to R \to R/I \to 0$). In particular,

$$N^i/N^{i+1} \otimes_R Q \cong N^i Q/N^{i+1}Q = N'^i/N'^{i+1}$$

for every i. Since $\rho_N(R)$ is finite and

$$\rho_N(R) = \sum_{i=0}^{n-1} \rho(N^i/N^{i+1})$$
$$= \sum_{i=0}^{n-1} \dim_Q(N^i/N^{i+1} \otimes_R Q) = \sum_{i=0}^{n-1} \dim_Q(N'^i/N'^{i+1}),$$

all the right Q-modules N'^i/N'^{i+1} have finite length, so that Q_Q is of finite length, that is, Q is right artinian.

Conversely, suppose that R has a right artinian classical right quotient ring Q. The prime radical N' of Q is equal to $J(Q)$ and is nilpotent (Proposition 1.9). Therefore the ideal $N' \cap R$ of R is nilpotent, so that $N' \cap R$ is contained in the prime radical N of R, and $Q/N' = Q/J(Q)$ is semisimple artinian. It is easily seen that Q/N' is the classical right quotient ring of $R/N' \cap R$, because:

(1) If $r + (N' \cap R)$ is a regular element of $R/N' \cap R$, left multiplication by r is a monomorphism $Q/N' \to Q/N'$ (for every element $xs^{-1} + N' \in Q/N'$, where $x \in R$ and $s \in \mathcal{C}_R(0)$, such that $rxs^{-1} \in N'$, one gets that

$$rx \in N' \cap R,$$

and therefore $x \in N' \cap R$). Since Q/N' is semisimple artinian, i.e., it is a Q-module of finite length, the left invertible element $r + (N' \cap R)$ of Q/N' is invertible in Q/N' (Proposition 2.38).

(2) Let $rs^{-1} + N' \in Q/N'$, $r \in R$ and $s \in \mathcal{C}_R(0)$, be an arbitrary element of Q/N'. In order to show that it can be written in the form

$$(r + N')(s + N')^{-1}$$

with $r, s \in R$ and $s + (N' \cap R)$ regular in $R/N' \cap R$, it is sufficient to show that $s \in \mathcal{C}_R(0)$ implies $s + (N' \cap R)$ regular in $R/N' \cap R$. Now if $s \in \mathcal{C}_R(0)$, s is invertible in Q, so that $s + N'$ is invertible in Q/N'. Therefore the element $s + (N' \cap R)$ is regular in the subring in $R/N' \cap R$ of Q/N'.

Since Q/N' is semisimple artinian and is the classical right quotient ring of $R/N' \cap R$, $R/N' \cap R$ is a semiprime right Goldie ring by Goldie's Theorem 6.6 . Hence $N' \cap R$ is an intersection of prime ideals of R, and therefore $N \subseteq N' \cap R$. This proves that $N = N' \cap R$. Hence (a) and (b) hold.

In the proof of Step 6 we have already seen that since the right quotient ring Q of R is a flat left R-module, $N^i/N^{i+1} \otimes_R Q \cong N^i Q/N^{i+1}Q$ for every i.

Now $N^i Q$ is a right ideal of the right artinian ring Q, and therefore it has finite length. It is clear, now, that $\rho_N(R_R)$ is finite.

Finally, if $s \in \mathcal{C}_R(N)$, then $s + N'$ is invertible in Q/N', so that $sQ + N' = Q$ and $Qs + N' = Q$. By Nakayama's Lemma 1.4 we have $sQ = Q$ and $Qs = Q$. Thus s is invertible in Q, whence $s \in \mathcal{C}_R(0)$. This shows that $\mathcal{C}_R(N) \subseteq \mathcal{C}_R(0)$. From Lemma 6.15(1) we know that $\mathcal{C}_R(0) \subseteq \mathcal{C}_R(N)$. This proves (d). $\square$

6.4 Localization in chain rings

In this section we shall see how it is possible to localize a chain ring at an arbitrary completely prime ideal as in the case of commutative rings. We begin with two elementary lemmas used in the sequel.

Lemma 6.17 *Let R be a right chain ring and let I be a right ideal of R. Then*

(a) *for every $r \in R$ and $x \in I$, there exists an invertible element $v \in R$ such that $rx \in vI$;*

(b) *I is a two-sided ideal in R if and only if $uI \subseteq I$ for every invertible element $u \in R$.*

Proof. Suppose that (a) does not hold. Let $r \in R$ and $x \in I$ be elements such that $rx \notin vI$ for every invertible element $v \in R$. In particular, r is not invertible in R and $rx \notin I$. Thus $rx \notin xR$, so that if $J(R)$ denotes the Jacobson radical of R we obtain $rxJ(R) \supseteq xR$. Therefore there exists an element $j \in J(R)$ such that $x = rxj$. Then both $j + 1$ and $1 + r$ are invertible elements in R, and

$$rx = rx(j+1)(j+1)^{-1} = (rxj + rx)(j+1)^{-1}$$
$$= (x + rx)(j+1)^{-1} = (1+r)x(j+1)^{-1} \in vI$$

with $v = 1 + r$. This is a contradiction.

(b) follows immediately from (a). $\square$

Lemma 6.18 *Let P be a subset of a right chain ring R. The following conditions are equivalent:*

(a) *$R \setminus P$ is a saturated multiplicatively closed subset of R;*

(b) *P is either the empty set or a completely prime ideal of R.*

Proof. (a) $\Rightarrow$ (b). Suppose that $R \setminus P$ is a saturated multiplicatively closed subset of R and that P is not empty.

If P is not additively closed, then there are elements $x, y \in P$ with $x + y \notin P$. Now either $xR \subseteq yR$ or $yR \subseteq xR$. If for instance $xR \subseteq yR$, then $x = yr$ for some $r \in R$, so that $x + y = y(r + 1)$. Since $x + y \in R \setminus P$ and $R \setminus P$ is saturated, it follows that $y \in R \setminus P$, a contradiction. Thus P is additively closed.

The proof that P is a two-sided completely prime ideal now follows immediately from the fact that $R \setminus P$ is saturated and multiplicatively closed.

(b) $\Rightarrow$ (a). This is trivial. $\qquad\qquad\square$

If P a completely prime ideal of a chain ring R, set

$$S = R \setminus P,$$

$$N_r(S) = \{\, x \in R \mid sx = 0 \text{ for some } s \in S \,\}$$

and

$$N_l(S) = \{\, x \in R \mid xs = 0 \text{ for some } s \in S \,\}.$$

Lemma 6.19 *Let P be a completely prime ideal in a chain ring R. Set $S = R \setminus P$ and $I = \{\, x \in R \mid sxs' = 0 \text{ for some } s, s' \in S \,\}$. Then:*

(a) $N_r(S)$ and $N_l(S)$ are two-sided ideals in R;

(b) $I = N_r(S) \cup N_l(S)$, in particular I is a two-sided ideal;

(c) $S \subseteq C_R(I)$;

(d) the set $\overline{S}$ of the images in $\overline{R} = R/I$ of the elements of S is a left and right denominator set in $\overline{R}$.

Proof. (a) Suppose $x, y \in N_r(S)$. Then $sx = 0$ and $ty = 0$ for some $s, t \in S$. Since R is a chain ring, either $Rs \subseteq Rt$ or $Rt \subseteq Rs$. If, for instance, $Rs \subseteq Rt$, then $Rs(x + y) \subseteq Rsx + Rty = 0$, so that $s(x + y) = 0$. This shows that $N_r(S)$ is additively closed.

It is easily seen that $N_r(S)$ is a right ideal of R. By Lemma 6.17(b), in order to show that $N_r(S)$ is two-sided, it is sufficient to prove that $ux \in N_r(S)$ for every $x \in N_r(S)$ and every invertible element $u \in R$. Now if $x \in N_r(S)$ and u is invertible, then $sx = 0$ for some element $s \in S$. But $u^{-1} \in S$, so that $su^{-1} \in S$, and $(su^{-1})(ux) = 0$. Thus $ux \in N_r(S)$. This shows that $N_r(S)$ is a two-sided ideal in R. By symmetry, the same holds true for $N_l(S)$.

(b) It is obvious that both $N_r(S)$ and $N_l(S)$ are contained in I. Conversely, let x be an element in I. Then $sxs' = 0$ for some $s, s' \in S$. Suppose

$$N_r(S) \subseteq N_l(S).$$

Then $xs' \in N_r(S)$, so that $xs' \in N_l(S)$, i.e., $xs't = 0$ for some $t \in S$, from which $x \in N_l(S)$. Similarly, from $N_r(S) \supseteq N_l(S)$ it follows that $x \in N_r(S)$. Thus, in both cases, $x \in N_r(S) \cup N_l(S)$.

(c) Let s be an element of S and suppose that $rs \in I$ for some $r \in R$. Then $trst' = 0$ for suitable $t, t' \in S$, so that $r \in I$. Similarly, $sr \in I$ implies $r \in I$. Hence $s \in C_R(I)$.

(d) By (c) we need only show that the multiplicatively closed subset S of R is a left and right Ore set in R. We must prove that for each $r \in R$ and $s \in S$, there exist $r' \in R$, $s' \in S$ such that $rs' = sr'$. Given $r \in R$ and $s \in S$, then either $rR \supseteq sR$ or $rR \subseteq sR$. If $rR \supseteq sR$, then $s = rs'$ for some $s' \in R$, so that $s' \in S$ and $rs' = s1_R$. If $rR \subseteq sR$, then $r = sr'$ for some $r' \in R$, so that $r1_R = sr'$ with $1_R \in S$. This proves that S is a right Ore set in R. Similarly, S is a left Ore set. $\qquad\qquad\qquad\qquad\qquad\qquad\qquad\qquad\qquad\qquad\qquad\qquad\qquad\quad\square$

By Theorem 6.2 and Proposition 6.3 the left and right quotient ring of $\overline{R} = R/I$ with respect to $\overline{S}$ exists. We denote this quotient ring $\overline{R}[\overline{S}^{-1}]$ by R_P and call it the *localization* of R at P. Observe that R_P is a left and right quotient ring of $\overline{R}$ with respect to $\overline{S}$, not of R with respect to S. More precisely, the kernel of the canonical ring homomorphism $\varphi \colon R \to R_P$ is $I = N_r(S) \cup N_l(S)$. If $N_r(S) \subseteq N_l(S)$, then $\ker \varphi = N_l(S)$, so that in this case R_P is a right quotient ring of R with respect to S. If $N_l(S) \subseteq N_r(S)$, then $\ker \varphi = N_r(S)$, and R_P is a left quotient ring of R with respect to S.

The localization R_P is a chain ring by Proposition 6.5(b).

6.5 Localizable systems in a serial ring

In the previous section we saw how it is possible to localize a chain ring at a completely prime ideal. In this section we shall extend this idea to serial rings.

Let R be a serial ring and let $\{e_1, \ldots, e_n\}$ be a fixed complete set of orthogonal primitive idempotents of R. An n-tuple $\mathcal{P} = (P_1, \ldots, P_n)$ is a *localizable system* for the serial ring R if:

1) for every $i = 1, 2, \ldots, n$, P_i is either the empty set or a completely prime ideal in the ring $e_i R e_i$;

2) there exists $i = 1, 2, \ldots, n$ such that $P_i \neq \emptyset$;

3) for every $r, s \in R$ and every $i, j = 1, \ldots, n$, if $P_i \neq \emptyset$ and

$$e_i r e_j \cdot e_j s e_i \notin P_i,$$

then $P_j \neq \emptyset$ and $e_j s e_j \cdot e_i r e_j \notin P_j$.

Let R be a serial ring, let S be the subring $e_1 R e_1 \times \cdots \times e_n R e_n$ of R, let $\mathcal{P} = (P_1, \ldots, P_n)$ be a localizable system and let $X_{\mathcal{P}}$ be the subset

$$X_{\mathcal{P}} = (e_1 R e_1 \setminus P_1) \times (e_2 R e_2 \setminus P_2) \times \cdots \times (e_n R e_n \setminus P_n)$$

of S. Note that the idempotents e_i commute with the elements of S. In particular, $e_i x = x e_i$ for every $i = 1, \ldots, n$ and every $x \in X_{\mathcal{P}}$.

Define a sequence of two-sided ideals I_t of R, $t = 0, 1, 2, \ldots$, as follows. The definition is by induction on t. Set $I_0 = 0$. For every $t \geq 0$ let I_{t+1} be the two-sided ideal of R generated by all the elements r of R such that either $xr \in I_t$ or $rx \in I_t$ for some $x \in X_{\mathcal{P}}$.

Since $1 \in X_{\mathcal{P}}$, the sequence $I_0 \subseteq I_1 \subseteq I_2 \subseteq \ldots$ is an ascending chain of ideals. Let I be the union $\bigcup_{t \geq 0} I_t$ of this chain of ideals.

Lemma 6.20 *Let R be a serial ring, let $\mathcal{P} = (P_1, \ldots, P_n)$ be a localizable system, and $i = 1, 2, \ldots, n$. The following conditions are equivalent:*

(a) $e_i I e_i \subseteq P_i$.

(b) $e_i \notin I$.

(c) $P_i \neq \emptyset$.

Proof. (a) $\Rightarrow$ (b). If $e_i I e_i \subseteq P_i$ and $e_i \in I$, then $e_i \in P_i$. This is a contradiction, because P_i is either the empty set or a proper ideal of $e_i R e_i$.

(b) $\Rightarrow$ (c). If $P_i = \emptyset$, then $1 - e_i \in X_\mathcal{P}$, so that $e_i \in I_1 \subseteq I$.

(c) $\Rightarrow$ (a). We shall prove that $e_j I e_j \subseteq P_j$ for every index j such that $P_j \neq \emptyset$. Suppose the contrary. Since the ideals $e_j I_t e_j$, $t \geq 0$, form an ascending chain of ideals in $e_j R e_j$ for every j, and $e_j I_0 e_j = 0 \subseteq P_j$ for every index j with $P_j \neq \emptyset$, there exists a greatest non-negative integer t such that $e_j I_t e_j \subseteq P_j$ for every j with $P_j \neq \emptyset$. Hence $e_k I_{t+1} e_k \not\subseteq P_k$ for some index $k = 1, \ldots, n$ with $P_k \neq \emptyset$. Since I_{t+1} is the two-sided ideal generated by

$$G = \{\, r \in R \mid xr \in I_t \text{ or } rx \in I_t \text{ for some } x \in X_\mathcal{P} \,\},$$

there exists an element $r \in G$ such that $e_k R r R e_k \not\subseteq P_k$. By symmetry, we may suppose that $xr \in I_t$ for a suitable $x \in X_\mathcal{P}$. From $e_k R r R e_k \not\subseteq P_k$ it follows that there are indices ℓ, m such that $e_k R e_\ell r e_m R e_k \not\subseteq P_k$. Hence there are $s, s' \in R$ with $e_k s e_\ell r e_m s' e_k \notin P_k$. Since $\mathcal{P}$ is a localizable system, we have that $P_\ell \neq \emptyset$ and $e_\ell r e_m s' e_k s e_\ell \notin P_\ell$. By the choice of t we know that $e_\ell I_t e_\ell \subseteq P_\ell$. Since $xr \in I_t$, we have that $e_\ell x r e_m s' e_k s e_\ell \in P_\ell$. But $e_\ell x = e_\ell x e_\ell \in e_\ell R e_\ell \setminus P_\ell$ because $x \in X_\mathcal{P}$. Hence $e_\ell r e_m s' e_k s e_\ell \in P_\ell$. As $\mathcal{P}$ is a localizable system, $e_\ell r e_m s' e_k \cdot e_k s e_\ell \in P_\ell$ forces $e_k s e_\ell r e_m s' e_k \in P_k$. This is a contradiction. $\square$

Observe that in the localizable system $\mathcal{P} = (P_1, \ldots, P_n)$ at least one of the completely prime ideals P_i is nonempty. From Lemma 6.20 it follows that $e_i \notin I$, hence I is a proper ideal of R. Set $\overline{R} = R/I$. For every element $r \in R$ let $\overline{r}$ denote the image of r in $\overline{R} = R/I$, and let $\overline{X_\mathcal{P}} = \{\, \overline{x} \mid x \in X_\mathcal{P} \,\}$. Clearly, the elements of $\overline{X_\mathcal{P}}$ are regular elements of $\overline{R}$. In particular, $\overline{X_\mathcal{P}}$ is a proper subset of $\overline{R}$.

Lemma 6.21 *The set $\overline{X_\mathcal{P}}$ is a left and right denominator set in $\overline{R}$.*

Proof. Since the subset $X_\mathcal{P}$ of S is a multiplicatively closed subset of R, it is obvious that $\overline{X_\mathcal{P}}$ is a multiplicatively closed subset of $\overline{R}$. In order to prove that $\overline{X_\mathcal{P}}$ is a right Ore set in $\overline{R}$, fix two elements $r \in R$ and $x \in X_\mathcal{P}$. We must prove that there exists $t \in R$ and $y \in X_\mathcal{P}$ such that $\overline{r} \overline{y} = \overline{x} \overline{t}$.

First of all we shall consider the case in which $x = e_i u e_i + (1 - e_i)$ for some index i and some $e_i u e_i \in e_i R e_i \setminus P_i$. If $P_i = \emptyset$, then $e_i \in I$ by Lemma 6.20, so $\overline{x} = \overline{1}$. Hence $y = 1 \in X_\mathcal{P}$ and $t = r \in R$ have the property that $\overline{r} \overline{y} = \overline{x} \overline{t}$. Thus we may assume $P_i \neq \emptyset$.

We claim that for every $j = 1, 2, \ldots, n$ there exist $t_j \in R$ and

$$y_j = e_j y_j e_j \in e_j R e_j \setminus P_j$$

such that $e_i r e_j \cdot e_j y_j e_j = e_i u e_i \cdot e_i t_j e_j$. In order to prove this claim, fix an index $j = 1, 2, \ldots, n$. Since the module $e_i R$ is uniserial, either $e_i r e_j \in e_i u e_i R$ or $e_i u e_i \in e_i r e_j R$.

If $e_i r e_j \in e_i u e_i R$, there exists $t_j \in R$ such that $e_i r e_j = e_i u e_i t_j$. Then $e_i r e_j \cdot e_j = e_i u e_i \cdot e_i t_j e_j$, so that t_j and $y_j = e_j$ satisfy the claim.

If $e_i u e_i \in e_i r e_j R$, there exists $h_j \in R$ such that $e_i u e_i = e_i r e_j h_j$. Hence $e_i r e_j \cdot e_j h_j e_i = e_i u e_i \in e_i R e_i \setminus P_i$. But $P_i \neq \emptyset$ and $\mathcal{P}$ is a localizable system, so that $e_j h_j e_i \cdot e_i r e_j \in e_j R e_j \setminus P_j$. Then $t_j = r$ and $y_j = e_j h_j e_i \cdot e_i r e_j$ satisfy the claim, because $e_i r e_j \cdot e_j y_j e_j = e_i r e_j h_j e_i r e_j = e_i u e_i r e_j = e_i u e_i \cdot e_i t_j e_j$.

This proves the claim. Now set

$$t = e_i \left(\sum_{j=1}^{n} t_j e_j \right) + (1 - e_i) r y \quad \text{and} \quad y = \sum_{j=1}^{n} e_j y_j e_j,$$

so that $y \in X_\mathcal{P}$. Then

$$
\begin{aligned}
xt &= (e_i u e_i + (1 - e_i)) \left(e_i \left(\sum_{j=1}^{n} t_j e_j \right) + (1 - e_i) r y \right) \\
&= \left(e_i u e_i \sum_{j=1}^{n} t_j e_j \right) + (1 - e_i) r y = \left(\sum_{j=1}^{n} e_i r e_j y_j e_j \right) + (1 - e_i) r y \\
&= e_i r y + (1 - e_i) r y = r y.
\end{aligned}
$$

Hence $\overline{xt} = \overline{ry}$. Thus we have verified the right Ore condition for the elements $\overline{x}$ in the case in which x is of the form $x = e_i u e_i + (1 - e_i)$ for some index i and some $e_i u e_i \in e_i R e_i \setminus P_i$.

Now let $r \in R$ and $x \in X_\mathcal{P}$ be arbitrary elements. Then

$$x = \prod_{i=1}^{n} (e_i x e_i + (1 - e_i)).$$

By induction on $i = 1, 2, \ldots, n$ define $t_i \in R$ and $y_i \in X_\mathcal{P}$ as follows. By the previous case, there exists $t_1 \in R$ and $y_1 \in X_\mathcal{P}$ such that

$$\overline{(e_1 x e_1 + (1 - e_1)) t_1} = \overline{r y_1}.$$

Suppose t_{i-1} and y_{i-1} have been defined. Again by the previous case there exists $t_i \in R$ and $y_i \in X_\mathcal{P}$ such that

$$\overline{(e_i x e_i + (1 - e_i)) t_i} = \overline{t_{i-1} y_i}.$$

Then $t = t_n$ and $y = y_1 y_2 \ldots y_n$ have the property that

$$
\begin{aligned}
\overline{ry} &= \overline{r y_1 y_2 \ldots y_n} = \overline{(e_1 x e_1 + (1 - e_1)) t_1 y_2 \ldots y_n} \\
&= \overline{(e_1 x e_1 + (1 - e_1))(e_2 x e_2 + (1 - e_2)) t_2 y_3 \ldots y_n} = \cdots = \overline{xt_n} = \overline{xt}.
\end{aligned}
$$

Hence $\overline{X_\mathcal{P}}$ is a right Ore set in $\overline{R}$. By symmetry, $\overline{X_\mathcal{P}}$ is a left Ore set also. $\square$

From Lemma 6.21, Theorem 6.2 and Proposition 6.3 it follows that the right and left quotient ring of $\overline{R}$ with respect to $\overline{X_\mathcal{P}}$ exists and is unique up to isomorphism. We shall denote this right and left quotient ring by $R_\mathcal{P}$ and call the ring $R_\mathcal{P}$ with the canonical mapping $\varphi\colon R \to R_\mathcal{P}$ the *localization of R at* $\mathcal{P}$. Often we shall call just $R_\mathcal{P}$ or φ the *localization* of R at $\mathcal{P}$. Note that the kernel of $\varphi\colon R \to R_\mathcal{P}$ is the ideal I. From this and Proposition 6.1 we obtain

Proposition 6.22 *Let* $\mathcal{P} = (P_1, \dots, P_n)$ *be a localizable system for a serial ring R. Then there exists a ring* $R_\mathcal{P}$ *and a ring homomorphism* $\varphi\colon R \to R_\mathcal{P}$ *with the following properties:*

(a) *The elements* $\varphi(x)$ *are invertible in* $R_\mathcal{P}$ *for every* $x \in X_\mathcal{P}$.

(b) *If* $\psi\colon R \to T$ *is a ring homomorphism such that the elements* $\psi(x)$ *are invertible in T for every* $x \in X_\mathcal{P}$, *then there is a unique ring homomorphism* $\omega\colon R_\mathcal{P} \to T$ *such that* $\psi = \omega\varphi$. $\square$

Every element of $R_\mathcal{P}$ can be written in the form $\varphi(x)^{-1}\varphi(r)$ and the form $\varphi(s)\varphi(y)^{-1}$ for suitable $r, s \in R$, $x, y \in X_\mathcal{P}$. As usual, since it is rather cumbersome to use this notation, we shall write elements of $R_\mathcal{P}$ in the form $x^{-1}r$ or the form sy^{-1}. Note that $x^{-1}e_i = e_i x^{-1}$ for every $x \in X_\mathcal{P}$ and every index i, because the elements of $X_\mathcal{P}$ and the idempotents e_i commute.

In Proposition 6.24 we shall study the structure of the localization $R_\mathcal{P}$. First we prove an easy lemma.

Lemma 6.23 *Let R be a serial ring, let* $\{e_1, \dots, e_n\}$ *be a complete set of orthogonal primitive idempotents in R and let* $\mathcal{P} = (P_1, \dots, P_n)$ *be a localizable system for R. Then:*

(a) *If* $i = 1, \dots, n$, *the elements of* $e_i R_\mathcal{P}$ *can be written in the form*

$$e_i x^{-1} r = x^{-1} e_i r$$

for suitable $r \in R$, $x \in X_\mathcal{P}$.

(b) *If* $i, j = 1, \dots, n$, *the elements of* $e_i R_\mathcal{P} e_j$ *can be written in the form* $x^{-1} e_i r e_j$ *and in the form* $e_i s e_j y^{-1}$ *for suitable* $r, s \in R$, $x, y \in X_\mathcal{P}$.

Proof. (a) This follows from the fact that an element of $e_i R_\mathcal{P}$ is of the form $e_i x^{-1} r$, with $r \in R$ and $x \in X_\mathcal{P}$, and $x^{-1} e_i = e_i x^{-1}$.

(b) The first form is an immediate consequence of (a). The second form is an immediate consequence of the left-right dual of (a). $\square$

In the next proposition we describe a complete set of primitive orthogonal idempotents for the localization of a serial ring R at a localizable system.

Proposition 6.24 *Let R be a serial ring, let $\{e_1, \ldots, e_n\}$ be a complete set of orthogonal primitive idempotents in R, let $\mathcal{P} = (P_1, \ldots, P_n)$ be a localizable system for R, and let $\varphi \colon R \to R_{\mathcal{P}}$ be the localization of R at $\mathcal{P}$. Then:*

(a) *The right ideals $e_i R_{\mathcal{P}}$ of $R_{\mathcal{P}}$ are uniserial for every $i = 1, \ldots, n$.*

(b) *The ring $R_{\mathcal{P}}$ is serial.*

(c) *$\{\, e_i \mid P_i \neq \emptyset, \ i = 1, 2, \ldots, n \,\}$ is a complete set of primitive orthogonal idempotents for $R_{\mathcal{P}}$.*

(d) *The Jacobson radical of $R_{\mathcal{P}}$ is contained in $\varphi(R)$.*

Proof. For (a) and (b) see Proposition 6.5 and its proof.

(c) Obviously $\{\, e_i \mid i = 1, 2, \ldots, n \,\}$ is a complete set of orthogonal idempotents for $R_{\mathcal{P}}$. From (a) it follows that each e_i is either zero or primitive in $R_{\mathcal{P}}$. Also $e_i = 0$ in $R_{\mathcal{P}}$ if and only if $e_i \in I$, that is, if and only if $P_i = \emptyset$.

(d) By (c) it is sufficient to show that $e_i J(R_{\mathcal{P}}) \subseteq \varphi(R)$ for every index i such that $P_i \neq \emptyset$. Let $x^{-1}r$ be an element of $J(R_{\mathcal{P}})$, with $r \in R$ and $x \in X_{\mathcal{P}}$. Since $e_i R$ is uniserial, either $e_i x R \subseteq e_i r R$ or $e_i x R \supseteq e_i r R$.

If $e_i x R \subseteq e_i r R$, then $e_i x = e_i r s$ for some $s \in R$, so that $x e_i = e_i r s$. Hence in the ring $R_{\mathcal{P}}$ we have $e_i = x^{-1} e_i r s = e_i x^{-1} r s \in J(R_{\mathcal{P}})$, and this is a contradiction, because the Jacobson radical cannot contain the non-zero idempotent e_i.

Therefore $e_i x R \supseteq e_i r R$, so that $e_i r = e_i x t$ for some $t \in R$. Then

$$e_i x^{-1} r = x^{-1} e_i r = x^{-1} e_i x t = e_i x^{-1} x t = e_i t \in \varphi(R). \qquad \square$$

As a corollary we obtain

Corollary 6.25 *Let R be a serial ring, let $\{e_1, \ldots, e_n\}$ be a complete set of orthogonal primitive idempotents in R, let $\mathcal{P} = (P_1, \ldots, P_n)$ be a localizable system for R and let $R_{\mathcal{P}}$ be the localization of R. Then every uniserial $R_{\mathcal{P}}$-module is uniserial as an R-module.*

Proof. Let M be a uniserial right $R_{\mathcal{P}}$-module. If $x, y \in M$, then $x R_{\mathcal{P}} + y R_{\mathcal{P}}$ is a uniserial cyclic $R_{\mathcal{P}}$-module, hence it is a homomorphic image of $e_i R_{\mathcal{P}}$ for some index $i = 1, 2, \ldots, n$. Thus in order to show that M is uniserial as an R-module, it is sufficient to show that each $e_i R_{\mathcal{P}}$ is a uniserial right R-module. In other words, we may suppose $M = e_i R_{\mathcal{P}}$.

We claim that for every $j, k = 1, 2, \ldots, n$, every $a \in e_i R_{\mathcal{P}} e_j$ and every $b \in e_i R_{\mathcal{P}} e_k$, either $a R \subseteq b R$ or $b R \subseteq a R$. Suppose we have proved the claim. Then if $c, d \in e_i R_{\mathcal{P}}$, then $cR = c e_1 R + \cdots + c e_n R$, and $c e_\ell \in e_i R_{\mathcal{P}} e_\ell$ for every

ℓ, so that by the claim there exists an index j such that $cR = ce_j R$. Similarly, there exists an index k such that $dR = de_k R$. By the claim, the submodules $cR = ce_j R$ and $dR = de_k R$ are comparable, which proves the corollary.

In order to prove the claim, fix two indices j, k and two elements

$$a \in e_i R_{\mathcal{P}} e_j \quad \text{and} \quad b \in e_i R_{\mathcal{P}} e_k.$$

As $e_i R_{\mathcal{P}}$ is a uniserial right ideal of $R_{\mathcal{P}}$, either $aR_{\mathcal{P}} \subseteq bR_{\mathcal{P}}$ or $bR_{\mathcal{P}} \subseteq aR_{\mathcal{P}}$. By symmetry we may suppose $bR_{\mathcal{P}} \subseteq aR_{\mathcal{P}}$, so that there exists $r' \in R_{\mathcal{P}}$ with $ar' = b$, from which $e_i a e_j \cdot e_j r' e_k = e_i b e_k$. By Lemma 6.23(b) the element $e_j r' e_k$ of $e_j R_{\mathcal{P}} e_k$ can be written in the form $e_j r e_k y^{-1}$ for some $r \in R$, $y \in X_{\mathcal{P}}$. Thus $e_i a e_j \cdot e_j r e_k = e_i b e_k \cdot e_k y e_k$.

Now the left ideal $R e_k$ is uniserial, hence either $R e_j r e_k \subseteq R e_k y e_k$ or $R e_j r e_k \supseteq R e_k y e_k$.

If $R e_j r e_k \subseteq R e_k y e_k$, there exists $s \in R$ with $e_j r e_k = e_j s e_k \cdot e_k y e_k$, so that $e_i a e_j \cdot e_j s e_k \cdot e_k y e_k = e_i b e_k \cdot e_k y e_k$. Since $y \in X_{\mathcal{P}} \subseteq S$, it follows that $e_i a e_j \cdot e_j s e_k \cdot y = e_i b e_k \cdot y$. But y is invertible in the ring $R_{\mathcal{P}}$, and thus in the module $e_i R_{\mathcal{P}}$ we have $e_i a e_j \cdot e_j s e_k = e_i b e_k$, and thus $aR \supseteq bR$.

If $R e_j r e_k \supseteq R e_k y e_k$, there exists $t \in R$ such that $e_k t e_j \cdot e_j r e_k = e_k y e_k$. Thus $e_k t e_j \cdot e_j r e_k \notin P_k$. If $P_k = \emptyset$, then $e_i \in I$ by Lemma 6.20, hence $e_k = 0$ in $R_{\mathcal{P}}$, so that the element $b \in e_i R_{\mathcal{P}} e_k$ is zero, and $bR \subseteq aR$. Thus we may assume $P_k \neq \emptyset$. This assumption and the relation $e_k t e_j \cdot e_j r e_k \notin P_k$ yield that $P_j \neq \emptyset$ and $e_j r e_k \cdot e_k t e_j \notin P_j$. Thus $x = e_j r e_k \cdot e_k t e_j + 1 - e_j \in X_{\mathcal{P}}$ is invertible in $R_{\mathcal{P}}$, so that

$$\begin{aligned} ax &= e_i a e_j \cdot x = e_i a e_j \cdot e_j r e_k \cdot e_k t e_j = e_i b e_k \cdot e_k y e_k \cdot e_k t e_j \\ &= e_i b e_k \cdot e_k t e_j \cdot e_j r e_k \cdot e_k t e_j = b \cdot e_k t e_j \cdot x \end{aligned}$$

forces $a = b \cdot e_k t e_j$. Hence $aR \subseteq bR$, which completes the proof of the claim. $\square$

6.6 An example

The reason why we have introduced localizable systems in the previous section is twofold. On the one hand we shall need localizable systems to study Ore subsets in serial rings. This will be done in the rest of this chapter, in particular in Section 6.10. On the other hand we shall use localizable systems to study the structure of Σ-pure-injective modules in Chapter 10. In order to study such modules the idea is the following: if M is the R-module we want to study and R is a serial ring, we shall construct a suitable localizable system $\mathcal{P}$ for R in such a way that M has a canonical structure as an $R_{\mathcal{P}}$-module. Proposition 6.27 will show how it is possible to construct localizable systems starting from modules with suitable properties.

The next elementary lemma will be useful not only in the proof of Proposition 6.27 but also in the study of direct sum decompositions of serial modules over arbitrary rings in Chapter 9.

Lemma 6.26 *Let R be an arbitrary ring, let A, B, C be non-zero R-modules and let $\alpha\colon A \to B$, $\beta\colon B \to C$ be homomorphisms. Then*

(a) *If B is uniform, the composite $\beta\alpha$ is a monomorphism if and only if β and α are both monomorphisms;*

(b) *If B is couniform, the composite $\beta\alpha$ is an epimorphism if and only if β and α are both epimorphisms.*

Proof. (a) It is sufficient to prove that if B is uniform and $\beta\alpha$ is a monomorphism, then β is also a monomorphism. Now if $\beta\alpha$ is a monomorphism, then $\alpha(A) \cap \ker(\beta) = 0$. Since B is uniform, either $\alpha(A) = 0$ or $\ker(\beta) = 0$. Now α is a monomorphism and $A \neq 0$, so that $\alpha(A) \neq 0$. Hence $\ker(\beta) = 0$, and β is a monomorphism.

(b) If B is couniform and $\beta\alpha$ is an epimorphism, then $\ker(\beta) + \alpha(A) = B$. Since B is couniform, either $\ker(\beta) = B$ or $\alpha(A) = B$. If $\ker(\beta) = B$, then $\beta = 0$, so that $\beta\alpha = 0$, and this yields a contradiction because $\beta\alpha$ is an epimorphism and $C \neq 0$. Therefore $\alpha(A) = B$, that is, α is an epimorphism. $\square$

In particular, Lemma 6.26 can be applied to any uniserial module $B \neq 0$.

For the next proposition, observe that if $_SM_R$ is a bimodule and $e \in R$ is idempotent, then $_SMe$ is a left S-module, and right multiplication by an element of eRe is an endomorphism of the left S-module $_SMe$. More generally, if $e, f \in R$ are idempotents, then right multiplication by an element of eRf is a homomorphism $_SMe \to _SMf$.

Proposition 6.27 *Let R be a serial ring, let $\{e_1, \ldots, e_n\}$ be a complete set of orthogonal primitive idempotents for R, let S be an arbitrary ring and let $_SM_R$ be a non-zero S-R-bimodule. Suppose that each $_SMe_i$ is a uniserial S-module. For every index $i = 1, 2, \ldots, n$ let P_i be the set of all $r \in e_iRe_i$ such that right multiplication by e_ire_i is not an injective endomorphism of $_SMe_i$, let P_i' be the set of all $r \in e_iRe_i$ such that right multiplication by e_ire_i is not a surjective endomorphism of $_SMe_i$, and let P_i'' be the set of all $r \in e_iRe_i$ such that right multiplication by e_ire_i is not an automorphism of $_SMe_i$. Then $\mathcal{P} = (P_1, \ldots, P_n)$, $\mathcal{P}' = (P_1', \ldots, P_n')$ and $\mathcal{P}'' = (P_1'', \ldots, P_n'')$ are localizable systems for R.*

Proof. It is easily seen that $P_i = \emptyset$ if and only if $Me_i = 0$. Hence at least one of the P_i is nonempty. Suppose $P_i \neq \emptyset$, so that $Me_i \neq 0$. By Lemma 6.26(a) the set $e_iRe_i \setminus P_i$ is a saturated multiplicatively closed subset of e_iRe_i, so that P_i is a completely prime two-sided ideal of e_iRe_i by Lemma 6.18.

In order to show that $\mathcal{P} = (P_1, \ldots, P_n)$ is a localizable system, fix two elements $r, s \in R$, two indices $i, j = 1, \ldots, n$ and suppose $P_i \neq \emptyset$ and

$$e_ire_j \cdot e_jse_i \notin P_i.$$

Then $Me_i \neq 0$ and the composite of the mapping $Me_i \to Me_j$ given by right multiplication by $e_i r e_j$ and the mapping $Me_j \to Me_i$ given by right multiplication by $e_j s e_i$ is injective. In particular $Me_j \neq 0$, so that $P_j \neq \emptyset$. By Lemma 6.26(a) both mappings $Me_i \to Me_j$ and $Me_j \to Me_i$ are injective. Hence their composite $Me_j \to Me_i \to Me_j$ is injective, that is,

$$e_j s e_i \cdot e_i r e_j \notin P_j.$$

The proof that $\mathcal{P}'$ and $\mathcal{P}''$ are localizable systems is similar. $\qquad\square$

In the notation of Proposition 6.27 it is easily seen that M turns out to be an S-$R_{\mathcal{P}''}$-bimodule in a natural way (use the universal property 6.22(b)).

As a particular case of Proposition 6.27 consider an arbitrary proper two-sided ideal I of a serial ring R and the bimodule $_R M_R = R/I$. Then every $Me_i \cong Re_i/Ie_i$ is a uniserial left R-module.

In Section 6.10 we shall give further examples of constructions of localizable systems.

6.7   Prime ideals in serial rings

In this section we describe the prime spectrum of a serial ring, that is, the set of all its prime ideals partially ordered by set inclusion.

Let R be a serial ring. In order to study the localizable systems for R introduced in Section 6.5, we must know the completely prime ideals in the chain rings $e_i R e_i$. There is a one-to-one correspondence, described in Theorem 6.28, between the prime ideals of the ring $e_i R e_i$ and the prime ideals of R that do not contain e_i.

Theorem 6.28 *Let R be an arbitrary ring and let e be a non-zero idempotent of R. Then there is an order preserving one-to-one correspondence between the set of all prime ideals P of the ring eRe and the set of all the prime ideals P' of the ring R not containing c. The correspondence is given by*

$$\begin{aligned} P &\mapsto P' = \{\, r \in R \mid eRrRe \subseteq P \,\} \quad and \\ P' &\mapsto P = eP'e. \end{aligned}$$

Proof. Let P be a prime ideal of eRe. It is easy to see that

$$P' = \{\, r \in R \mid eRrRe \subseteq P \,\}$$

is a two-sided ideal in R not containing e. In particular, P' is a proper ideal of R. Moreover, $eP'e = P$, because if $x \in P$, then $x = exe$ and $eRxRe = eRexeRe \subseteq P$, so that $x \in P'$; and if $y \in P'$, then $eRyRe \subseteq P$, so that $eye \in P$. In order to see that P' is a prime ideal, let I and J be two ideals of R with $IJ \subseteq P'$. Then eIe and eJe are two two-sided ideals of eRe, and

$(eIe)(eJe) \subseteq eP'e = P$. Since P is prime, either $eIe \subseteq P$ or $eJe \subseteq P$. If, for instance, $eIe \subseteq P$, then $eRxRe \subseteq P$ for every $x \in I$, and thus $I \subseteq P'$. Hence P' is a prime ideal in R.

Conversely, let P' be a prime ideal in R not containing e, and set $P = eP'e$. Then P is a two-sided ideal in eRe. Moreover, $e \notin P'$ and $P \subseteq P'$, so that $e \notin P$ and the ideal P of eRe is proper. If $x, y \in R$ and

$$exe, eye \notin P,$$

then $exe, eye \notin P'$, so that there exists $r \in R$ with $exereye \notin P'$. Thus

$$exe \cdot eRe \cdot eye \nsubseteq P,$$

i.e., P is a prime ideal in eRe. In order to conclude the proof it remains to show that

$$P' = \{ r \in R \mid eRrRe \subseteq eP'e \}.$$

The inclusion $\subseteq$ is obvious. For the other inclusion, fix an element $r \in R$ with $eRrRe \subseteq eP'e$. Then for every $x \in R$ one has $eRrxe \subseteq eP'e \subseteq P'$. Since P' is prime and $e \notin P'$, it follows that $rxe \in P'$, i.e., $rRe \subseteq P'$. Thus $r \in P'$. $\square$

Observe that $P' \cap eRe = P$ in Theorem 6.28.

As a first application of Theorem 6.28 we shall study prime ideals in right serial rings. Let R be a right serial ring, so that there exists a complete set $\{e_1, \ldots, e_n\}$ of orthogonal idempotents for R with each $e_i R$ uniserial. If P' is a prime ideal in R, then there is an idempotent e_i that does not belong to P'. Hence P' correspond to a prime ideal of the right chain ring $e_i Re_i$. Let Spec R be the set of all prime ideals of R and $\mathrm{Spec}_i R = \{ P \in \mathrm{Spec}\, R \mid e_i \notin P \}$. Then Spec R is a set partially ordered by inclusion, and $\mathrm{Spec}\, R = \bigcup_{i=1}^n \mathrm{Spec}_i R$. The one-to-one correspondence given in Theorem 6.28 is order preserving, so that every $\mathrm{Spec}_i R$ is a linearly ordered subset of Spec R. Note that if $P \subseteq Q$ are prime ideals of R and $Q \in \mathrm{Spec}_i R$, then $P \in \mathrm{Spec}_i R$. Let min R denote the set of all the minimal elements of Spec R. Then the elements of min R are exactly the least elements of the subsets $\mathrm{Spec}_i R$. Thus min R has at most n elements.

Lemma 6.29 *If P, Q are two incomparable prime ideals of a right serial ring R, then P and Q are comaximal, that is, $P + Q = R$.*

Proof. For every $i = 1, 2, \ldots, n$ we have either $e_i P \subseteq e_i Q$ or $e_i Q \subseteq e_i P$. If $e_i P \subseteq e_i Q$, then $e_i P \subseteq Q$, so that either $e_i \in Q$ or $P \subseteq Q$. Since P and Q are incomparable, $e_i \in Q$. Similarly, $e_i Q \subseteq e_i P$ implies $e_i \in P$. In both cases, $e_i \in P + Q$. This holds for every i, so that $P + Q = R$. $\square$

In particular, any two elements of $\min R$ are comaximal ideals. For every $P \in \min R$ set $V(P) = \{Q \in \operatorname{Spec} R \mid Q \supseteq P\}$. Then $\operatorname{Spec} R$ is the disjoint union of these subsets $V(P)$, and each $V(P)$ is a *tree*, that is, a partially ordered set with a least element and with the property that the subset of all the elements below a fixed element Q is linearly ordered for every $Q \in V(P)$. Hence $\operatorname{Spec} R$ is the disjoint union of a finite number of trees.

Proposition 6.30 *Every semiprime right serial ring is the direct product of a finite number of prime right serial rings.*

Proof. We have just seen that the ring has finitely many minimal prime ideals $P_1, \dots, P_m$. And $P_1 \cap \cdots \cap P_m = 0$ because R is semiprime. Finally, $P_i + P_j = R$ for every $i \neq j$ (Lemma 6.29). Therefore $R \cong R/P_1 \times \cdots \times R/P_m$ by the Chinese Remainder Theorem. $\square$

As a second application of Theorem 6.28 we shall give a description of the localizable systems in a serial ring R in terms of the prime ideals of R. Let $\{e_1, \dots, e_n\}$ be a complete set of orthogonal primitive idempotents of R. We have defined a localizable system in R as a suitable n-tuple $\mathcal{P} = (P_1, \dots, P_n)$, where each P_i is either a completely prime ideal of $e_i R e_i$ or is the empty set.

Define a localizable *prime* system as follows. An n-tuple $\mathcal{P}' = (P_1', \dots, P_n')$ is a *localizable prime system* for the serial ring R if:

1) for every $i = 1, 2, \dots, n$, P_i' is either the empty set or a prime ideal of R;

2) there exists $i = 1, 2, \dots, n$ such that $P_i' \neq \emptyset$;

3) $e_i \notin P_i'$ for every $i = 1, 2, \dots, n$;

4) for every $r, s \in R$ and every $i, j = 1, \dots, n$, if $P_i' \neq \emptyset$ and

$$e_i r e_j \cdot e_j s e_i \notin P_i',$$

then $P_j' \neq \emptyset$ and $e_j s e_i \cdot e_i r e_j \notin P_j'$.

By Theorem 6.28 there is a one-to-one correspondence between the set of all prime ideals P in the ring $e_i R e_i$ and the set of the prime ideals P' in the ring R with $e_i \notin P'$. The prime ideal of $e_i R e_i$ that corresponds to the prime P' of R with $e_i \notin P'$ is $e_i P' e_i$. For $P' = \emptyset$ set $e_i P' e_i = \emptyset$. Hence Theorem 6.28 induces a one-to-one correspondence between the set of all the n-tuples $(P_1, \dots, P_n)$ of prime ideals of $e_i R e_i$ and empty sets, and the set of all the n-tuples $(P_1', \dots, P_n')$ of prime ideals of R with $e_i \notin P_i'$ and empty sets.

Lemma 6.31 *Let R be a serial ring and let $\{e_1, \dots, e_n\}$ be a complete set of orthogonal primitive idempotents of R. Let $\mathcal{P}' = (P_1', \dots, P_n')$ be an n-tuple in which every P_i' is either a prime ideal of R or the empty set. Then $\mathcal{P}'$ is a localizable prime system for R if and only if $(e_1 P_1' e_1, \dots, e_n P_n' e_n)$ is a localizable system for R.*

Proof. Suppose that $\mathcal{P}' = (P_1', \ldots, P_n')$ is a localizable prime system for R. Let us show that if $P_i' \neq \emptyset$, the prime ideal $e_i P_i' e_i$ of $e_i R e_i$ is completely prime. If $a, b \in e_i R e_i$ and $ab \in e_i P_i' e_i$, then $ab \in P_i'$, so that $ba \in P_i'$. Hence $bar \in P_i'$ for every $r \in e_i R e_i$. Therefore $arb \in P_i'$, i.e., $ae_i R e_i b \subseteq e_i P_i' e_i$. Since $e_i P_i' e_i$ is a prime ideal of $e_i R e_i$, it follows that either $a \in e_i P_i' e_i$ or $b \in e_i P_i' e_i$. Now it is easily seen that $(e_1 P_1' e_1, \ldots, e_n P_n' e_n)$ is a localizable system for R.

The converse is an easy exercise. $\qquad\square$

Hence the one-to-one correspondence between the set of all prime ideals P in the ring $e_i R e_i$ and the set of all the prime ideals P' in the ring R with $e_i \notin P'$ defined in Theorem 6.28 induces a one-to-one correspondence between localizable systems and localizable prime systems.

Lemma 6.32 *Let P_i', P_j' be two prime ideals in a localizable prime system*

$$\mathcal{P}' = (P_1', \ldots, P_n')$$

for a serial ring R. The following statements hold:

(a) *If $P_i' \subseteq P_j'$, then $P_i' = P_j'$.*

(b) *If $P_i' \neq P_j'$, then P_i', P_j' are incomparable, $e_j \in P_i'$ and $e_i \in P_j'$.*

Proof. (a) Suppose $P_i' \subseteq P_j'$. For every $r \in R$ and every $x \in P_j'$ we have that $e_j x e_i r e_j \in P_j'$. Since $\mathcal{P}'$ is a localizable prime system, it follows that $e_i r e_j x e_i \in P_i'$. Hence $e_i R e_j x e_i \subseteq P_i'$. As P_i' is prime and $e_i \notin P_i'$, we obtain that $e_j x e_i \in P_i'$. Thus $e_j P_j' e_i \subseteq P_i'$. But $P_i' \subseteq P_j'$ implies that neither e_i nor e_j are in P_i'. Hence $P_j' \subseteq P_i'$.

(b) If $P_i' \neq P_j'$, then P_i', P_j' are incomparable by (a). But if, for instance, $e_j \notin P_i'$, then P_i', P_j' are comparable (Theorem 6.28). Therefore $e_j \in P_i'$ and $e_i \in P_j'$. $\qquad\square$

6.8   Goldie semiprime ideals

As a further application of Theorem 6.28 we describe the prime ideals of a serial ring R that correspond to completely prime ideals in the rings $e_i R e_i$. Note that by Goldie's Theorem 6.6 a semiprime serial ring is a right Goldie ring if and only if it is right nonsingular. But a serial ring is right nonsingular if and only if it is left nonsingular (Corollary 5.12). Hence a semiprime serial ring is right Goldie if and only if it is left Goldie. We shall say that a semiprime ideal K of a serial ring R is a *Goldie* semiprime ideal if R/K is a right (equivalently, left) Goldie ring.

Proposition 6.33 *Let R be a serial ring, let $\{e_1, \ldots, e_n\}$ be a complete set of orthogonal primitive idempotents of R, and let P' be a prime ideal of R. The following conditions are equivalent:*

(a) P' *is a Goldie ideal.*

(b) *For every $i = 1, \ldots, n$ either $e_i \in P'$ or $e_i P' e_i$ is a completely prime ideal of $e_i R e_i$.*

(c) *There exists an $i = 1, \ldots, n$ such that $e_i P' e_i$ is a completely prime ideal of $e_i R e_i$.*

Proof. (a) $\Rightarrow$ (b). Let i be an index such that $e_i \notin P'$. By Theorem 6.28 $e_i P' e_i$ is a prime ideal in $e_i R e_i$. Let $r, s \in e_i R e_i$ be two elements such that $r, s \notin e_i P' e_i$, so that $r, s \notin P'$. By Theorem 6.6 the prime right Goldie ring R/P' is right nonsingular, and we know that

$$\{ e_j + P' \mid j = 1, \ldots, n, \ e_j \notin P' \}$$

is a complete set of pairwise orthogonal primitive idempotents in R/P' (remark after the proof of Lemma 1.20). Since $r + P', s + P'$ are non-zero elements of

$$(e_i + P')(R/P')(e_i + P'),$$

it follows that their product is non-zero (Proposition 5.11). Hence $rs \notin P'$, so that $rs \notin e_i P' e_i$.

(b) $\Rightarrow$ (c). This is trivial.

(c) $\Rightarrow$ (a). Suppose that P' is a prime ideal of R for which (c) holds, but P' is not a Goldie ideal. By Theorem 6.6 the ring $\overline{R} = R/P'$ is not right nonsingular. For every $x \in R$ let $\overline{x}$ denote the element $x + P'$ of $\overline{R}$. The set $\{ \overline{e_j} \mid j = 1, \ldots, n, \ e_j \notin P' \}$ is a complete set of pairwise orthogonal primitive idempotents in R/P'. By Proposition 5.11 there exist $j, k, \ell = 1, 2, \ldots, n$ with $e_j, e_k, e_\ell \notin P'$ and $r, s \in R$ such that $\overline{e_j r e_k} \neq 0$, $\overline{e_k s e_\ell} \neq 0$ and $\overline{e_j r e_k s e_\ell} = 0$. Since $e_i P' e_i$ is a completely prime ideal of $e_i R e_i$, we know that $e_i \notin P'$. Set

$$P = e_i P' e_i,$$

so that P is the prime ideal of $e_i R e_i$ corresponding to P' in the one-to-one correspondence of Theorem 6.28. In particular,

$$P' = \{ x \in R \mid e_i R x R e_i \subseteq P \}.$$

As $e_j r e_k \notin P'$, we obtain that $e_i R e_j r e_k R e_i \not\subseteq P$. Hence $e_i t e_j r e_k u e_i \notin P$ for suitable elements $t, u \in R$. Similarly, from $e_k s e_\ell \notin P'$ it follows that $e_i v e_k s e_\ell w e_i \notin P$ for suitable $v, w \in R$. Consider the elements $e_k u e_i, e_k s e_\ell$ of the uniserial module $e_k R$. If $e_k u e_i \in e_k s e_\ell R$, then $e_i t e_j r e_k u e_i \in e_i t e_j r e_k s e_\ell R \subseteq P'$. This is a contradiction, because $e_i t e_j r e_k u e_i \notin P$. Hence $e_k s e_\ell \in e_k u e_i R$, so that $e_k s e_\ell = e_k u e_i s' e_\ell$ for some $s' \in R$. Then

$$e_i t e_j r e_k u e_i s' e_\ell w e_i = e_i t e_j r e_k s e_\ell w e_i \in P'.$$

But $e_i t e_j r e_k u e_i \notin P$ and P is completely prime, whence $e_i s' e_\ell w e_i \in P$. Then $e_i v e_k s e_\ell w e_i = e_i v e_k u e_i s' e_\ell w e_i \in P \subseteq P'$, contradiction. This contradiction shows that (c) $\Rightarrow$ (a). $\qquad \square$

Therefore in a localizable prime system $\mathcal{P}' = (P_1', \ldots, P_n')$ all the non-empty P_i' are Goldie prime ideals.

Proposition 6.33 has a generalization to semiprime ideals. Note that if K is a semiprime ideal of a serial ring R, then R/K is a finite direct product of prime rings (Proposition 6.30). Hence there are a finite number of pairwise incomparable prime ideals $P_1, \ldots, P_m$ containing K such that the canonical homomorphism $R/K \to R/P_1 \times \cdots \times R/P_m$ is an isomorphism. In particular $K = P_1 \cap \cdots \cap P_m$, and K is Goldie if and only if every P_j is Goldie. Thus we have proved

Lemma 6.34 *If K is a semiprime ideal of a serial ring R, then K is the intersection $K = P_1 \cap \cdots \cap P_m$ of a finite number of pairwise incomparable prime ideals $P_1, \ldots, P_m$. Such a representation of K is unique, and K is a Goldie ideal if and only if every P_j is a Goldie ideal.* □

As a corollary of Proposition 6.33 we have that

Corollary 6.35 *If K is a semiprime ideal of a serial ring R, then K is a Goldie ideal if and only if for every $i = 1, \ldots, n$ either $e_i \in K$ or $e_i K e_i$ is completely prime in $e_i R e_i$.*

Proof. Let K be a semiprime Goldie ideal and let i be an index such that $e_i \notin K$. Then $K = P_1 \cap \cdots \cap P_m$, where $P_1, \ldots, P_m$ are pairwise incomparable prime Goldie ideals. Since $e_i \notin K$, $e_i \notin P_k$ for a some P_k, $k = 1, \ldots, m$. As the prime ideals of R not containing e_i are comparable, it follows that $e_i \in P_j$ for every $j \neq k$, so that $e_i P_j e_i = e_i R e_i$. Then

$$e_i K e_i = e_i P_1 e_i \cap \cdots \cap e_i P_m e_i = e_i P_k e_i,$$

which is completely prime by Proposition 6.33.

Conversely, suppose that for every $i = 1, \ldots, n$ either $e_i \in K$ or $e_i K e_i$ is completely prime. By the previous lemma $K = P_1 \cap \cdots \cap P_m$ with the P_j pairwise incomparable prime ideals, and we must prove that every P_j is Goldie. Fix one of these prime ideals P_j, say P_k. Since P_k is proper, there is an index i such that $e_i \notin P_k$. In particular $e_i \notin K$, hence $e_i K e_i$ is completely prime. As in the previous paragraph $e_i \in P_j$ for every $j \neq k$, and $e_i K e_i = e_i P_k e_i$. By Proposition 6.33 the prime ideal P_k is a Goldie ideal. □

We conclude this section by showing that for a serial ring R there is a one-to-one correspondence between the set of all localizable prime systems for R and the set of all semiprime Goldie ideals of R.

Theorem 6.36 *Let R be a serial ring and let $\{e_1, \ldots, e_n\}$ be a complete set of orthogonal primitive idempotents of R. There is a one-to-one correspondence between the set of all localizable prime systems for R and the set of all semiprime Goldie ideals of R given by $\mathcal{P}' \mapsto \bigcap\{P' \mid P' \in \mathcal{P}', \ P' \neq \emptyset\}$ for every localizable prime system $\mathcal{P}'$.*

Proof. The prime ideals in any localizable prime system $\mathcal{P}'$ are Goldie prime ideals of R (Proposition 6.33). And any finite intersection of Goldie prime ideals is a Goldie semiprime ideal (Lemma 6.34).

Define another correspondence between the set of all semiprime Goldie ideals of R and the set of all localizable prime systems for R by

$$Q_1 \cap \cdots \cap Q_m \mapsto (P_1', \ldots, P_n'),$$

where $Q_1, \ldots, Q_m$ are finitely many pairwise incomparable prime Goldie ideals of R, $P_i' = Q_j$ if $e_i \notin Q_j$ for some index $j = 1, 2, \ldots, m$, and $P_i' = \emptyset$ otherwise, that is, if $e_i \in Q_1 \cap \cdots \cap Q_m$. Observe that if there exists an index j such that $e_i \notin Q_j$, then such an index j is necessarily unique, because if $e_i \notin Q_j$ and $e_i \notin Q_k$, then Q_j, Q_k are comparable, hence $Q_j = Q_k$ and $j = k$. In order to prove that this defines a localizable prime system, suppose that $r, s \in R$, $i, j = 1, \ldots, n$, $P_i' \neq \emptyset$ and $e_i r e_j \cdot e_j s e_i \notin P_i'$. Then $e_j \notin P_i'$. Hence $P_j' = P_i'$. Since P_i' is a Goldie ideal, the ring R/P_i' is a nonsingular by Theorem 6.6. As $e_i r e_j, e_j s e_i \notin P_i'$, it follows that $e_j s e_i r e_j \notin P_i' = P_j'$ (Proposition 5.11). Hence this second correspondence is well-defined also.

The verification that the compositions of the two correspondences are identities is an exercise that follows from Lemma 6.32 and is left to the reader. $\square$

In the last two sections we showed that in a serial ring there are one-to-one correspondences between localizable systems, localizable prime systems (Lemma 6.31), semiprime Goldie ideals (Theorem 6.36), and finite nonempty sets of pairwise incomparable prime Goldie ideals (Lemma 6.34).

Note that in the proof of Theorem 6.36 we saw that if K is a semiprime Goldie ideal, $(P_1', \ldots, P_n')$ is the corresponding localizable prime system and $i = 1, 2, \ldots, n$, then $e_i \notin K$ if and only if $P_i' \neq \emptyset$.

6.9 Diagonalization of matrices

Let R be a serial ring and let $\{e_1, \ldots, e_n\}$ be a fixed complete set of orthogonal primitive idempotents of R. Then the modules $e_i R$ and $R e_i$ are uniserial (Example 3.11) and the rings $e_i R e_i$ are chain rings (Proposition 1.11). Let

$$\mathbf{M}_n(e_i R e_j) = \begin{pmatrix} e_1 R e_1 & e_1 R e_2 & \cdots & e_1 R e_n \\ e_2 R e_1 & e_2 R e_2 & \cdots & e_2 R e_n \\ \vdots & \vdots & \vdots & \\ e_n R e_1 & e_n R e_2 & \cdots & e_n R e_n \end{pmatrix}$$

denote the ring of $n \times n$ matrices whose (i, j) entry is an arbitrary element of $e_i R e_j$. There is a canonical ring isomorphism $A \colon R \to \mathbf{M}_n(e_i R e_j)$, where for every element $r = \sum_{i,j=1}^n e_i r e_j$ of the ring R, the matrix $A(r)$ has $e_i r e_j$ in (i, j)-position.

Theorem 6.37 (Drozd) *For every element a of a serial ring R there are invertible elements $u, v \in R$ such that the matrix $A(uav)$ has at most one non-zero entry in every row and every column.*

Proof. Let a be an element of a serial ring R. If we apply Theorem 3.29 to the projective module R_R and its finitely generated submodule aR, we see that there is a decomposition $R_R = P_1 \oplus P_2 \oplus \cdots \oplus P_n$ of R_R into uniserial projective modules such that $aR = (aR \cap P_1) \oplus (aR \cap P_2) \oplus \cdots \oplus (aR \cap P_n)$. In the two decompositions $R_R = e_1 R \oplus e_2 R \oplus \cdots \oplus e_n R = P_1 \oplus P_2 \oplus \cdots \oplus P_n$ every direct summand has a local endomorphism ring, so that by the Krull-Schmidt-Remak-Azumaya Theorem there is an automorphism ω of R_R such that $\omega(P_i) = e_{\sigma(i)} R$ for a suitable permutation σ of $\{1, 2, \ldots, n\}$. Now endomorphisms of R_R are given by left multiplication by elements of R, and automorphisms of R_R are given by left multiplication by invertible elements of R. Thus there is an invertible element $u \in R$ such that $u P_i = e_{\sigma(i)} R$ for every $i = 1, 2, \ldots, n$.

Since $aR \cap P_i$ is a direct summand of aR contained in P_i, it is a cyclic uniserial module. The ring R is serial, and so semiperfect. Thus for every $i = 1, 2, \ldots, n$ such that $aR \cap P_i \neq 0$ the simple module $aR \cap P_i/\mathrm{rad}(aR \cap P_i)$ has a projective cover $C_i \to aR \cap P_i/\mathrm{rad}(aR \cap P_i)$, and C_i is cyclic and uniserial. As C_i is projective, there is an epimorphism $\varphi_i \colon C_i \to aR \cap P_i$, which is necessarily a projective cover of $aR \cap P_i$. For every $i = 1, 2, \ldots, n$ such that $aR \cap P_i = 0$, set $C_i = 0$ and $\varphi_i = 0$, so that $\varphi_i \colon C_i \to aR \cap P_i$ is a projective cover of $aR \cap P_i$ also in this case. By Proposition 3.2

$$\varphi_1 \oplus \cdots \oplus \varphi_n \colon C_1 \oplus \cdots \oplus C_n \to (aR \cap P_1) \oplus (aR \cap P_2) \oplus \cdots \oplus (aR \cap P_n)$$

is a projective cover of $(aR \cap P_1) \oplus (aR \cap P_2) \oplus \cdots \oplus (aR \cap P_n) = aR$. Apply Proposition 3.1(a) to the epimorphism $\psi \colon R_R \to aR$ given by left multiplication by a. Then there is a direct sum decomposition $R_R = Q' \oplus Q''$ of R_R with $\psi|_{Q'} \colon Q' \to aR$ a projective cover, and $\psi|_{Q''} \colon Q'' \to aR$ the zero homomorphism. Moreover, by the uniqueness of projective cover (Proposition 3.1(b)), there exists an isomorphism $f \colon C_1 \oplus \cdots \oplus C_n \to Q'$ such that $\varphi_1 \oplus \cdots \oplus \varphi_n = \psi|_{Q'} \circ f$. Thus $R_R = Q' \oplus Q'' = f(C_1) \oplus \cdots \oplus f(C_n) \oplus Q''$. Since the modules C_i are cyclic and uniserial, and so indecomposable or zero, the last decomposition has a refinement $R_R = f(C_1) \oplus \cdots \oplus f(C_n) \oplus Q_1'' \oplus \cdots, \oplus Q_m''$ into direct summands that are either indecomposable or zero.

If we apply Theorem 2.12 to the decomposition $R_R = e_1 R \oplus e_2 R \oplus \cdots \oplus e_n R$ into direct summands with local endomorphism rings and the decomposition $R_R = f(C_1) \oplus \cdots \oplus f(C_n) \oplus Q_1'' \oplus \cdots \oplus Q_m''$ into indecomposable or zero modules, we find that there is an automorphism α of R_R such that for every index j either $\alpha(e_j R) = f(C_k)$ for a uniquely determined index $k = 1, 2, \ldots, n$ or $\alpha(e_j R) = Q_\ell''$ for a uniquely determined index $\ell = 1, 2, \ldots, m$. Since automorphisms of R_R are given by left multiplication by invertible elements of R, there exists an invertible element $v \in R$ such that either $ve_j R = f(C_k)$ for some k or $ve_j R \subseteq Q''$.

We are ready to show that the matrix $A(uav)$ has the required property. In order to prove this, we must compute $e_i uave_j$. Fix an index j. If $ve_j R \subseteq Q''$, then $ave_j R \subseteq aQ'' = \psi(Q'') = 0$. Hence in this case $e_i uave_j = 0$ for every i. Thus we may assume $ve_j R = f(C_k)$ for an index $k = 1, 2, \ldots, n$ uniquely determined. Then

$$uave_j R = uaf(C_k) = u\psi|_{Q'} f(C_k) = u\varphi_k(C_k) = u(aR \cap P_k) \subseteq uP_k = e_{\sigma(k)} R.$$

Hence $e_i uave_j \neq 0$ for at most $i = \sigma(k)$. This shows that for every j there is at most one index i with $e_i uave_j \neq 0$. And if j, j' are two indices with $e_i uave_j \neq 0$ and $e_i uave_{j'} \neq 0$, then $i = \sigma(k)$, where $ve_j R = f(C_k)$ and $ve_{j'} R = f(C_k)$, so that $j = j'$. This proves that for every i there is at most one index j with $e_i uave_j \neq 0$. $\square$

We need the following proposition of a combinatorial nature.

Proposition 6.38 *Let A be a $n \times n$ matrix with entries in a ring R. Suppose that A has at most one non-zero entry in every row and every column. Then there exists a positive integer m such that the matrix A^m is diagonal.*

Proof. Let $A = (a_{ij})$ be a matrix with at most one non-zero entry in every row and every column. Consider the directed graph G with vertices the positive integers $1, 2, \ldots, n$ and with an arrow from the vertex i to the vertex j if and only if $a_{ij} \neq 0$. Thus G is a directed graph with n vertices, the number of arrows of which is equal to the number of non-zero entries in A. Note that in G for every vertex i there is at most one arrow that leaves that vertex and there is at most one arrow that arrives at that vertex.

If the entry (i, j) in the matrix A^2 is non-zero, then in the graph G there is an oriented path of length 2 from the vertex i to the vertex j. More generally, if the entry (i, j) in the matrix A^t is non-zero, then in G there is an oriented path of length t from the vertex i to the vertex j.

Since in G for each vertex i there is at most one arrow that leaves i and there is at most one arrow that arrives at i, the connected components of G are of two possible types: they are either oriented cycles of length t on t distinct vertices, or they are open oriented paths of length t on $t + 1$ distinct vertices. Let m be a positive integer that is at the same time greater than n and a common multiple of the length of all the oriented cycles that are connected components of G. Then in G every oriented path of length m is necessarily a closed path from a vertex i to the same vertex i. Hence in A^m only the entries on the diagonal are possibly non-zero. $\square$

From the previous two results we obtain the following corollary.

Corollary 6.39 *Let $\{e_1, \ldots, e_n\}$ be a complete set of orthogonal primitive idempotents of a serial ring R. For every $a \in R$ there exist invertible elements $u, v \in R$ and a positive integer m such that $(uav)^m \in e_1 R e_1 \times \cdots \times e_n R e_n$.* $\square$

6.10   Ore sets in serial rings

Let R be a serial ring and let $\mathcal{P} = (P_1, \ldots, P_n)$ be a localizable system for R. On page 142 we constructed an ascending chain $I_0 \subseteq I_1 \subseteq I_2 \subseteq \ldots$ of two-sided ideals with the property that the union $I = \bigcup_{t \geq 0} I_t$ is the smallest ideal of R such that the images of the elements of the subset $X_{\mathcal{P}}$ of R are regular elements of R/I. This ideal I turned out to be the kernel of the localization $\varphi \colon R \to R_{\mathcal{P}}$. Clearly, the construction of the ideal I is possible not only for $X_{\mathcal{P}}$, but for any subset X of a ring R; without loss of generality we may always suppose that $1 \in X$. The construction is as follows.

Let R be an arbitrary ring. Fix a subset X of R with $1 \in X$. Define a sequence of two-sided ideals I_t of R, $t = 0, 1, 2, \ldots$, by induction on the integer $t \geq 0$ in the following way. Set $I_0 = 0$. For every $t \geq 0$ let I_{t+1} be the two-sided ideal of R generated by the subset $\{r \in R \mid xr \in I_t \text{ or } rx \in I_t \text{ for some } x \in X\}$. Since $1 \in X$, the ideals I_t form an ascending chain $I_0 \subseteq I_1 \subseteq I_2 \subseteq \ldots$. Set $I(X) = \bigcup_{t \geq 0} I_t$. If $r \in R$ let $\bar{r} = r + I(X)$ be the image of r in $\overline{R} = R/I(X)$. Set $\overline{X} = \{\bar{x} \mid x \in X\}$. Then, obviously, the elements of $\overline{X}$ are regular elements of $\overline{R}$.

If $\overline{X}$ is a right and left Ore set in $\overline{R}$, let $\varphi \colon \overline{R} \to Q = \overline{R}[\overline{X}^{-1}]$ denote the quotient ring of $\overline{R}$ with respect to $\overline{X}$. Note that if R is serial, then Q is serial by Proposition 6.5.

A subset X of R is called a *saturated Ore set* in R if X is a saturated multiplicatively closed subset of R and $\overline{X}$ is a right and left Ore set in the ring $\overline{R}$. Note that with this definition a saturated Ore set in R is not necessarily an Ore subset of R in the sense defined on page 129, but this will not cause problems. A subset X of a serial ring R is called a *maximal Ore set* if $1 \in X$, $\overline{X}$ is a right and left Ore set in the ring $\overline{R} = R/I(X)$ and $X = \varphi^{-1}(U(\overline{R}[\overline{X}^{-1}]))$. Hence X is the set of all the elements that are mapped onto units of $\overline{R}[\overline{X}^{-1}]$ via the canonical homomorphism $\varphi \colon \overline{R} \to \overline{R}[\overline{X}^{-1}]$. Obviously, every maximal Ore set is a saturated Ore set.

In the next lemma we shall see that for every maximal Ore set X in a serial ring R it is possible to associate a localizable system to X.

Lemma 6.40 *Let R be a serial ring, let $\{e_1, \ldots, e_n\}$ be a complete set of orthogonal primitive idempotents of R and let X be a maximal Ore set. If $P_i = \{r \in e_i R e_i \mid 1 - e_i + r \notin X\}$ for every $i = 1, 2, \ldots, n$, then $\mathcal{P} = (P_1, \ldots, P_n)$ is a localizable system for R.*

Proof. Set $Q = \overline{R}[\overline{X}^{-1}]$, so that by Proposition 6.5 and its proof every Qe_i is a uniserial left ideal of Q. By Proposition 6.27 applied to the Q-R-bimodule ${}_QQ_R$, if $P_i'' = \{r \in e_i R e_i \mid \text{right multiplication by } e_i R e_i \text{ is not an automorphism of } {}_QQe_i\}$ for every i, then $\mathcal{P}'' = (P_1'', \ldots, P_n'')$ is a localizable system for R. Hence it suffices to show that $P_i = P_i''$ for every index i.

Let r be an element of $e_i R e_i$. Then $r \notin P_i$ if and only if $1 - e_i + r \in X$, that is, if and only if $\overline{1 - e_i + r}$ is invertible in Q, or, equivalently, if and only if right multiplication by $1 - e_i + r$ is an automorphism of $_Q Q$. This happens if and only if right multiplication by r is an automorphism of $_Q Q e_i$, i.e., if and only if $r \notin P_i''$. $\qquad \square$

Given a maximal Ore set X, the localizable system associated to X defined in Lemma 6.40 will be denoted $\mathcal{P}(X)$. Now our aim is to show that if X is a maximal Ore set and $\mathcal{P}(X)$ is its associated localizable system, then the quotient ring $\overline{R}[\overline{X}^{-1}]$ of $\overline{R} = R/I(X)$ with respect to $\overline{X}$ and the localization $R_{\mathcal{P}(X)}$ of the serial ring R at $\mathcal{P}(X)$ coincide. This will be done in Proposition 6.43. First we need two preparatory lemmas. Recall that in Section 6.5 we defined $X_{\mathcal{P}} = (e_1 R e_1 \setminus P_1) \times (e_2 R e_2 \setminus P_2) \times \cdots \times (e_n R e_n \setminus P_n)$ for any localizable system $\mathcal{P} = (P_1, \ldots, P_n)$.

Lemma 6.41 *Let X be a maximal Ore set of a serial ring R. Then*

$$X_{\mathcal{P}(X)} = X \cap (e_1 R e_1 \times \cdots \times e_n R e_n).$$

Proof. Let r be an element of $e_1 R e_1 \times \cdots \times e_n R e_n$, so that

$$r = (1 - e_1 + e_1 r e_1) \cdot (1 - e_2 + e_2 r e_2) \cdots (1 - e_n + e_n r e_n).$$

Then $r \in X$ if and only if $1 - e_i + e_i r e_i \in X$ for every i, that is, if and only if $e_i r e_i \notin P_i$ for every i, i.e., if and only if $r \in X_{\mathcal{P}(X)}$. $\qquad \square$

Let X be a maximal Ore set and let $\mathcal{P}(X)$ be its associated localizable system. We will now show that the ideal $I(X)$, defined at the beginning of this section, and which is the kernel of the canonical mapping $R \to \overline{R}[\overline{X}^{-1}]$, coincides with the ideal $I(X_{\mathcal{P}(X)})$ defined at the beginning of Section 6.5, and which is the kernel of the localization mapping $R \to R_{\mathcal{P}(X)}$.

Lemma 6.42 *Let X be a maximal Ore set of a serial ring R and let $\mathcal{P}(X)$ be the localizable system associated to X. Let $I(X)$ be the kernel of the canonical mapping $R \to \overline{R}[\overline{X}^{-1}]$ and $I(X_{\mathcal{P}(X)})$ the kernel of the localization mapping $R \to R_{\mathcal{P}(X)}$. Then $I(X) = I(X_{\mathcal{P}(X)})$.*

Proof. We shall show that $I_t(X) = I_t(X_{\mathcal{P}(X)})$ for every $t \geq 0$ by induction on t.

Suppose $I_t(X) = I_t(X_{\mathcal{P}(X)})$. Since $X_{\mathcal{P}(X)} \subseteq X$ (Lemma 6.41), we have that $I_{t+1}(X_{\mathcal{P}(X)}) \subseteq I_{t+1}(X)$. Conversely, suppose that

$$rx \in I_t(X) = I_t(X_{\mathcal{P}(X)})$$

for some $r \in R$, $x \in X$. By Corollary 6.39 there exist $u, v \in U(R)$ and $m > 0$ such that $y = (uxv)^m \in e_1 R e_1 \times \cdots \times e_n R e_n$. Hence $y \in X$. From

Lemma 6.41 it follows that $y \in X_{\mathcal{P}(X)}$. Now $ru^{-1}y \in I_t(X_{\mathcal{P}(X)})$ implies that $ru^{-1} \in I_{t+1}(X_{\mathcal{P}(X)})$, hence $r \in I_{t+1}(X_{\mathcal{P}(X)})$. Similarly, from

$$xr \in I_t(X) = I_t(X_{\mathcal{P}(X)})$$

for some $r \in R$ and $x \in X$, it is easily seen that $r \in I_{t+1}(X_{\mathcal{P}(X)})$. Therefore $I_{t+1}(X) \subseteq I_{t+1}(X_{\mathcal{P}(X)})$. $\qquad\qquad\qquad\qquad\qquad\qquad\qquad\qquad\qquad\qquad\qquad\square$

Proposition 6.43 *If X is a maximal Ore set of a serial ring R, $\mathcal{P}(X)$ is the localizable system associated to X, and $\overline{R} = R/I(X)$, then the mapping $\overline{R} \to R_{\mathcal{P}(X)}$ induced by the localization $\varphi\colon R \to R_{\mathcal{P}(X)}$ of R at $\mathcal{P}(X)$ is the quotient ring of $\overline{R}$ with respect to $\overline{X}$.*

Proof. By Lemma 6.42 the localization $\varphi\colon R \to R_{\mathcal{P}(X)}$ induces an injective ring homomorphism $R/I(X) \to R_{\mathcal{P}(X)}$. This injective ring homomorphism is the quotient ring of $\overline{R} = R/I(X)$ with respect to $\overline{X}$ as follows from two remarks:

(a) If $x \in X$, by Corollary 6.39 there exist $u, v \in U(R)$ and $m > 0$ such that $y = (uxv)^m \in e_1 R e_1 \times \cdots \times e_n R e_n$. Then $y \in X$, so that from Lemma 6.41 it follows that $y \in X_{\mathcal{P}(X)}$. Thus $\overline{y}$ is invertible in $R_{\mathcal{P}(X)}$, and so $\overline{x}$ is invertible in $R_{\mathcal{P}(X)}$.

(b) Since $X_{\mathcal{P}(X)} \subseteq X$ (Lemma 6.41), every element of $R_{\mathcal{P}(X)}$ can be written as a quotient of an element $\overline{r} \in \overline{R}$ and an element $\overline{x} \in \overline{X}$. $\qquad\square$

Our aim is now to prove the final result of this section: that the assignement $X \mapsto \mathcal{P}(X)$ is a one-to-one correspondence between maximal Ore sets and localizable systems.

Theorem 6.44 *Let R be a serial ring. Then there is a one-to-one correspondence between the set of all maximal Ore sets X in R and the set of all localizable systems for R given by $X \mapsto \mathcal{P}(X)$.*

Proof. For a localizable system $\mathcal{P}$, let $\varphi\colon R \to R_{\mathcal{P}}$ denote the canonical homomorphism of R into its localization at $\mathcal{P}$ (Section 6.5). Let $X(\mathcal{P})$ be the inverse image of the group of units $U(R_{\mathcal{P}})$ of $R_{\mathcal{P}}$ via φ. Then $X(\mathcal{P})$ is a maximal Ore set.

Let us show that the two correspondences $X \mapsto \mathcal{P}(X)$ and $\mathcal{P} \mapsto X(\mathcal{P})$ are mutually inverse. Let X be a maximal Ore set. In order to see that

$$X = X(\mathcal{P}(X)),$$

observe that X is the inverse image of the group of units $U(\overline{R}[\overline{X}]^{-1})$ via

$$R \to \overline{R}[\overline{X}]^{-1},$$

and $X(\mathcal{P}(X))$ is the inverse image of the group of units $U(R_{\mathcal{P}(X)})$ via

$$\varphi\colon R \to R_{\mathcal{P}(X)}.$$

From Proposition 6.43 we obtain that $X = X(\mathcal{P}(X))$.

Conversely, let $Q = (Q_1, \ldots, Q_n)$ be a localizable system and $\mathcal{P}(X(Q)) = (P_1, \ldots, P_n)$. In order to show that $Q = \mathcal{P}(X(Q))$ we must prove that $Q_i = P_i$ for every i.

Let r be an element in $e_i R e_i$. On the one hand, if $r \notin Q_i$, then

$$1 - e_i + r \in X_Q,$$

so that $\varphi(1 - e_i + r)$ is invertible in R_Q. Hence $1 - e_i + r \in X(Q)$, and so $r \notin P_i$. This proves that $P_i \subseteq Q_i$. To prove the reverse inclusion, suppose that $r \notin P_i$. If $Q_i = \emptyset$, then $r \notin Q_i$. Assume that $Q_i \neq \emptyset$. Then $1 - e_i + r \in X(Q)$, so that $\varphi(1 - e_i + r)$ is invertible in R_Q. Hence $\varphi(1 - e_i + r)\varphi(a)\varphi(b)^{-1} = 1$ for some $a \in R$ and $b \in X_Q$. Then $(1 - e_i + r)a - b \in \ker \varphi = I(X_Q)$, so that $e_i r a e_i - e_i b e_i \in I(X_Q)$, i.e., $e_i r a e_i - e_i b e_i = e_i c e_i$ for some $c \in I(X_Q)$. From $b \in X_Q$ we get $e_i b e_i \notin Q_i$, and from $c \in I(X_Q)$ and Lemma 6.20(c) $\Rightarrow$ (a) we obtain that $e_i c e_i \in Q_i$. Thus $e_i r a e_i = r e_i a e_i \notin Q_i$. Hence $r \notin Q_i$. This shows that $Q_i = P_i$. $\qquad\square$

6.11   Goldie semiprime ideals and maximal Ore sets

Theorem 6.44 and the results of Sections 6.7 and 6.8 say that in a serial ring there are one-to-one correspondences between localizable systems, localizable prime systems, semiprime Goldie ideals, finite nonempty sets of pairwise incomparable prime Goldie ideals, and maximal Ore sets. The one-to-one correspondence from semiprime Goldie ideals to maximal Ore sets is described directly in the following proposition.

Proposition 6.45 *Let Q be a semiprime Goldie ideal of a serial ring R, let $\{e_1, \ldots, e_n\}$ be a complete set of orthogonal primitive idempotents of R, let $\mathcal{P}' = (P_1', \ldots, P_n')$ be the localizable prime system corresponding to Q, let $\mathcal{P} = (e_1 P_1' e_1, \ldots, e_n P_n' e_n)$ be the corresponding localizable system, and let X be the corresponding maximal Ore set. Then $X = \mathcal{C}(Q)$.*

Proof. Put $J = \{ i \mid i = 1, \ldots, n, \ P_i' \neq \emptyset \}$. By Theorem 6.36

$$Q = \bigcap_{i \in J} P_i',$$

and by Lemma 6.32 the distinct prime ideals that appear in this intersection are pairwise incomparable, and so comaximal (Lemma 6.29). Thus R/Q is canonically isomorphic to the direct product of the rings R modulo the distinct P_i''s. Hence $\mathcal{C}(Q) = \bigcap_{i \in J} \mathcal{C}(P_i')$. We shall first show that $X_{\mathcal{P}} \subseteq \mathcal{C}(Q)$.

Put $S = e_1 R e_1 \times \cdots \times e_n R e_n$, so that $X_{\mathcal{P}} \subseteq S$. In order to prove that $X_{\mathcal{P}} \subseteq \mathcal{C}(Q)$ we shall show that if $x \in S$ and $x \notin \mathcal{C}(Q)$, then $x \notin X_{\mathcal{P}}$. Let x be an element of $S \setminus \mathcal{C}(Q)$. Then there is an index $i \in J$ such that $x \notin \mathcal{C}(P_i')$.

Hence there exists $r \in R \setminus P'_i$ such that either $xr \in P'_i$ or $rx \in P'_i$. Suppose that $xr \in P'_i$. Then $e_j r e_k \notin P'_i$ for suitable $j, k = 1, \dots, n$, but

$$e_j x e_j r e_k = e_j x r e_k \in P'_i.$$

As $e_j r e_k \notin P'_i$, there exist $s, t \in R$ such that $e_i s e_j r e_k t e_i \notin e_i P'_i e_i$ (Theorem 6.28). Hence $e_i s e_j r e_k t e_i \notin P'_i$. Since $\mathcal{P}'$ is a localizable prime system, P'_i is not empty, that is, it is a prime ideal. From $e_j r e_k \notin P'_i$ we know that $e_j \notin P'_i$. Hence $P'_i = P'_j$ by Lemma 6.32(b). Thus $e_j r e_k \notin P'_j$. From Theorem 6.28 it follows that there exists $u \in R$ such that $e_j r e_k u e_j \notin P'_j$. As $e_j x e_j r e_k \in P'_i = P'_j$, we get that $e_j x e_j r e_k u e_j \in e_j P'_j e_j$. But $e_j P'_j e_j$ is a completely prime ideal, so that $e_j x e_j \in e_j P'_j e_j$. Thus $x \notin X_{\mathcal{P}}$. This proves that $X_{\mathcal{P}} \subseteq \mathcal{C}(Q)$.

In order to prove that $X \subseteq \mathcal{C}(Q)$ fix an element $x \in X$. By Corollary 6.39 there are invertible elements $u, v \in R$ and a positive integer m such that

$$y = (uxv)^m \in S.$$

Then $y \in X_{\mathcal{P}}$ (Lemma 6.41), so that $y \in \mathcal{C}(Q)$. Now it is easily seen that $x \in \mathcal{C}(Q)$.

To prove the reverse inclusion $\mathcal{C}(Q) \subseteq X$, suppose that $x \in \mathcal{C}(Q)$. By Corollary 6.39 there are invertible elements $u, v \in R$ and a positive integer m such that $y = (uxv)^m \in S$. Clearly $y = (uxv)^m \in \mathcal{C}(Q)$. Suppose $i \in J$. Then $y \in \mathcal{C}(P'_i)$ and $e_i \notin P'_i$, so that $y e_i \notin P'_i$, that is, $e_i y e_i \notin e_i P'_i e_i$. Also if $i \notin J$, then $e_i P'_i e_i = \emptyset$, so that $e_i y e_i \notin e_i P'_i e_i$. Hence $y \in X_{\mathcal{P}} \subseteq X$. But X is a saturated multiplicatively closed set, and so $x \in X$. $\qquad\square$

6.12   Classical quotient ring of a serial ring

In this section we apply localizable systems to the study of the classical quotient ring of a serial ring.

Lemma 6.46 *Let* $\{e_1, \dots, e_n\}$ *be a complete set of orthogonal primitive idempotents of a serial ring* R. *For every* $i = 1, 2, \dots, n$ *set*

$$P_i = (Z(R_R) \cap e_i R e_i) \cup (Z(_R R) \cap e_i R e_i).$$

Then $\mathcal{P} = (P_1, \dots, P_n)$ *is a localizable system for* R.

Proof. Since every $e_i R e_i$ is a chain ring, it is immediate that every P_i is a proper ideal of $e_i R e_i$. In particular, all the P_i are non-empty. Let a and b be two elements of $e_i R e_i$ such that $ab \in P_i$. By symmetry, we can suppose that $abI = 0$ for some essential right ideal I of R. If $bI = 0$ then $b \in P_i$. Otherwise bI is essential in $e_i R$, so that $bI + (1 - e_i)R$ is an essential right ideal of R that annihilates a. Hence $a \in P_i$, and P_i is a completely prime ideal in $e_i R e_i$.

Suppose that $r, s \in R$ and $i, j = 1, \ldots, n$ are such that $e_j s e_i r e_j \in P_j$. Then either $e_j s e_i r e_j \in Z(R_R)$ or $e_j s e_i r e_j \in Z(_R R)$. Suppose that $e_j s e_i r e_j \in Z(R_R)$, so that $e_j s e_i r e_j I = 0$ for an essential right ideal I. If $e_i r e_j I \neq 0$, then $e_i r e_j I + (1 - e_i) R$ is an essential right ideal of R that annihilates $e_i r e_j s e_i$, so that $e_i r e_j s e_i \in P_i$. If $e_i r e_j I = 0$, then $e_i r e_j \in Z(R_R)$. Hence $e_i r e_j s e_i \in Z(R_R)$, i.e., $e_i r e_j s e_i \in P_i$. Similarly, if $e_j s e_i r e_j \in Z(_R R)$, then there is an essential left ideal I' of R such that $I' e_j s e_i r e_j = 0$, and this yields $e_i r e_j s e_i \in P_i$. $\square$

The localizable system defined in Lemma 6.46 is used to prove the following theorem.

Theorem 6.47 *Every serial ring has a classical two-sided quotient ring.*

Proof. Put $S = e_1 R e_1 \times \cdots \times e_n R e_n$. If $\mathcal{P}$ is the localizable system defined in Lemma 6.46 and X is the maximal Ore set corresponding to $\mathcal{P}$, it suffices to show that $X = \mathcal{C}(0)$. Let x be an element of $\mathcal{C}(0)$. By Corollary 6.39 there exist $u, v \in U(R)$ and $m \geq 1$ such that $y = (uxv)^m \in S$. Then $y \in \mathcal{C}(0)$, i.e., y is regular. If I is a right ideal of R and $y e_i I = 0$, then $e_i I = 0$, and thus I is not essential in R_R. Similarly, if I' is a left ideal and $I' e_i y = 0$, then I' is not essential in $_R R$. This proves that $y e_i = e_i y \notin P_i$ for every i. Thus $y \in X_{\mathcal{P}} \subseteq X$. But X is saturated, so that $x \in X$.

Conversely, let $x \in X$. There exist $u, v \in U(R)$ and $m \geq 1$ such that $y = (uxv)^m \in S$. Then $y \in X_{\mathcal{P}}$ by Lemma 6.41, so that $e_i y e_i \notin Z(R_R)$ and $e_i y e_i \notin Z(_R R)$. If $r \in R$ satisfies $yr = 0$, then for every index i we have $e_i y r = y e_i r = 0$, from which $e_i y e_i (e_i r R + (1 - e_i) R) = 0$. It follows that $e_i r = 0$ for every i, hence $r = 0$. Similarly, $ry = 0$ implies $r = 0$. Thus y, hence x, is regular. $\square$

As an application of the description of the structure of noetherian serial rings R obtained in Chapter 5, we prove a result about the right R-module $Q/J(R)$, where Q is the classical (two-sided) quotient ring of R and $J(R)$ is the Jacobson radical.

Proposition 6.48 *If R is a noetherian serial ring, $J(R)$ is the Jacobson radical of R, and Q is the classical quotient ring of R, then $Q/J(R)$ is an artinian right R-module and $R/J(R)$ is its socle.*

Proof. By Theorem 5.29 we may suppose that the noetherian serial ring R is either artinian or a hereditary ring with no simple left or right ideals. Let $\{e_1, \ldots, e_n\}$ be a complete set of orthogonal primitive idempotents for R, so that $\{e_1, \ldots, e_n\}$ is also a complete set of orthogonal primitive idempotents for the classical quotient ring Q of R. Then $e_i Q_Q$ is a uniserial module for every i, so that $e_i Q_R$ is uniserial by Corollary 6.25. If R is artinian, every cyclic R-module is artinian, so that every uniserial R-module is artinian. Hence $e_i Q / e_i J(R)$ is an artinian R-module in this case. If R is a hereditary serial ring with no simple left or right ideals, then $e_i J(R) \neq 0$ (otherwise $e_i R$ would be a simple right ideal of R), so that $e_i Q / e_i J(R)$ is a proper homomorphic image of a uniserial R-module. In order to prove that $e_i Q / e_i J(R)$ is artinian, it is therefore sufficient to show that every proper homomorphic image of a uniserial R-module is artinian. For this it suffices to show that every proper homomorphic image of a cyclic uniserial R-module is artinian. Thus we are reduced to proving that every proper homomorphic image of an $e_j R$ is artinian. We have seen in Theorem 5.28 and in the remark after the proof of Lemma 5.20 that either $e_j R$ is a uniserial module of finite composition length or $e_j R$ is a noetherian non-artinian uniserial module whose lattice of its submodules is order antiisomorphic to $\omega + 1$. Thus every proper homomorphic image of $e_j R$ has finite composition length. This proves that every $e_i Q / e_i J(R)$ is artinian. Since $e_i R / e_i J(R)$ is a simple submodule of the uniserial module $e_i Q / e_i J(R)$, it follows that $e_i R / e_i J(R)$ is the socle of $e_i Q / e_i J(R)$. Hence $Q / J(R) = \oplus_{i=1}^{n} e_i Q / e_i J(R)$ is an artinian R-module and $R / J(R) = \oplus_{i=1}^{n} e_i R / e_i J(R)$ is its socle. $\square$

The aim of the second part of this section it to prove Warfield-Chatters' Theorem 6.54. We split the proof into some preliminary partial results.

Theorem 6.49 *If R is a ring with the a.c.c. on right annihilators, then $Z(R_R)$ is nilpotent.*

Proof. Put $Z = Z(R_R)$ and suppose that Z is not nilpotent. Since R is a ring with the a.c.c. on right annihilators, there exists $n > 0$ such that $\text{r.ann}(Z^n) = \text{r.ann}(Z^{n+1})$. Now $Z^{n+1} \neq 0$. Hence the set

$$ T = \{ t \in Z \mid Z^n t \neq 0 \} $$

is non-empty. Let $z \in T$ be an element such that $\text{r.ann}(z)$ is maximal in the set $\{ \text{r.ann}(t) \mid t \in T \}$. If x in an arbitrary element of Z, then $\text{r.ann}(x)$ is essential in R_R, so $\text{r.ann}(x) \cap zR \neq 0$. Hence there is an element $r \in R$ such that $zr \neq 0$ and $xzr = 0$. It follows that $\text{r.ann}(xz) \supset \text{r.ann}(z)$, so $xz \notin T$. Hence $Z^n xz = 0$. This shows that $Z^{n+1} z = 0$, i.e., $z \in \text{r.ann}(Z^{n+1}) = \text{r.ann}(Z^n)$, hence $Z^n z = 0$. This is a contradiction because $z \in T$. $\square$

Proposition 6.50 *If R is a serial ring with the a.c.c. on right annihilators, then $R / N(R)$ is a semiprime Goldie ring.*

Proof. Let $\{ e_1, \ldots, e_n \}$ be a complete set of orthogonal primitive idempotents of R. For every $i = 1, 2, \ldots, n$ put $P_i = Z(R_R) \cap e_i R e_i$. A proof identical to that of Lemma 6.46 shows that $\mathcal{P} = (P_1, \ldots, P_n)$ is a localizable system for R. Since $P_i \subseteq Z(R_R)$, P_i is nilpotent by Theorem 6.49. Therefore P_i is the minimal prime ideal of $e_i R e_i$. By Theorem 6.28 the prime Goldie ideal P_i' corresponding to P_i is a minimal prime ideal of R. Hence $P_1', \ldots, P_n'$ are all the minimal prime ideals of R, so that $N(R) = P_1' \cap \cdots \cap P_n'$ and $N(R)$ is a Goldie semiprime ideal. $\square$

The following lemma is taken from [Chatters, Lemma 3.2].

Lemma 6.51 *Let R be a serial ring, let $\{e_1, \ldots, e_n\}$ be a complete set of orthogonal primitive idempotents of R, and let K be a semiprime Goldie ideal of R such that $e_i \notin K$ for all $i = 1, 2, \ldots, n$. Then $K = Kc = cK$ for every $c \in \mathcal{C}(K)$.*

Proof. Fix an element $c \in \mathcal{C}(K)$. By Corollary 6.39 there exist invertible elements $u, v \in R$ and a positive integer m such that

$$y = (ucv)^m \in e_1 Re_1 \times \cdots \times e_n Re_n.$$

Then $yR = e_1 yR \oplus \cdots \oplus e_n yR$ and $y \in \mathcal{C}(K)$. Now $K = e_1 K \oplus \cdots \oplus e_n K$, and $R/K \cong e_1 R/e_1 K \oplus \cdots \oplus e_n R/e_n K$ is a direct sum of n non-zero uniserial modules. Since R/K is a semiprime Goldie ring, its right ideal $yR + K/K$ is essential (Theorem 6.6). In particular, $yR + K/K$ has Goldie dimension n. But $yR + K/K \cong e_1 yR + e_1 K/e_1 K \oplus \cdots \oplus e_n yR + e_n K/e_n K$. Hence every $e_i yR + e_i K/e_i K$ is non-zero, so that $e_i K \subset e_i yR$ for every i. Thus $K \subset yR$. Then $K \subseteq yK$, because for every $x \in K$ there exists $r \in R$ with $x = yr$, so that $y \in \mathcal{C}(K)$ and $yr \in K$ imply $r \in K$. Thus $K = yK$, and so $K = ucvK$, i.e., $K = cK$. By symmetry, $K = Kc$. □

Lemma 6.52 *Let R be a serial ring with the a.c.c. on right annihilators, let $\{e_1, \ldots, e_n\}$ be a complete set of orthogonal primitive idempotents of R, and let K be a semiprime Goldie ideal of R such that $e_i \notin K$ for all $i = 1, 2, \ldots, n$. Then $\mathrm{r.ann}(c) = 0$ for every $c \in \mathcal{C}(K)$.*

Proof. Fix an element $c \in \mathcal{C}(K)$. Then $\mathrm{r.ann}(c^n) \subseteq K$ for every $n \geq 0$. From Lemma 6.51 we know that $K = cK$, so that the endomorphism f of the right R-module K given by left multiplication by c is surjective. Since $\mathrm{r.ann}(c^n) \subseteq K$, it follows that $\ker f^n = \mathrm{r.ann}(c^n)$ for every $n \geq 0$. But the ascending chain $\mathrm{r.ann}(c) \subseteq \mathrm{r.ann}(c^2) \subseteq \mathrm{r.ann}(c^3) \subseteq \ldots$ is stationary. Thus $\ker f^n = \ker f^{n+1}$ for some n. Hence $\ker f^n \cap f^n(K) = 0$ (Lemma 2.17). Therefore $\ker f^n = 0$, i.e., $\mathrm{r.ann}(c^n) = 0$, from which $\mathrm{r.ann}(c) = 0$. □

The next lemma is a key result.

Lemma 6.53 *Every serial ring with the a.c.c. on right annihilators has the a.c.c. on left annihilators.*

Proof. If R is a serial ring with the a.c.c. on right annihilators, then the prime radical N of R is nilpotent (Corollary 6.9) and R/N is a semiprime Goldie ring (Proposition 6.50). Hence the reduced rank of any right or left R-module is defined.

We prove that if A and B are left annihilators in a serial ring R with the a.c.c. on right annihilators, $A \supseteq B$ and the reduced rank $\rho_N(A/B)$ as a left

R-module is zero, then $A = B$. If $a \in A$, then there exists $c \in \mathcal{C}_R(N)$ such that $ca \in B$ by the left-right dual of Proposition 6.13(c). Hence $ca\,\mathrm{r.ann}(B) = 0$. From Lemma 6.52 it follows that $a\,\mathrm{r.ann}(B) = 0$, so that

$$a \in \mathrm{l.ann}(\mathrm{r.ann}(B)) = B.$$

As $\rho_N(_RR)$ is finite (Example 6.14), we get that every strictly ascending chain of left annihilators in R must have finite length $\leq \rho_N(_RR)$. $\qquad\square$

We are now in a position to prove Warfield-Chatters' Theorem.

Theorem 6.54 *The following conditions are equivalent for a serial ring R:*

(a) *The classical two-sided quotient ring of R is artinian;*

(b) *R has the a.c.c. on right annihilators;*

(c) *R has the a.c.c. on left annihilators.*

Proof. Every ring contained in an artinian ring has the a.c.c. on right and left annihilators, so that (a) implies (b) and (c). By Lemma 6.53 and its left-right dual, a serial ring R has the a. c. c. on right annihilators if and only if it has the a. c. c. on left annihilators. Hence (b) and (c) are equivalent. To conclude the proof of the theorem it suffices to see that (b) and (c) imply (a). Thus we must show that if a serial ring R has the a. c. c. on right and left annihilators, then its classical two-sided quotient ring, which exists by Theorem 6.47, is artinian.

Put $N = N(R)$. We shall apply Theorem 6.16. Note that R/N is a Goldie ring (Proposition 6.50), N is nilpotent (Corollary 6.9), $\rho_N(R_R)$ and $\rho_N(_RR)$ are finite (Example 6.14), and $\mathcal{C}_R(0) \supseteq \mathcal{C}_R(N)$ (Lemma 6.52 and its left-right dual). Hence we need only prove that $\mathcal{C}_R(0) \subseteq \mathcal{C}_R(N)$.

If $c \in \mathcal{C}_R(0)$, left multiplication by c is an isomorphism $R_R \to cR_R$. From Proposition 6.13(b) it follows that $\rho_N(R_R/cR_R) = 0$, so that there exists an element $d \in \mathcal{C}_R(N)$ such that $(1 + cR)d = 0$ (Proposition 6.13(c)), that is, $d \in cR$. Hence $d = cr$ for a suitable $r \in R$. Let $\overline{x}$ denote the image in $\overline{R} = R/N$ of any element $x \in R$. Then $\overline{cr}$ is a regular element of $\overline{R}$, hence

$$\overline{r}\overline{R} \cap \mathrm{r.ann}_{\overline{R}}(\overline{c}) = 0.$$

Since $\mathrm{r.ann}_{\overline{R}}(\overline{r}) \subseteq \mathrm{r.ann}_{\overline{R}}(\overline{cr}) = 0$, we have that $\mathrm{r.ann}_{\overline{R}}(\overline{r}) = 0$. Set $I = \mathrm{r.ann}_{\overline{R}}(\overline{c})$. From $\overline{r}\overline{R} \cap I = 0$ it follows that the sum $I + \overline{r}I + \overline{r}^2I + \ldots$ is direct. As $\overline{R}$ has finite Goldie dimension, we must have $\overline{r}^nI = 0$ for some n, so that $I = 0$. Thus $\mathrm{r.ann}_{\overline{R}}(\overline{c}) = 0$. By symmetry, $\mathrm{l.ann}_{\overline{R}}(\overline{c}) = 0$. Hence $c \in \mathcal{C}_R(N)$. $\qquad\square$

By Proposition 6.50 and Lemma 6.51, if R is a serial ring with the a.c.c. on right annihilators and $c \in \mathcal{C}(N(R))$, then $N(R) = N(R)c = cN(R)$. Since $\mathcal{C}_R(0) = \mathcal{C}_R(N(R))$ (proof of Theorem 6.54), $N(R)$ is an ideal in the classical quotient ring Q of R. But $N(R)$ is nilpotent by Corollary 6.9. Hence $N(R) \subseteq N(Q) = J(Q)$ (Proposition 1.9). From $\mathcal{C}_R(0) = \mathcal{C}_R(N(R))$ it follows that $Q/N(R)$ is the classical quotient ring of the semiprime Goldie ring $R/N(R)$, so that $Q/N(R)$ is semisimple artinian by Goldie's Theorem 6.6. Thus its Jacobson radical $J(Q/N(R)) = J(Q)/N(R)$ is zero. This proves that

Proposition 6.55 *If Q is the classical quotient ring of a serial ring R with the a.c.c. on right annihilators, then $J(Q) = N(R)$.* $\qquad\square$

Examples of (commutative) chain rings with or without the a.c.c. on (right) annihilators are given easily. For an example of a commutative chain ring with the a.c.c. on annihilators take any commutative valuation domain. For an example of a commutative chain ring R without the a.c.c. on annihilators take the *trivial extension* $R = \mathbf{Z}_p \propto \mathbf{Z}(p^\infty)$ of the discrete valuation ring $\mathbf{Z}_p$ via the Prüfer group $\mathbf{Z}(p^\infty)$. Here p is a prime number, the valuation ring $\mathbf{Z}_p$ is the localization of the ring of integers $\mathbf{Z}$ at its maximal ideal (p), and the trivial extension $\mathbf{Z}_p \propto \mathbf{Z}(p^\infty)$ is the direct sum $\mathbf{Z}_p \oplus \mathbf{Z}(p^\infty)$ of the abelian groups $\mathbf{Z}_p$ and $\mathbf{Z}(p^\infty)$ with the multiplication defined by $(a, x)(b, y) = (ab, ay + xb)$ for every $(a, x), (b, y) \in \mathbf{Z}_p \oplus \mathbf{Z}(p^\infty)$.

Let us go back to the general case, that is, localizing a serial ring at an arbitrary semiprime Goldie ideal K, i.e., with respect to $\mathcal{C}(K)$.

Lemma 6.56 *Let R be a serial ring, let $\{e_1, \ldots, e_n\}$ be a complete set of orthogonal primitive idempotents of R and let K be a semiprime Goldie ideal of R such that $e_i \notin K$ for all $i = 1, 2, \ldots, n$. Suppose that at least one of the following three conditions holds:*

(a) *R is prime;*

(b) *R has the a.c.c. on right annihilators;*

(c) *R is nonsingular.*

Then $\mathcal{C}(K) \subseteq \mathcal{C}(0)$.

Proof. Suppose condition (a) holds. Then the statement is trivial for $K = 0$. Assume that $K \neq 0$. Fix an element $c \in \mathcal{C}(K)$. If $a \in R$ has the property that $ca = 0$ or $ac = 0$, then $Kca = 0$ or $acK = 0$, so that $Ka = 0$ or $aK = 0$ by Lemma 6.51. Since R is prime and $K \neq 0$, we get that $a = 0$, i.e., c is regular.

If condition (b) holds, the statement follows by Lemmas 6.52 and 6.53.

Suppose condition (c) holds. If $(P_1', \ldots, P_n')$ is the localizable prime system for R associated to K, $\mathcal{P} = (e_1 P_1' e_1, \ldots, e_n P_n' e_n)$ is the corresponding localizable system and X is the maximal Ore set corresponding to K, then $X = \mathcal{C}(K)$

by Proposition 6.45. In order to prove that $X \subseteq \mathcal{C}(0)$ we need only show that the ideal $I(X)$, which is the kernel of the canonical mapping $R \to \overline{R[\overline{X}^{-1}]}$, is zero. From Lemma 6.42 we know that $I(X) = I(X_{\mathcal{P}(X)})$. Hence it suffices to show that if $x \in X_{\mathcal{P}}$, $r \in R$ and either $xr = 0$ or $rx = 0$, then $r = 0$. By symmetry we may assume $xr = 0$, so that $e_i x e_i \cdot e_i r e_j = e_i x r e_j = 0$ for every i, j. In the last paragraph of Section 6.8 we observed that $e_i \notin K$ implies $P_i' \neq \emptyset$ for every i, from which $0 \in e_i P_i' e_i$. Hence $x \in X_{\mathcal{P}}$ forces $e_i x e_i \neq 0$. By Proposition 5.11 we get $e_i r e_j = 0$ for every i and every j. Thus $r = 0$. $\qquad\square$

The last result of this chapter will be used in an induction argument when we describe serial rings of finite Krull dimension in the next chapter.

Proposition 6.57 *Let R be a serial ring, let $\{e_1, \ldots, e_n\}$ be a complete set of orthogonal primitive idempotents of R and let K be a semiprime Goldie ideal of R such that $e_i \notin K$ for all $i = 1, 2, \ldots, n$. Suppose that R either has the a.c.c. on right annihilators or is nonsingular. Let T be the quotient ring of R with respect to the maximal Ore set $\mathcal{C}_R(K)$, (so that $R \subseteq T$ by Lemma 6.56). Then T is a serial ring, K is its Jacobson radical, T/K is a semisimple artinian ring and is the classical two-sided quotient ring of R/K. Moreover*

(a) *If R has the a.c.c. on right annihilators, then T has the a.c.c. on right annihilators.*

(b) *If R is nonsingular, then T is nonsingular.*

Proof. The quotient ring T is serial by Proposition 6.5(a), and $K \subseteq R \subseteq T$ is an ideal of T by Lemma 6.51. If $x \in K$, then right and left multiplication by $1 - x$ are the identity mapping of R/K, so that $1 - x \in \mathcal{C}_R(K)$. Therefore $1 - x$ is invertible in T for every $x \in K$. Hence the ideal K of T is contained in the Jacobson radical $J(T)$ of T.

As $K \subseteq R \subseteq T$, the inclusion $R \to T$ induces an injective ring homomorphism $R/K \to T/K$. For every $t \in T$ there exist $r, r' \in R$ and $c, c' \in \mathcal{C}_R(K)$ such that $tc = r$ and $c't = r'$. Thus $(t + K)(c + K) = (r + K)$ and $(c' + K)(t + K) = (r' + K)$, that is, every element $t + K$ of T/K can be written in the form $(r + K)(c + K)^{-1}$ and in the form $(c' + K)^{-1}(r' + K)$ with $c + K, c' + K \in \mathcal{C}_{R/K}(0)$. This shows that T/K is the classical two-sided quotient ring of R/K. Since R/K is a semiprime Goldie ring, its classical quotient ring T/K is semisimple artinian (Theorem 6.6).

As $K \subseteq J(T)$, the Jacobson radical of T/K is $J(T)/K$. But T/K is semisimple artinian. Hence its Jacobson radical $J(T)/K$ is zero, and so

$$K = J(T).$$

Suppose, now, that R has the a.c.c. on right annihilators. Then R has an artinian classical quotient ring Q. From Lemma 6.56 we know that $R \subseteq T \subseteq Q$. Hence T has the a.c.c. on right annihilators.

Finally, suppose that R is nonsingular. Since $\{e_1, \ldots, e_n\}$ is a complete set of orthogonal primitive idempotents of T, it suffices by Proposition 5.11 to show that for every $i, j, k = 1, 2, \ldots, n$, if $t \in e_i T e_j$ and $t' \in e_j T e_k$ are non-zero, then tt' is non-zero. Let $\mathcal{P} = (P_1, \ldots, P_n)$ be the localizable system corresponding to K, so that $T = R_{\mathcal{P}}$. By Lemma 6.23(b) t and t' can be written in the form $t = x^{-1} e_i r e_j$ and $t' = e_j r' e_k y^{-1}$ for suitable $r, r' \in R$, $x, y \in X_{\mathcal{P}}$. It is now clear that t, t' non-zero implies tt' non-zero because R is nonsingular. $\qquad \square$

6.13 Notes on Chapter 6

The account of the material presented in Section 6.1 is standard. For the proofs of Proposition 6.1, Theorem 6.2, Propositions 6.3 and 6.4 and Goldie's Theorem 6.6 see, for instance, [Goodearl and Warfield 89, Proposition 9.4, Theorem 9.7, Corollary 9.5, Proposition 9.8, Theorem 9.17 (a) and Proposition 9.19(b)] and [McConnell and Robson, Theorem 2.3.6]. E. Steinitz was the first mathematician to construct the field of fractions of an arbitrary commutative integral domain in 1910. [Ore 31,33] proved that a domain has a classical right quotient ring that is a division ring if and only if the set of its regular (= non-zero) elements is a right Ore set. [Malcev] gave an example of a domain that cannot be embedded in a division ring. For the history of the existence of quotient rings and Goldie's Theorem see [Goodearl and Warfield 89, p. 98 and p. 162]. Levitzki proved that in a right noetherian ring each nil right ideal is nilpotent. [Lanski] extended Levitzki's result by proving that in a right Goldie ring each nil subring is nilpotent (Corollary 6.9). Corollary 6.10 is due to [Herstein and Small]. The way we have presented the results of Section 6.2, in particular Propositions 6.7 and 6.8, is due to [Fisher 70].

The reduced rank was introduced by [Goldie 64]. Theorem 6.16 is due to [Warfield 79, Theorem 3] and is a variant of Small's Theorem proved in [Small] (the necessity of the condition $C_R(0) = C_R(N)$ was proved by [Talintyre]). The results about localization in chain rings presented in Section 6.4 are taken from [Bessenrodt, Brungs and Törner]. The notion of a localizable system for a serial ring and the results in Section 6.5 are due to [Facchini and Puninski 95, 96].

Lemma 6.26 and Proposition 6.27 are generalizations of [Facchini 96, Lemma 1.1] and [Facchini and Puninski 95, Proposition 3.6] respectively.

Theorem 6.28 is taken from [McConnell and Robson, Theorem 3.6.2]. The study of the prime spectrum of a serial ring, in particular Lemma 6.29 and Proposition 6.30, is due to [Müller and Singh]. The remaining part of Section 6.7 and Section 6.8 is due to [Facchini and Puninski 96]. Theorem 6.37 is due to [Drozd].

The material in Sections 6.10 and 6.11 is due to [Facchini and Puninski 96]. Previously, [Müller 92] and [Chatters] had used localization with respect to a semiprime Goldie ideal in a serial ring to study the structure of some classes of serial rings. Theorem 6.47 is due to [Gregul' and Kirichenko], but the proof presented here is taken essentially from [Facchini and Puninski 96]. Proposition 6.48 comes from [Camps, Facchini and Puninski, Lemma 3.2] and Theorem 6.49 is due to Mewborn and Winton. The proof of Lemma 6.52 given here follows a hint of [Chatters and Hajarnavis, p. 97], and Lemma 6.53 is taken from [Chatters and Hajarnavis, p. 88].

The important Theorem 6.54 is due to Chatters ([Chatters and Hajarnavis, Theorem 6.10]) and [Warfield 79, Theorem 5]. The proof we have given here is a simplification of the proof presented in [Chatters and Hajarnavis, Chapter 6]. The simplification is possible thanks to the theory of localizable systems developed in this chapter. Proposition 6.55 is taken from [Warfield 79, Theorem 5]. Lemma 6.56 is due to [Facchini and Puninski 96, Lemma 4.5], and Proposition 6.57 is essentially taken from [Camps, Facchini and Puninski, Theorem 3.1].

Chapter 7

Krull Dimension and Serial Rings

7.1 Deviation of a poset

Let A be a poset. If $a, b \in A$ and $b \le a$, the *factor* of a by b is the subposet a/b of A defined by $a/b = \{ x \in A \mid b \le x \le a \}$, that is, the interval $[b, a]$. Given a countable descending chain $a_0 \ge a_1 \ge a_2 \ge \ldots$ of elements of A, the *factors of the chain* are the factors a_i/a_{i+1}.

For $\alpha = -1$ or an ordinal number, we define a class $\mathcal{D}_\alpha$ of posets recursively as follows:

for $\alpha = -1$ let $\mathcal{D}_{-1}$ be the class of all *trivial* posets, that is, the posets A in which $a \le b$ implies $a = b$ for every $a, b \in A$. If α is an ordinal and the class $\mathcal{D}_\beta$ has been defined for every $\beta < \alpha$, let $\mathcal{D}_\alpha$ be the class of all posets A such that

(a) $A \notin \bigcup_{\beta < \alpha} \mathcal{D}_\beta$;

(b) for every countable descending chain $a_0 \ge a_1 \ge a_2 \ge \ldots$ of elements of A, there exists an index n such that the factors a_i/a_{i+1} belong to $\bigcup_{\beta < \alpha} \mathcal{D}_\beta$ for every $i \ge n$.

The classes $\mathcal{D}_\alpha$ are pairwise disjoint, that is, $\mathcal{D}_\alpha \cap \mathcal{D}_\beta = \emptyset$ whenever $\alpha \ne \beta$. If A is a poset and $A \in \mathcal{D}_\alpha$, the poset A is said to have *deviation* α, and we shall write $\operatorname{dev} A = \alpha$. Note that a poset A can fail to have a deviation, i.e., $A \notin \mathcal{D}_\alpha$ for every α.

Example 7.1 *A poset A has deviation 0 if and only if it is artinian and non-trivial.* $\qquad\square$

Example 7.2 *Let* **Z** *be the poset of integers with the usual order. Then*

$$\text{dev } \mathbf{Z} = 1.$$

In fact, dev $\mathbf{Z} \neq -1$ because $\mathbf{Z}$ is not trivial, and dev $\mathbf{Z} \neq 0$ because $\mathbf{Z}$ is not artinian. If $a_0 \geq a_1 \geq a_2 \geq \ldots$ is a descending chain of elements of $\mathbf{Z}$, then all the factors a_i/a_{i+1} have finite length, so that they all have deviation ≤ 0. Thus dev $\mathbf{Z} = 1$. □

In this section we shall review the first elementary properties of the deviation of a poset.

Proposition 7.3 *If A is a poset with deviation α and B is a subposet of A, then B has deviation $\beta \leq \alpha$.*

Proof. We prove this by transfinite induction on α. The case $\alpha = -1$ is trivial. Suppose that the result is true for all $\gamma < \alpha$. Let A be a poset with dev $A = \alpha$ and let $B \subseteq A$. If $b_0 \geq b_1 \geq b_2 \geq \ldots$ is a descending chain of elements of B, then the factors of this chain in B are subposets of the factors of the chain in A. The conclusion follows immediately from the induction hypothesis. □

Example 7.4 *Let* **R** *be the poset of real numbers with the natural order. Then* **R** *fails to have a deviation.*

In fact, suppose that dev $\mathbf{R} = \alpha$ for some ordinal α and note that dev $\mathbf{R} \neq -1$ because $\mathbf{R}$ is not trivial. Consider the descending chain

$$0 > -1 > -2 > -3 > \ldots$$

in $\mathbf{R}$. Since dev $\mathbf{R} = \alpha$ and the factors $i/i+1$ are all isomorphic, they all have the same deviation $\beta < \alpha$. But $\mathbf{R}$ is order isomorphic to a subposet of $i/i+1$, and thus $\alpha \leq \beta$ by Proposition 7.3. This contradiction shows that $\mathbf{R}$ fails to have a deviation. □

A poset A is *noetherian* if every ascending chain $a_0 \leq a_1 \leq a_2 \leq \ldots$ of elements of A is stationary.

Proposition 7.5 *Every noetherian poset has a deviation.*

Proof. It is sufficient to show that every noetherian poset with a greatest element and a least element has a deviation, because it is easily seen that every noetherian poset is a subposet of a noetherian poset with a greatest element and a least element. Assume there exists a noetherian poset A with a greatest element 1 and a least element 0 that fails to have a deviation. Let B be the set of all elements $a \in A$ such that $1/a$ fails to have a deviation. Since $0 \in B$, the set B is non-empty. Let b be a maximal element in B. Put

$\alpha = \sup\{\,\mathrm{dev}(1/a) \mid a \in A,\ a > b\,\}$. We will show that $1/b$ has deviation $\le \alpha + 1$, and this contradiction will prove the proposition.

Let $a_0 \ge a_1 \ge a_2 \ge \dots$ be a descending chain of elements of $1/b$. If $a_n = b$ for some n, then $a_i = b$ for every $i \ge n$. Thus $\mathrm{dev}(a_i/a_{i+1}) = -1$ for every $i \ge n$. If $a_n > b$ for every n, then $\mathrm{dev}(a_i/a_{i+1}) \le \mathrm{dev}(1/a_{i+1}) \le \alpha$ for every i. Therefore $\mathrm{dev}(1/b) \le \alpha + 1$. $\qquad\square$

There is a certain freedom in constructing countable descending chains of elements in a poset of deviation α, as the next lemma shows.

Lemma 7.6 *If A is a poset of deviation α and $\beta < \alpha$, then there is a chain $a_0 \ge a_1 \ge a_2 \ge \dots$ in A such that $\mathrm{dev}(a_i/a_{i+1}) = \beta$ for every i.*

Proof. We prove this by transfinite induction on α. The case $\alpha = 0$ is trivial. Suppose that the result is true for posets of deviation $< \alpha$. Fix a poset A of deviation α and an ordinal $\beta < \alpha$. Then there exists a chain $a_0 \ge a_1 \ge a_2 \ge \dots$ in A such that for every index n there exists $i \ge n$ with $\mathrm{dev}(a_n/a_{i+1}) \ge \beta$, because otherwise for every chain $a_0 \ge a_1 \ge a_2 \ge \dots$ there would exist an index n such that $\mathrm{dev}(a_n/a_{i+1}) < \beta$ for every $i \ge n$, and this forces $\mathrm{dev}(A) \le \beta < \alpha$, contradiction.

Passing to a subchain we may suppose that there exists a chain

$$a_0 \ge a_1 \ge a_2 \ge \dots$$

in A such that $\mathrm{dev}(a_i/a_{i+1}) \ge \beta$ for every i. Since $\mathrm{dev}(A) = \alpha$, there exists an index n such that $\beta \le \mathrm{dev}(a_i/a_{i+1}) < \alpha$ for every $i \ge n$. Passing once more to a subchain we may suppose that $\beta \le \mathrm{dev}(a_i/a_{i+1}) < \alpha$ for every i.

If $\mathrm{dev}(a_i/a_{i+1}) = \beta$ for every i, we are done. Hence we may suppose that $\mathrm{dev}(a_m/a_{m+1}) > \beta$ for some index m. Since $\beta < \mathrm{dev}(a_m/a_{m+1}) < \alpha$, we can apply the induction hypothesis to the poset a_m/a_{m+1}, i.e., we can find a chain having all its factors of deviation β. $\qquad\square$

Let α be a non-zero ordinal, and consider its opposite poset α^{op}. For instance, 1^{op} is the trivial poset with one element. The poset α^{op} is a noetherian poset, and therefore it has a deviation $\mathrm{dev}(\alpha^{\mathrm{op}})$ (Proposition 7.5). Recall that any non-zero ordinal α can be written uniquely in the form

$$\alpha = \omega^{\beta_1} n_1 + \cdots + \omega^{\beta_t} n_t,$$

where $n_1, \dots, n_t$ are non-zero finite ordinals, and $\beta_1 > \cdots > \beta_t$.

Theorem 7.7 *If $n_1, \dots, n_t$ are non-zero finite ordinals, $\beta_1 > \cdots > \beta_t$ are ordinals and $\alpha = \omega^{\beta_1} n_1 + \cdots + \omega^{\beta_t} n_t > 1$, then $\mathrm{dev}(\alpha^{\mathrm{op}}) = \beta_1$.*

Proof. We prove this by transfinite induction on α. If $\alpha = n \ge 2$ is a finite ordinal, then $\alpha = \omega^0 \cdot n$ and $\mathrm{dev}(\alpha^{\mathrm{op}}) = 0$. Suppose that the theorem holds for ordinals less than a fixed ordinal $\alpha = \omega^{\beta_1} n_1 + \cdots + \omega^{\beta_t} n_t$. Let

$$a_0 \ge a_1 \ge a_2 \ge \dots$$

be a descending chain in α^{op}. Since

$$\alpha = \underbrace{\omega^{\beta_1} + \cdots + \omega^{\beta_1}}_{n_1 \text{ times}} + \cdots + \underbrace{\omega^{\beta_t} + \cdots + \omega^{\beta_t}}_{n_t \text{ times}}$$

is the sum of a finite number of ordinals of type ω^{β_j}, the set α is the disjoint union of n_1 posets isomorphic to ω^{β_1}, n_2 posets isomorphic to ω^{β_2}, ..., n_t posets isomorphic to ω^{β_t}. Hence there exist an index $j = 1, 2, \ldots, t$ and a positive integer n such that the terms a_i of the chain are in one of the summands ω^{β_j} for every $i \geq n$. Each factor a_i/a_{i+1}, $i \geq n$, is well-ordered with respect to its opposite order. More precisely, it is order antiisomorphic to an ordinal $< \omega^{\beta_j}$. By the induction hypothesis $\mathrm{dev}(a_i/a_{i+1}) < \beta_j$ for every $i \geq n$, so that $\mathrm{dev}(\alpha^{\mathrm{op}}) \leq \beta_j \leq \beta_1$.

If β_1 is a limit ordinal, then for every $\gamma < \beta_1$ one has

$$\gamma = \mathrm{dev}((\omega^\gamma)^{\mathrm{op}}) \leq \mathrm{dev}(\alpha^{\mathrm{op}}) \leq \beta_1,$$

so that

$$\beta_1 = \sup_{\gamma < \beta_1} \gamma \leq \mathrm{dev}(\alpha^{\mathrm{op}}) \leq \beta_1.$$

If β_1 is not a limit ordinal, $\beta_1 = \beta + 1$ say, then α^{op} has the descending chain $a_i = \omega^\beta i$. By the induction hypothesis the factors in this chain have deviation β. Therefore $\mathrm{dev}(\alpha^{\mathrm{op}}) \geq \beta + 1 = \beta_1$. Hence $\mathrm{dev}(\alpha^{\mathrm{op}}) = \beta_1$. $\qquad\square$

In the last proposition of this section we give a useful criterion to decide whether a poset has finite deviation.

Proposition 7.8 *Let A be a poset and n a finite ordinal. Then $\mathrm{dev}(A) \leq n$ if and only if there do not exist injective order-reversing mappings $\omega^{n+1} \to A$.*

Proof. We prove this by induction on n. The case $n = 0$ is immediate.

If there exists an injective order-reversing mapping $\omega^{n+1} \to A$, that is, if A has a subposet order isomorphic to $(\omega^{n+1})^{\mathrm{op}}$, then either A fails to have a deviation or A has deviation $\geq n + 1$ by Proposition 7.3 and Theorem 7.7.

Conversely, suppose that there do not exist injective order-reversing mappings $\omega^{n+1} \to A$. Consider a descending chain $a_0 \geq a_1 \geq a_2 \geq \ldots$ in A. Then there cannot be an injective order-reversing mapping $\omega^n \to a_i/a_{i+1}$ for infinitely many factors a_i/a_{i+1}, otherwise there would be an injective order-reversing mapping $\omega^{n+1} \to A$. By the induction hypothesis $\mathrm{dev}(a_i/a_{i+1}) < n$ for all but at most finitely many factors. Therefore $\mathrm{dev}(A) \leq n$. $\qquad\square$

Proposition 7.8 cannot be extended to infinite ordinals, because if α is an infinite ordinal, the cardinality of $\omega^{\alpha+1}$ is at least $2^{\aleph_0}$. Hence there do not exist injective order-reversing mappings of $\omega^{\alpha+1}$ into any countable poset A. But it is easy to construct countable posets A, like $\mathbf{Q}$ or the poset D described in the notes at the end of this chapter, that fail to have a deviation.

7.2   Krull dimension of arbitrary modules and rings

Let $\mathcal{L}(M_R)$ denote the lattice of all submodules of the module M_R. The *Krull dimension* of M_R is defined to be the deviation of $\mathcal{L}(M_R)$. It is denoted K.dim(M_R). Therefore K.dim$(M_R) = \text{dev}(\mathcal{L}(M_R))$. For instance K.dim$(M_R) = -1$ if and only if $M_R = 0$, whereas K.dim$(M_R) = 0$ if and only if M_R is a non-zero artinian module.

The class $\mathcal{K}_\alpha$ of modules of Krull dimension α is defined as follows. The class $\mathcal{K}_{-1}$ contains all modules $M = 0$. If the class $\mathcal{K}_\beta$ of modules of Krull dimension β has been defined for every $\beta < \alpha$, then $\mathcal{K}_\alpha$ is defined as the class of all modules M such that

(a) $M \notin \bigcup_{\beta < \alpha} \mathcal{K}_\beta$;

(b) for every countable descending chain $A_0 \supseteq A_1 \supseteq A_2 \supseteq \ldots$ of submodules of M, there exists an index n such that the factors A_i/A_{i+1} belong to $\bigcup_{\beta < \alpha} \mathcal{K}_\beta$ for every $i \geq n$.

If $M \notin \mathcal{K}_\alpha$ for every α one says that M *fails to have Krull dimension*, or that K.dim(M_R) *is not defined*.

Example 7.9 *If R is a commutative principal ideal domain which is not a field, then* K.dim$(R_R) = 1$.

Since R_R is not an artinian module, K.dim(R_R) is neither -1 nor 0. Let $A_0 \supseteq A_1 \supseteq A_2 \supseteq \ldots$ be a descending chain of submodules of R_R. If $A_n = 0$ for some n, then K.dim$(A_i/A_{i+1}) = -1$ for every $i \geq n$. If $A_n \neq 0$ for every n, then A_n/A_{n+1} is a cyclic artinian module for every n, and therefore K.dim$(A_n/A_{n+1}) \leq 0$. Thus K.dim$(R_R) = 1$. $\square$

Example 7.10 *Every noetherian module has Krull dimension.*

This follows immediately from Proposition 7.5. $\square$

Krull dimension behaves properly with respect to short exact sequences, as the next proposition shows.

Proposition 7.11 *If N is a submodule of a module M, then M has Krull dimension if and only if both N and M/N have Krull dimension. In this case*

$$\text{K.dim}(M) = \max\{\text{K.dim}(N), \text{ K.dim}(M/N)\}.$$

Proof. Observe that $\mathcal{L}(N)$ and $\mathcal{L}(M/N)$ are subposets of $\mathcal{L}(M)$. Thus if M has Krull dimension, then N and M/N have Krull dimension and

$$\text{K.dim}(M) \geq \max\{\text{K.dim}(N), \text{ K.dim}(M/N)\}$$

by Proposition 7.3. In order to prove the theorem, we suppose that both N and M/N have Krull dimension and set $\alpha = \max\{\text{K.dim}(N), \text{K.dim}(M/N)\}$. We shall show that M has Krull dimension $\leq \alpha$ by induction on α. The case $\alpha = -1$ is trivial.

Let

$$A_0 \supseteq A_1 \supseteq A_2 \supseteq \ldots$$

be a descending chain of submodules of M_R. In order to show that M has Krull dimension $\leq \alpha$ we must prove that $\text{K.dim}(A_i/A_{i+1})$ is strictly less than α for all but finitely many indices i. Now $A_0 \cap N \supseteq A_1 \cap N \supseteq A_2 \cap N \supseteq \ldots$ is a descending chain of submodules of N and $A_0 + N/N \supseteq A_1 + N/N \supseteq A_2 + N/N \supseteq \ldots$ is a descending chain of submodules of M/N, so that

$$\text{K.dim}(A_i \cap N/A_{i+1} \cap N) < \text{K.dim}(N)$$

for all but finitely many indices i and $\text{K.dim}(A_i + N/A_{i+1} + N) < \text{K.dim}(M/N)$ for all but finitely many indices i. There is a natural epimorphism

$$A_i/A_{i+1} \to A_i + N/A_{i+1} + N,$$

whose kernel is $(A_i \cap (A_{i+1} + N))/A_{i+1}$. It is easily seen that the natural mapping $A_i \cap N \to (A_i \cap (A_{i+1} + N))/A_{i+1}$ is surjective and has $A_{i+1} \cap N$ as its kernel. Therefore $A_i \cap N/A_{i+1} \cap N \cong (A_i \cap (A_{i+1} + N))/A_{i+1}$. This shows that there is an exact sequence

$$0 \to A_i \cap N/A_{i+1} \cap N \to A_i/A_{i+1} \to A_i + N/A_{i+1} + N \to 0.$$

By the induction hypothesis $\text{K.dim}(A_i/A_{i+1})$ is equal to

$$\max\{\text{K.dim}(A_i \cap N/A_{i+1} \cap N), \text{K.dim}(A_i + N/A_{i+1} + N)\} < \alpha$$

for all but finitely many indices i. This proves that M has Krull dimension $\leq \alpha$. $\qquad\square$

As an immediate corollary we have

Corollary 7.12 *If $M_1, M_2, \ldots, M_n$ are modules, then $M_1 \oplus M_2 \oplus \cdots \oplus M_n$ has Krull dimension if and only if $M_1, M_2, \ldots, M_n$ have Krull dimension, and in this case*

$$\text{K.dim}(M_1 \oplus M_2 \oplus \cdots \oplus M_n) = \max\{\text{K.dim}(M_1), \text{K.dim}(M_2), \ldots, \text{K.dim}(M_n)\}.$$
$$\square$$

In the next proposition we give a criterion to decide whether a module fails to have left Krull dimension.

Proposition 7.13 *Every module with Krull dimension has finite Goldie dimension.*

Proof. Let M_R be a module with infinite Goldie dimension. Then

$$M \supseteq \oplus_{i \in I} N_i$$

for non-zero submodules N_i and an infinite index set I. Without loss of generality we can suppose $I = \mathbf{Q}$, the set of rational numbers. Define an injective order-preserving mapping $\mathbf{R} \rightarrow \mathcal{L}(M_R)$, via $x \mapsto M_x = \oplus_{i \in \mathbf{Q}, i \leq x} N_i$. If M_R has Krull dimension, then $\mathcal{L}(M_R)$ has a deviation, and therefore $\mathbf{R}$ has a deviation, which contradicts Example 7.4. $\quad\square$

Hence, if a module M contains a submodule N such that M/N has infinite Goldie dimension, then M fails to have left Krull dimension (Proposition 7.11).

For any ring R, the Krull dimension of the right module R_R (if it exists) is called the *right Krull dimension of R* and is denoted r.K.dim(R). Similarly, the *left Krull dimension of R* is denoted l.K.dim(R) and is K.dim$(_R R)$. It is clear that every finitely generated right module M over a ring R with right Krull dimension has Krull dimension $\leq$ r.K.dim(R), and every homomorphic image of a ring with right Krull dimension has right Krull dimension $\leq$ r.K.dim(R).

Example 7.14 *There exist right serial rings with right Krull dimension which fail to have left Krull dimension.*

Let R be the ring constructed in Example 3.13, where K and F are fields and K is an extension of infinite degree of F. We saw in Example 3.13 that R is a right artinian, right serial ring that is not left serial, and that the Goldie dimension of $_R R$ is infinite. Thus r.K.dim$(R) = 0$, but l.K.dim(R) is not defined by Proposition 7.13. $\quad\square$

We shall see in Corollary 7.30 that for every serial ring the right Krull dimension and the left Krull dimension coincide.

7.3 Nil subrings of rings with right Krull dimension

The central aim of this section is to prove that nil subrings of a ring with right Krull dimension are nilpotent. The proof we give makes use of the notion of critical modules. A critical module may be viewed as the analogue of a simple module.

Let α be an ordinal. A non-zero module M is α-*critical* if K.dim$(M) = \alpha$ and K.dim$(M/N) < \alpha$ for each non-zero submodule N of M. A *critical* module is one which is α-critical for some ordinal α. For example, a module if 0-critical if and only if it is simple.

Theorem 7.15 *Every non-zero module with Krull dimension has a critical submodule.*

Proof. Assume the contrary, i.e., that there exist non-zero modules with Krull dimension and without critical submodules. Amongst all such modules choose one, M say, of minimal possible Krull dimension. Set $\alpha = \mathrm{K.dim}(M)$, so that $\alpha \geq 0$. Then every non-zero submodule of M has Krull dimension α and has no critical submodules. By induction on n define a strictly descending chain of non-zero submodules M_n of M such that $\mathrm{K.dim}(M_n/M_{n+1}) = \alpha$ for every n as follows. Put $M_0 = M$. If $M_n \neq 0$ has been defined, then $\mathrm{K.dim}(M_n) = \alpha$ and M_n has no critical submodules. In particular, M_n itself is not α-critical, so that there exists a non-zero submodule $M_{n+1} \subseteq M_n$ with $\mathrm{K.dim}(M_n/M_{n+1}) = \alpha$, and therefore this descending chain of submodules of M is strictly descending. The existence of the chain shows that $\mathrm{K.dim}(M) > \alpha$, a contradiction.　　□

Next it is possible to define the analogue of the socle of a module. Let M be an R-module with Krull dimension. Put

$$\beta = \min\{\, \mathrm{K.dim}(N) \mid N \subseteq M,\ N \neq 0 \,\}.$$

By Theorem 7.15 M has a β-critical submodule. The sum of all the β-critical submodules of M is called the *critical socle* of M, denoted $S(M)$. We now have the analogue of the Loewy series of a module (Section 2.11). Define a sequence of submodules of M as follows:

$S_0(M) = 0$,
$S_{\alpha+1}(M)/S_\alpha(M) = S(M/S_\alpha(M))$ for every ordinal α,
$S_\beta(M) = \bigcup_{\alpha < \beta} S_\alpha(M)$ for every limit ordinal β.

The *critical socle series* of M is the chain

$$S_0(M) \subseteq S_1(M) \subseteq S_2(M) \subseteq \cdots \subseteq S_\alpha(M) \subseteq \cdots.$$

The following result is an immediate consequence of Theorem 7.15.

Lemma 7.16 *If M has Krull dimension, then $M = S_\lambda(M)$ for some ordinal λ.*
　　　　　　　　　　　　　　　　　　　　　　　　　　　　　　　　□

The proof of the main theorem of this section (Theorem 7.21) is subdivided into a number of partial results.

Lemma 7.17 *The Jacobson radical of a ring with right Krull dimension does not contain non-zero idempotent right ideals.*

Proof. Let R be a ring with right Krull dimension, let J be its Jacobson radical, and let $I = I^2$ be a right ideal contained in J. In order to prove the lemma we shall prove that $MI = 0$ for every right R-module M with Krull dimension α by induction on α. The case $\alpha = -1$ is trivial. Suppose $\alpha \geq 0$. Since $M = S_\lambda(M)$ for some ordinal λ and $I = I^2$, it is sufficient to prove that I annihilates all β-critical modules with $\beta \leq \alpha$. Let C be a β-critical module. If $\beta < \alpha$, then $CI = 0$ by the induction hypothesis. Suppose that $\beta = \alpha$ and let D be any non-zero submodule of C. Then K.dim$(C/D) < \alpha$, so that $(C/D)I = 0$, that is, $CI \subseteq D$. Hence CI is contained in every non-zero submodule of C, so that either CI is simple or $CI = 0$. But if CI is simple, then $CI = CI^2 = (CI)I \subseteq (CI)J = 0$, a contradiction. Thus $CI = 0$, as desired. □

Lemma 7.18 *Let R be a ring with right Krull dimension and let C be a critical right ideal of R.*

(a) *If $C^2 \neq 0$, then there exists $c \in C$ such that* r.ann$_R(c) \cap C = 0$.

(b) *If C is nil, then $C^2 = 0$.*

Proof. (a) If $C^2 \neq 0$, there exists $c \in C$ such that $cC \neq 0$. Consider the homomorphism $\varphi \colon C \to C$ given by left multiplication by c. If φ is not injective, then cC is a proper homomorphic image of C, so that K.dim$(cC) <$ K.dim(C) because C is critical. But the Krull dimension of a non-zero submodule of a critical module C is equal to the Krull dimension of C. Hence

$$\text{K.dim}(cC) = \text{K.dim}(C),$$

contradiction. Thus φ is injective, that is, r.ann$_R(c) \cap C = 0$.

(b) If C is nil and $C^2 \neq 0$, then there exists $c \in C$ such that

$$\text{r.ann}_R(c) \cap C = 0$$

by (a), and there exists $n \geq 0$ such that $c^n \neq 0$ and $c^{n+1} = 0$. Then $c^n \in$ r.ann$_R(c) \cap C = 0$, contradiction. □

Corollary 7.19 *Every semiprime ring with right Krull dimension is a right Goldie ring.*

Proof. By characterization (c) in Theorem 6.6 and Proposition 7.13 it suffices to show that $Z(R_R) = 0$ for a semiprime ring R with right Krull dimension. If $Z(R_R) \neq 0$, $Z(R_R)$ contains a critical right ideal C of R (Theorem 7.15). Since r.ann$_R(c)$ is essential for every $c \in C$, $C^2 = 0$ by Lemma 7.18(a). But R is semiprime, so that $C = 0$, contradiction. □

The hypothesis that the ring is semiprime is necessary in Corollary 7.19, even if the ring is a commutative chain ring. For instance, let R be the trivial extension $\mathbf{Z}_p \propto \mathbf{Z}(p^\infty)$ (page 167). Then R is a commutative chain ring without the a.c.c. on annihilators, and from Proposition 7.8 it is easily seen that R has Krull dimension 1.

Proposition 7.20 *Let M_R be a right module over an arbitrary ring R. Suppose that there exist an ascending chain of submodules $0 = A_0 \subsetneq A_1 \subsetneq A_2 \subsetneq \ldots$ and a descending chain of submodules $M = M_0 \supseteq M_1 \supseteq M_2 \supseteq \ldots$ such that $A_{i+1} \cap M_i \not\subseteq A_i + M_{i+1}$ for all i. Then M fails to have Krull dimension.*

Proof. Put $B = \sum_{i=1}^{\infty} A_i \cap M_i$, so that $B \subseteq A_i + M_{i+1}$ for all i. Then $A_{i+1} \cap M_i \not\subseteq B$. Hence $(A_{i+1} \cap M_i) + B/B$ is a non-zero submodule of M/B for all i. From $A_i \subseteq A_{i+1}$ it follows that $B \subseteq \sum_{i=1}^{\infty} A_{i+1} \cap M_i$. We claim that $\left(\sum_{i=1}^{\infty} A_{i+1} \cap M_i\right)/B$ is the direct sum of its submodules $(A_{i+1} \cap M_i) + B/B$, $i \geq 1$. From this we shall obtain that M/B has infinite Goldie dimension, so that M does not have Krull dimension (Propositions 7.11 and 7.13).

In order to prove the claim it is sufficient to show that

$$\left(\sum_{i=1}^{n}(A_{i+1} \cap M_i) + B/B\right) \cap ((A_{n+2} \cap M_{n+1}) + B/B) = 0$$

for every $n \geq 1$, or, equivalently, that

$$\left(\sum_{i=1}^{n}(A_{i+1} \cap M_i) + B\right) \cap (A_{n+2} \cap M_{n+1}) \subseteq B.$$

Suppose that

$$x \in \left(\sum_{i=1}^{n}(A_{i+1} \cap M_i) + B\right) \cap (A_{n+2} \cap M_{n+1}).$$

Then

$$x \in \sum_{i=1}^{n}(A_{i+1} \cap M_i) + B$$

$$= \sum_{i=1}^{n}(A_{i+1} \cap M_i) + \sum_{i=n+1}^{\infty}(A_i \cap M_i) \subseteq A_{n+1} + \sum_{i=n+1}^{\infty}(A_i \cap M_i),$$

so that $x = y + z$, where $y \in A_{n+1}$ and $z \in \sum_{i=n+1}^{\infty}(A_i \cap M_i) \subseteq B$. From $x \in A_{n+2} \cap M_{n+1}$ it follows that $x - z \in M_{n+1} + \sum_{i=n+1}^{\infty}(A_i \cap M_i) \subseteq M_{n+1}$. Thus $y = x - z \in A_{n+1} \cap M_{n+1} \subseteq B$. Hence $x \in B$, as desired. $\qquad\square$

We are ready to prove the main result of this section.

Theorem 7.21 *The prime radical $N(R)$ of a ring R with right Krull dimension is nilpotent.*

Proof. If A is any ideal of R contained in $N = N(R)$, by Zorn's Lemma there exists an ideal $\overline{A} \subseteq N$ such that $A \subseteq \overline{A}$ and $\overline{A}$ is maximal with respect to

$\overline{A}^2 \subseteq A$. Hence it is possible to construct a chain $0 = A_0 \subseteq A_1 \subseteq A_2 \subseteq \ldots$ of ideals of R contained in N in which A_{i+1} is maximal with respect to $A_{i+1}^2 \subseteq A_i$ for every i. Put $I = \bigcup_{i=0}^{\infty} A_i$.

Suppose that $I^n \not\subseteq I^{n+1} + A_i$ for every $i, n \geq 1$. Define a sequence of non-negative integers $t_0 < t_1 < t_2 < \ldots$ such that $I^n \cap A_{t_{n+1}} \not\subseteq I^{n+1} + A_{t_n}$ for every $n \geq 0$ by induction in the following way. Put $t_0 = 0$. Assume that t_n has been defined for an index n. As $I^n \not\subseteq I^{n+1} + A_{t_n}$, there exists an element $x_n \in I^n$, $x_n \notin I^{n+1} + A_{t_n}$. Let t_{n+1} denote an integer such that $t_{n+1} > t_n$ and $x_n \in A_{t_{n+1}}$. Then t_{n+1} has the required property, because $x_n \in I^n \cap A_{t_{n+1}}$ and $x_n \notin I^{n+1} + A_{t_n}$. If we apply Proposition 7.20 to the chains of submodules $0 = A_{t_0} \subseteq A_{t_1} \subseteq A_{t_2} \subseteq \ldots$ and $R_R = I^0 \supseteq I^1 \supseteq I^2 \supseteq \ldots$, we see that R_R does not have Krull dimension, contradiction.

Hence there exist indices $i, n \geq 1$ such that $I^n \subseteq I^{n+1} + A_i$. In particular $(I^n + A_i)/A_i$ is an idempotent ideal of the ring R/A_i. From $I \subseteq N$ it follows that $(I^n + A_i)/A_i \subseteq N(R/A_i) \subseteq J(R/A_i)$. By Lemma 7.17 we obtain that $I^n \subseteq A_i$. Let us prove by induction on $j = 0, 1, \ldots, n-1$ that $I^{n-j} \subseteq A_{i+j}$. Suppose $j \leq n-1$ and $I^{n-j+1} \subseteq A_{i+j-1}$. Then

$$(I^{n-j} + A_{i+j})^2 \subseteq I^{2(n-j)} + I^{n-j} \cdot I + A_{i+j}^2 \subseteq A_{i+j-1}$$

and $I^{n-j} + A_{i+j} \supseteq A_{i+j}$, so that $I^{n-j} + A_{i+j} = A_{i+j}$ by the maximality of A_{i+j}. Thus $I^{n-j} \subseteq A_{i+j}$, which completes the induction. In particular, for $j = n-1$, we obtain that $I \subseteq A_{i+n-1}$. Hence $A_{i+n} = A_{i+n-1}$, that is, $\overline{R} = R/A_{i+n-1}$ contains no non-zero ideal whose square is zero. As $A_{i+n-1} \subseteq N$, the prime radical of $\overline{R} = R/A_{i+n-1}$ is $\overline{N} = N/A_{i+n-1}$. If $\overline{N} \neq 0$, then $\overline{N}$ contains a critical right ideal $\overline{C} \neq 0$ by Theorem 7.15. Since every ideal contained in the prime radical is nil, $\overline{C}$ is nil. From Lemma 7.18(b) we obtain that $\overline{C}^2 = 0$. Then $\overline{R}\,\overline{C}$ is a non-zero ideal of $\overline{R}$ whose square is zero, contradiction. Thus $\overline{N} = 0$, that is, $N = A_{i+n-1}$, and thus N is nilpotent. $\qquad\square$

As a generalization of Theorem 7.21 we obtain

Corollary 7.22 *Nil subrings of a ring with right Krull dimension are nilpotent.*

Proof. Let S be a nil subring of a ring R with right Krull dimension. Then $S/S \cap N(R)$ is a nil subring of $R/N(R)$. By Corollary 7.19 the ring $R/N(R)$ is right Goldie, so that $S/S \cap N(R)$ is nilpotent (Corollary 6.9). By Theorem 7.21 $N(R)$, and so $S \cap N(R)$, is nilpotent. Thus S is nilpotent. $\qquad\square$

As an application of Theorem 7.21 we obtain a characterization of the rings with right Krull dimension that have a right artinian right quotient ring.

Theorem 7.23 *Let R be a ring with right Krull dimension and let N be its prime radical. Then R has a right artinian classical right quotient ring if and only if $\mathcal{C}_R(0) = \mathcal{C}_R(N)$.*

Proof. By Theorem 6.16 it is sufficient to show that if R has right Krull dimension, then R/N is right Goldie, N is nilpotent and $\rho_N(R_R)$ is finite. Now R/N is a right Goldie ring by Corollary 7.19 and N is nilpotent by Theorem 7.21. Finally, every $\rho_N(N^i/N^{i+1})$ is finite by Proposition 6.11(g) and Proposition 7.13, so that $\rho_N(R_R)$ is finite. $\qquad\square$

7.4   Transfinite powers of the Jacobson radical

The following result and its Corollary 7.25 are useful in proofs by induction on the Krull dimension.

Proposition 7.24 (a) *If M_R is a module with Krull dimension, then*

$$\mathrm{K.dim}(M) \leq \sup\{\,\mathrm{K.dim}(M/E) + 1 \mid E \text{ is an essential submodule of } M_R\,\}.$$

(b) *If R is a semiprime ring with right Krull dimension, then*

$$\mathrm{r.K.dim}(R) = \sup\{\,\mathrm{K.dim}(R/E) + 1 \mid E \text{ is a right ideal of } R \text{ essential in } R_R\,\}.$$

Proof. (a) Let

$$A_0 \supseteq A_1 \supseteq A_2 \supseteq \cdots$$

be a descending chain of submodules of M. By Proposition 7.13 the module M has finite Goldie dimension $\dim(M)$, so that there exists an index n such that $\dim(A_i) = \dim(A_{i+1})$ for every $i \geq n$. Let C be a submodule of M maximal with respect to $C \cap A_n = 0$. Then $C \cap A_i = 0$ for every $i \geq n$, and $C \oplus A_i$ is essential in M. Set

$$\alpha = \sup\{\,\mathrm{K.dim}(M/E) + 1 \mid E \text{ is an essential submodule of } M_R\,\}.$$

From $A_{i+1}/A_i \cong (C \oplus A_{i+1})/(C \oplus A_i) \subseteq M/(C \oplus A_i)$, it follows that

$$\mathrm{K.dim}(A_{i+1}/A_i) \leq \mathrm{K.dim}(M/(C \oplus A_i)) < \mathrm{K.dim}(M/(C \oplus A_i)) + 1 \leq \alpha.$$

By the definition of Krull dimension we get that $\mathrm{K.dim}(M) \leq \alpha$.

(b) Let α denote the supremum, so that $\mathrm{r.K.dim}(R) \leq \alpha$ by part (a). It is sufficient to prove that $\mathrm{r.K.dim}(R) \geq \mathrm{K.dim}(R/E) + 1$ for every right ideal E essential in R_R. The ring R is right Goldie by Corollary 7.19, so that every essential right ideal E contains a regular element c (Theorem 6.6). In the descending chain $R \supseteq cR \supseteq c^2R \supseteq c^3R \supseteq \cdots$, all the quotients $c^nR/c^{n+1}R$ are isomorphic to R/cR. Hence $\mathrm{r.K.dim}(R) > \mathrm{K.dim}(R/cR)$ by the definition of Krull dimension. Therefore

$$\mathrm{r.K.dim}(R) \geq \mathrm{K.dim}(R/cR) + 1 \geq \mathrm{K.dim}(R/E) + 1. \qquad\square$$

Corollary 7.25 *If R is a prime ring with right Krull dimension and I is a non-zero ideal of R, then* r.K.dim$(R/I) <$ r.K.dim(R).

Proof. If K is a non-zero right ideal in the prime ring R, then

$$0 \neq KI \subseteq K \cap I.$$

Therefore the non-zero ideal I is essential in R_R. The conclusion follows from Proposition 7.24(b). □

Let I be an ideal of a ring R. The *transfinite powers* I_α of I are defined inductively by

$$I_0 = I,$$
$$I_{\alpha+1} = \bigcap_{n=1}^{\infty} I_\alpha^n \text{ for every ordinal } \alpha,$$
$$I_\alpha = \bigcap_{\beta < \alpha} I_\beta \text{ for every limit ordinal } \alpha.$$

Theorem 7.26 *If R is a ring with right Krull dimension α and J is the Jacobson radical of R, then J_α is nilpotent.*

Proof. We prove the result by transfinite induction on α. If $\alpha = 0$, then R is right artinian, so that its Jacobson radical is nilpotent. Assume the result is true for all rings with right Krull dimension $< \alpha$. If R is a prime ring with right Krull dimension α, Jacobson radical J and $J_\alpha \neq 0$, then

$$\text{r.K.dim}(R/J_\alpha) = \beta < \alpha$$

by Corollary 7.25. By the induction hypothesis $J(R/J_\alpha)_\beta = J_\beta/J_\alpha$ is nilpotent, so that $J_\beta^n \subseteq J_\alpha$ for some n. Then $J_\beta^i \subseteq J_\alpha$ for every $i \geq n$. Thus

$$J_\alpha \subseteq J_{\beta+1} \subseteq J_\beta^i$$

forces $J_\beta^i = J_\alpha$ for every $i \geq n$. In particular, $J_\alpha^2 = J_\alpha$. This contradicts Lemma 7.17. Hence $J_\alpha = 0$ for a prime ring with right Krull dimension α.

Now let R be an arbitrary ring with right Krull dimension α. Then we know that for every prime ideal P of R we have $J(R/P)_\alpha = 0$. Since

$$J(R) + P/P \subseteq J(R/P),$$

by induction on γ it is easily seen that $J(R)_\gamma + P/P \subseteq J(R/P)_\gamma$ for every ordinal γ. Thus $J(R)_\alpha \subseteq P$. This shows that $J(R)_\alpha \subseteq N(R)$, which is nilpotent by Theorem 7.21. □

We shall now apply transfinite powers of the Jacobson radical to compute the right Krull dimension of a right serial ring. Observe that we have already encountered the transfinite powers J_α of the Jacobson radical $J = J(R)$ of a serial ring R. For instance, Theorem 5.28 says that a serial ring R is noetherian if and only if $J_1 = 0$. We shall prove that the nilpotency of J_α determines the right Krull dimension of a right serial ring (Proposition 7.29). First we need two lemmas.

Lemma 7.27 *Let R be a right serial ring, let M_R be a cyclic uniserial module and let $\mathcal{F}$ be a family of ideals of R. Then $M\left(\bigcap_{I \in \mathcal{F}} I\right) = \bigcap_{I \in \mathcal{F}} MI$.*

Proof. Observe that $e\left(\bigcap_{I \in \mathcal{F}} I\right) = \bigcap_{I \in \mathcal{F}} eI$ for any idempotent e and any family of ideals in any ring.

Let $R_R = e_1 R \oplus \cdots \oplus e_n R$ be a direct sum decomposition of R_R with the e_j idempotents and the $e_j R$ uniserial right ideals. Let

$$f: R_R = e_1 R \oplus \dots e_n R \to M_R$$

be an epimorphism. Then $\sum_j f(e_j R) = M$, so that $f(e_i R) = M$ for some index i. Hence $M_R \cong e_i R/K$ for a right ideal K of R contained in $e_i R$. Set

$$A = \bigcap_{I \in \mathcal{F}} I,$$

so that we only have to prove that $MA \supseteq \bigcap_{I \in \mathcal{F}} MI$, or, equivalently, that

$$e_i A + K \supseteq \bigcap_{I \in \mathcal{F}} (e_i I + K). \tag{7.1}$$

If $K \subseteq e_i A$, then $K \subseteq e_i I$ for every $I \in \mathcal{F}$, so that (7.1) holds by the observation in the first paragraph of this proof. If $K \supset e_i A$, from the same observation we get that $K \supset e_i I$ for some $I \in \mathcal{F}$, and so (7.1) holds in any case. $\qquad\square$

Lemma 7.28 *If M_R is a uniserial module over a right serial ring R and $M(J_\alpha)^n = 0$ for some ordinal α and some positive integer n, then*

$$\mathrm{K.dim}(M_R) \le \alpha.$$

Proof. We prove the result by transfinite induction on α. If $\alpha = 0$ and $MJ^n = 0$, then M_R is a uniserial module of finite length $\le n$, and thus $\mathrm{K.dim}(M_R) \le 0$. Let $\alpha > 0$ be an ordinal. By Proposition 7.11 we may assume $MJ_\alpha = 0$. Let $M = M_0 \supset M_1 \supset M_2 \supset \dots$ be a descending chain of submodules of M, and fix an element $x \in M \setminus M_1$. If α if a limit ordinal, then $0 = xJ_\alpha = \bigcap_{\beta < \alpha} xJ_\beta$ by Lemma 7.27 applied to the cyclic uniserial module xR. Hence for every $i \ge 0$ there exists $\beta < \alpha$ such that $xJ_\beta \subseteq M_{i+1}$. Then $(M_i/M_{i+1})J_\beta = 0$ for every $i \ge 1$, so that $\mathrm{K.dim}(M_i/M_{i+1}) \le \beta < \alpha$ for every $i \ge 1$ in this case. If α if a not limit ordinal, $\alpha = \beta + 1$ say, then $0 = xJ_\alpha = \bigcap_{n=1}^\infty xJ_\beta^n$ by Lemma 7.27, so that for every $i \geq 1$ there exists $n \geq 1$ such that $xJ_\beta^n \subseteq M_{i+1}$. Then $(M_i/M_{i+1})J_\beta^n = 0$. Hence by the induction hypothesis K.dim$(M_i/M_{i+1}) \leq \beta < \alpha$ for every $i \geq 1$ in any case. Therefore K.dim$(M_R) \leq \alpha$. $\qquad\square$

We are ready to prove a useful criterion to determine the right Krull dimension of a right serial ring.

Proposition 7.29 *A right serial ring has right Krull dimension α if and only if α is the least ordinal with J_α nilpotent.*

Proof. We shall prove that a right serial ring has right Krull dimension $\leq \alpha$ if and only if J_α nilpotent. If R is an arbitrary ring with right Krull dimension $\leq \alpha$, then J_α is nilpotent by Theorem 7.26. Conversely, if R is a right serial ring and J_α is nilpotent, write $R_R = e_1 R \oplus \ldots e_n R$ as a direct sum of uniserial right ideals $e_i R$. Then $e_i J_\alpha^n = 0$ for some $n > 0$, so that K.dim$(e_i R) \leq \alpha$ by Lemma 7.28. Thus r.K.dim$(R) \leq \alpha$. $\qquad\square$

Proposition 7.29 and its left-right dual give

Corollary 7.30 *A serial ring has right Krull dimension α if and only if it has left Krull dimension α, if and only if α is the least ordinal with J_α nilpotent.* $\qquad\square$

Hence the left Krull dimension and the right Krull dimension of a serial ring are equal. Example 7.14 shows that this is not true for right serial rings.

We conclude the section with a proposition that will be used in the sequel.

Proposition 7.31 *Let $\{e_1, \ldots, e_n\}$ be a complete set of orthogonal primitive idempotents for a serial ring R. Then:*

(a) *If R is right noetherian, then $e_i R e_i$ is noetherian for every $i = 1, 2, \ldots, n$.*

(b) *If R has Krull dimension, then $e_i R e_i$ has Krull dimension $\leq$ K.dim(R) for every $i = 1, 2, \ldots, n$.*

(c) *If $e_i R e_i$ has finite Krull dimension for every $i = 1, 2, \ldots, n$, then R has finite Krull dimension and*

$$\text{K.dim}(R) = \max\{\,\text{K.dim}(e_i R e_i) \mid i = 1, 2, \ldots, n\,\}.$$

Proof. Let R be a serial ring, $\mathcal{L}(e_i R e_i)$ the linearly ordered set of right ideals of $e_i R e_i$, and $\mathcal{L}(e_i R_R)$ the linearly ordered set of R-submodules of $e_i R_R$.

For every right ideal I of $e_i R e_i$ let $\varepsilon(I) = IR$ be the right ideal of R generated by I. This defines a mapping $\varepsilon \colon \mathcal{L}(e_i R e_i) \to \mathcal{L}(e_i R_R)$. We show that this mapping φ is order-preserving and injective. Fix two right ideals $I \subset J$

of $e_j Re_j$. Then $IR \subset JR$, otherwise $IR \supseteq JR$ implies $I = IRe_i \supseteq JRe_i = J$, contradiction.

Therefore, if R is right noetherian, then $\mathcal{L}(e_i R_R)$ has the a.c.c., so that $\mathcal{L}(e_i Re_i)$ has the a.c.c., that is, $e_i Re_i$ is right noetherian. By Proposition 5.3(b) the ring $e_i Re_i$ is left noetherian also, and this proves (a). Similarly, if R has Krull dimension, then $e_i R_R$ has Krull dimension, so that the deviation of $\mathcal{L}(e_i Re_i)$ exists and is less than or equal to the deviation of $\mathcal{L}(e_i R_R)$ by Proposition 7.3. Hence r.K.dim$(e_i Re_i) \leq$ K.dim$(e_i R_R)$ for every index i. This shows that (b) holds.

In view of (b), in order to prove (c) it is sufficient to show that if all the rings $e_i Re_i$ have finite Krull dimension $\leq k$ for a non-negative integer k, then R_R has finite Krull dimension $\leq k$. Suppose the contrary, i.e., assume that K.dim(R) is not defined or K.dim$(R) > k$. Then there is an index j for which the deviation of the linearly ordered set $\mathcal{L}(e_j R_R)$ of all the R-submodules of $e_j R_R$ is not defined or is $> k$. By Proposition 7.8 there is a strictly descending chain of submodules of $e_j R_R$ indexed in ω^{k+1}, that is, there is a family of R-submodules I_α, $\alpha < \omega^{k+1}$, of $e_j R_R$ such that $I_\alpha \supset I_\beta$ for every $\alpha < \beta < \omega^{k+1}$. For every $\alpha < \omega^{k+1}$ let $e_j r_\alpha$ be an element of $I_\alpha \setminus I_{\alpha+1}$. Then the chain $e_j r_\alpha R$, $\alpha < \omega^{k+1}$, is a strictly descending chain of submodules of $e_j R$. Now for every α there is an index $i(\alpha) = 1, 2, \ldots, n$ such that $e_j r_\alpha R = e_j r_\alpha e_{i(\alpha)} R$. Passing to a subchain we may assume without loss of generality that there exists an index i such that $r_\alpha = r_\alpha e_i$ for every $\alpha < \omega^{k+1}$. From $e_j r_0 e_i R \supseteq e_j r_\alpha e_i R$ we obtain that for every $\alpha < \omega^{k+1}$ there exists $s_\alpha \in R$ such that

$$e_j r_\alpha e_i = e_j r_0 e_i s_\alpha e_i.$$

Then the right ideals $e_i s_\alpha e_i Re_i$ of $e_i Re_i$, $\alpha < \omega^{k+1}$, form a strictly descending chain, because if $\alpha < \beta < \omega^{k+1}$ and $e_i s_\alpha e_i Re_i \subseteq e_i s_\beta e_i Re_i$, then

$$e_j r_\alpha e_i R \subseteq e_j r_0 e_i s_\alpha e_i Re_i R \subseteq e_j r_0 e_i s_\beta e_i Re_i R \subseteq e_j r_\beta e_i R,$$

contradiction. By Proposition 7.8 the deviation of the linearly ordered set $\mathcal{L}(e_i Re_i)$ of all the right ideals of $e_i Re_i$, if it exists, is greater than k, that is, K.dim$(e_i Re_i) > k$. $\qquad\square$

Note that the hypothesis "R is serial" is essential in Proposition 7.31(c). In order to see this consider the case of the ring

$$R = \begin{pmatrix} k & V \\ 0 & k \end{pmatrix},$$

where k is a field and V is an infinite dimensional vector space over k.

Corollary 7.32 *Every serial, right noetherian ring has (right and left) Krull dimension at most one.*

Proof. If R is such a ring, then $e_i R e_i$ is a noetherian chain ring for every $i = 1, 2, \ldots, n$ (Proposition 7.31(a)). By Proposition 5.3(b) and Corollary 7.30 all the rings $e_i R e_i$ have Krull dimension at most one. Proposition 7.31(c) allows us to conclude the proof. $\qquad\square$

7.5 Structure of serial rings of finite Krull dimension

The aim of this section is to describe serial rings of finite Krull dimension. As usual, the case of chain rings is simpler, so we begin with chain rings.

A right chain ring R is *discrete* if $P \neq P^2$ for every completely prime ideal $P \neq 0$ of R [Brungs].

Lemma 7.33 *In a discrete right chain ring all prime ideals are completely prime.*

Proof. Let R be a discrete right chain ring and let P be a prime ideal that is not completely prime. Let Q denote the intersection of all the completely prime ideals containing P. Then Q is a completely prime ideal, so that Q contains P properly. Thus $Q^n \supset P$ for all integers $n > 0$. In particular, Q is not nilpotent, so that $\bigcap_{n=1}^{\infty} Q^n$ is completely prime by Proposition 5.2(d). This is a contradiction because $Q \supset \bigcap_{n=1}^{\infty} Q^n \supseteq P$. $\qquad\square$

The reason why we have introduced discrete right chain rings is that every right chain ring with right Krull dimension is discrete (Lemma 7.17).

We are mainly interested in chain and serial rings with Krull dimension and the a.c.c. on right annihilators, because these rings appears naturally in the study of endomorphism rings of artinian modules, as we shall see in Chapter 10.

Lemma 7.34 *Let R be a discrete chain ring with the a.c.c. on right annihilators and with only a finite number $P_0 \subset P_1 \subset \cdots \subset P_n$ of prime ideals. Let $k = 1, 2, \ldots, n$ be a fixed index and let R_{P_k} be the localization of R at P_k. Then R_{P_k} is a discrete chain ring with the a.c.c. on right annihilators, and its prime ideals are $P_0 \subset P_1 \subset \cdots \subset P_k$.*

Proof. The ring R has a classical two-sided quotient ring Q (Theorem 6.47), which is an artinian chain ring (Theorem 6.54), and $J(Q) = N(R) = P_0$ (Proposition 6.55). Hence all the elements in $R \setminus P_0$ are regular, because if $r \in R \setminus P_0$, then $r \in Q \setminus J(Q) = U(Q)$.

Since P_k is a completely prime ideal in R and all the elements of $R \setminus P_k$ are regular, we may suppose $R \subseteq R_{P_k} \subseteq Q$. In particular, R_{P_k} has the a.c.c. on right annihilators.

If $x \in P_i$ for some $i \leq k$ and $s \in R \setminus P_k$, then $s^{-1}x, xs^{-1} \in P_i$ (Lemma 7.33 and Proposition 5.2(e)), so that $P_0, P_1, \ldots, P_k$ are ideals of R_{P_k} and P_k is the maximal ideal of R_{P_k}. In particular, every proper ideal of R_{P_k} is contained in P_k, and thus every proper ideal of R_{P_k} is an ideal of R. By Proposition 6.4(b) the ideals $P_0, P_1, \ldots, P_k$ of R_{P_k} are prime. If $i = 1, 2, \ldots, k$, then P_i is a completely prime ideal of R by Lemma 7.33, so that $P_i \neq P_i^2$. By Lemma 5.1 there is no prime ideal of R properly between P_i and $\bigcap_{t=1}^{\infty} P_i^t$, and $\bigcap_{t=1}^{\infty} P_i^t$ is a completely prime ideal of R by Proposition 5.2(d). Hence $\bigcap_{t=1}^{\infty} P_i^t = P_{i-1}$. By Lemma 5.1 there is no prime ideal of R_{P_k} properly between P_i and $\bigcap_{t=1}^{\infty} P_i^t = P_{i-1}$. Thus $P_0 \subset P_1 \subset \cdots \subset P_k$ are all the prime ideals of R_{P_k}. In particular, the chain ring R_{P_k} is discrete. $\square$

The next theorem characterizes the chain rings of finite Krull dimension with the a.c.c. on right annihilators.

Theorem 7.35 *Let R be a chain ring with the a.c.c. on right annihilators and let $n \geq 0$ be an integer. The following conditions are equivalent:*

(a) *The chain ring R has Krull dimension n.*

(b) *The chain ring R is discrete and has exactly $n + 1$ prime ideals;*

(c) *The chain ring R has exactly $n + 1$ completely prime ideals*

$$P_0 \subset P_1 \subset \cdots \subset P_n$$

and the localizations $(R/P_{i-1})_{P_i/P_{i-1}}$, $i = 1, 2, \ldots, n$, are noetherian rings.

Proof. (a) $\Rightarrow$ (b). Let R be a chain ring of Krull dimension n. Consider the transfinite powers $J_0 \supseteq J_1 \supseteq \cdots \supseteq J_n$ of the Jacobson radical J of R. By Corollary 7.30 the ideal J_n is nilpotent and $J_0, J_1, \ldots, J_{n-1}$ are not nilpotent, so that in particular $J_0 \supset J_1 \supset \cdots \supset J_n$. By Proposition 5.2(d) the ideals $J_1, \ldots, J_n$ are completely prime, and by Lemma 5.1 there is no prime ideal properly between J_i and J_{i+1} for every $i = 0, 1, \ldots, n-1$. It is now clear that R is a discrete chain ring and $J_0 \supset J_1 \supset \cdots \supset J_n$ are all its prime ideals.

(b) $\Rightarrow$ (c). Let R be a discrete chain ring with exactly $n + 1$ prime ideals $P_0 \subset P_1 \subset \cdots \subset P_n$. Fix an index $i = 1, 2, \ldots, n$. Then R/P_{i-1} is a discrete chain ring and $0 = P_{i-1}/P_{i-1} \subset P_i/P_{i-1} \subset \cdots \subset P_n/P_{i-1}$ are all its prime ideals. Note that R/P_{i-1} is an integral domain by Lemma 7.33. By Lemma 7.34 $(R/P_{i-1})_{P_i/P_{i-1}}$ is a discrete chain ring with exactly two prime ideals, its Jacobson radical and 0. If J is the Jacobson radical of a discrete chain ring whose only prime ideals are J and 0, then $\bigcap_{t=1}^{\infty} J^t = 0$ by Proposition 5.2(d). Hence the ring is noetherian by Proposition 5.3(b).

(c) $\Rightarrow$ (a). Let R be a ring for which (c) holds. For $i = 1, 2, \ldots, n$ the Jacobson radical of the ring $(R/P_{i-1})_{P_i/P_{i-1}}$ is P_i/P_{i-1}. Since $(R/P_{i-1})_{P_i/P_{i-1}}$ is noetherian, from Proposition 5.3(b) it follows that $\bigcap_{n=1}^{\infty} (P_i/P_{i-1})^n = 0$, so that $\bigcap_{n=1}^{\infty} P_i^n \subseteq P_{i-1}$. But $P_i^n \supseteq P_{i-1}$ for every n because P_{i-1} is prime. Hence $\bigcap_{n=1}^{\infty} P_i^n = P_{i-1}$. It follows that the transfinite powers of the Jacobson radical

J of R are $J_t = P_{n-t}$ for every $t = 0, 1, \ldots, n$. In particular J_t is not nilpotent for every $t = 0, 1, \ldots, n-1$.

The ring R has an artinian classical two-sided quotient ring Q and

$$J(Q) = N(R) = P_0$$

(Theorems 6.47 and 6.54 and Proposition 6.55). Hence $J_n = P_0 = J(Q)$ is nilpotent. Thus R has Krull dimension n by Corollary 7.30. □

Note that chain rings of finite Krull dimension do not necessarily have the a.c.c. on right annihilators. For instance, the trivial extension $R = \mathbf{Z}_p \propto \mathbf{Z}(p^\infty)$ of the discrete valuation ring $\mathbf{Z}_p$ via the Prüfer group $\mathbf{Z}(p^\infty)$ (p. 167) is a commutative chain ring of Krull dimension 1 without the a.c.c. on annihilators.

We now turn to consider serial rings.

Lemma 7.36 *If R is a serial ring, $N = N(R)$ is its prime radical, J is the Jacobson radical of R, $J_1 = \bigcap_{n=1}^{\infty} J^n$ and K is the ideal of R containing J_1 such that $N(R/J_1) = K/J_1$, then:*

(a) *K is the least semiprime ideal of R contained in J such that R/K is noetherian.*

(b) *K contains no non-zero idempotent of R.*

(c) *If N is nilpotent, then $N \subseteq K$.*

Proof. By Theorem 5.28 the ring R/J_1 is noetherian, so that R/K is obviously a semiprime noetherian serial ring. From

$$K/J_1 = N(R/J_1) \subseteq J(R/J_1) = J/J_1$$

it follows that $K \subseteq J$. In particular, K contains no non-zero idempotent.

If I is an arbitrary semiprime ideal of R such that $I \subseteq J$ and R/I is noetherian, then $J(R/I) = J/I$, so that

$$J_1 + I/I \subseteq \bigcap_{n=1}^{\infty} (J^n + I)/I = \bigcap_{n=1}^{\infty} J(R/I)^n = 0$$

(Theorem 5.28). Hence $J_1 \subseteq I$. In particular, $(R/J_1)/(I/J_1) \cong R/I$ is semiprime, so that I/J_1 is a semiprime ideal of R/J_1. Thus

$$K/J_1 = N(R/J_1) \subseteq I/J_1,$$

from which $K \subseteq I$. Finally, if N is nilpotent, then $N + J_1/J_1$ is a nilpotent ideal of R/J_1, and thus $N + J_1/J_1 \subseteq N(R/J_1) = K/J_1$, from which $N \subseteq K$. □

If R is a serial ring, the ideal K defined in the previous lemma will be called the *derived ideal* of R. Since R/K is a noetherian ring, the derived ideal K of R is a semiprime Goldie ideal of R.

Now let R be a serial ring, $\{e_1, \ldots, e_n\}$ a complete set of orthogonal primitive idempotents of R and K the derived ideal of R. Suppose that R either has the a.c.c. on right annihilators or is nonsingular. The maximal Ore set corresponding to K is $\mathcal{C}(K)$ (Proposition 6.45). By Lemma 6.56 the elements of $\mathcal{C}(K)$ are regular elements of R. Hence if T is the quotient ring of R with respect to $\mathcal{C}_R(K)$, we may suppose $R \subseteq T$. Then T is a serial ring, K is its Jacobson radical, T/K is a semisimple artinian ring, and T/K is the classical two-sided quotient ring of R/K (Proposition 6.57). Moreover T either has the a.c.c. on right annihilators or is nonsingular.

Theorem 7.37 *Let R be a serial ring of finite Krull dimension m. Suppose that R has the a.c.c. on right annihilators (or is nonsingular). Let K be the derived ideal of R and $T \supseteq R$ the quotient ring of R with respect to $\mathcal{C}_R(K)$. Then:*

(a) *If $m = 0$, then $T = R$.*

(b) *If $m > 0$, then T is a serial ring of finite Krull dimension $m - 1$ that has the a.c.c. on right annihilators (is nonsingular, respectively).*

Proof. Let J be the Jacobson radical of R, so that K/J_1 is the prime radical of the noetherian ring R/J_1.

If $m = 0$, the ring R is artinian, so that $K = N(R) = J$. Thus

$$\mathcal{C}_R(K) \subseteq \mathcal{C}_R(0) = U(R).$$

Hence $T = R$. If $m = 1$, then J_1 is nilpotent (Corollary 7.30). Since

$$N(R/J_1) = K/J_1$$

is the prime radical of the Noetherian ring R/J_1, K/J_1 is nilpotent, so that K is nilpotent. But K is the Jacobson radical of the serial ring T, so that T is artinian, that is, r.K.dim$(T) = 0$.

Suppose that $m \geq 2$. Since T is a serial ring, in order to show that r.K.dim$(T) = m - 1$ it suffices to compute the transfinite powers K_i of its Jacobson radical K, and show that K_{m-1} is nilpotent, but K_{m-2} is not nilpotent (Corollary 7.30). As K/J_1 is the prime radical of the noetherian ring R/J_1 and the prime radical of a noetherian ring is nilpotent, there exists a positive integer t such that $K^t \subseteq J_1$. Thus $K \supseteq J_1 \supseteq K^t$. Taking the intersections of the n-th powers of this inclusion one sees that $K_1 \supseteq J_2 \supseteq K_1$. Hence $K_1 = J_2$. Thus $K_i = J_{i+1}$ for every $i \geq 1$. From Corollary 7.30 it follows immediately that r.K.dim$(T) = $ r.K.dim$(R) - 1$. $\qquad\square$

Theorem 7.37 tells us that every serial ring of finite Krull dimension m which has the a.c.c. on right annihilators (or is nonsingular), can be obtained inductively in m steps from an artinian serial ring and m noetherian serial rings. The construction is the following. Let R be such a serial ring of finite Krull dimension m. If $m = 0$, then R is artinian. Suppose that $m \geq 1$. For every $i = 0, 1, \ldots, m$ define a serial ring T_i of Krull dimension $m - i$ in the following way. Put $T_0 = R$. Assume T_i has been defined. If K_i denotes the derived ideal of T_i, then T_i has a quotient ring T_{i+1} with respect to $\mathcal{C}_{T_i}(K_i)$, and T_{i+1} is a serial ring of Krull dimension $m - i - 1$. From Proposition 6.57 we know that $R = T_0 \subseteq T_1 \subseteq \cdots \subseteq T_m$, K_i is the Jacobson radical of T_{i+1}, the T_{i+1}/K_i are semisimple artinian rings, and every T_{i+1}/K_i is the classical two-sided quotient ring of T_i/K_i. Hence each T_i is the inverse image via the canonical projection $T_{i+1} \to T_{i+1}/K_i$ of the noetherian serial subring T_i/K_i of T_{i+1}/K_i. Thus T_i is obtained from the serial ring T_{i+1} and the noetherian serial ring T_i/K_i. By induction we find that R can be constructed in m steps from the artinian serial ring T_m and the m noetherian serial rings $T_0/K_0, T_1/K_1, \ldots, T_{m-1}/K_{m-1}$.

7.6 Notes on Chapter 7

The Krull dimension for objects of an abelian category was introduced by [Gabriel] using localizing subcategories. The definition we have given here is due to [Rentschler and Gabriel] (for finite ordinals) and [Krause] (for arbitrary ordinals). The fundamental references on Krull dimension are [Lemonnier 72a] and [Gordon and Robson 73a]. We have often followed Gordon and Robson's Memoir [Gordon and Robson 73a] and the book [McConnell and Robson] in the first three sections of this chapter. The deviation of an arbitrary poset A, that is, the "Krull dimension" of A, was introduced by [Lemonnier 72a].

Let D be the set of all rational numbers between 0 and 1 that can be written as a fraction whose denominator is a power of 2, that is,

$$D = \{\, m/2^n \mid m, n \text{ nonnegative integers}, 0 \leq m/2^n \leq 1 \,\}.$$

The set D, with the order induced from the usual order of $\mathbf{Q}$, is a linearly ordered set. [Lemonnier 72a] proved that a poset A fails to have a deviation if and only if A contains a poset order isomorphic to D.

Example 7.10 is due to [Gabriel]. Proposition 7.11 is a result of [Rentschler and Gabriel], and Proposition 7.13 is due to [Krause].

Critical modules were introduced by [Hart], who called them restricted modules. Lemma 7.17 is taken from [Krause and Lenagan, Lemma 7]. Corollary 7.19 appears in [Lemonnier 72b] and in [Gordon and Robson 73b], and Proposition 7.20, a variant of Proposition 4.5 in [Gordon and Robson 73a], is due to Lenagan ([Lenagan 77, Proposition 1] and [Lenagan 80, Lemma]). For the proof of Theorem 7.21, due to [Gordon and Robson 73a, Lemma 5.6] and

[Lenagan 73], we have essentially followed the nice proof given in [Lenagan 77].
Theorem 7.23 is taken from [Lenagan 77, Theorem 8].

Proposition 7.24(a) and Theorem 7.26 are due to [Krause] and [Krause and
Lenagan, Theorem 8], respectively. The results from Lemma 7.27 to Proposition 7.29 are due to [Müller and Singh]. Corollary 7.30 appears in [Wright 89].
For its proof we have followed [Müller and Singh]. Proposition 7.31 is proved
in [Facchini and Puninski 95, Lemma 5.5].

The description of serial rings of finite Krull dimension given in Section 7.5
is due to [Camps, Facchini and Puninski].

Chapter 8

Krull-Schmidt Fails for Finitely Generated Modules and Artinian Modules

In this chapter we show that the Krull-Schmidt Theorem fails for finitely generated modules even if the base ring is a commutative noetherian ring, and that it can fail for artinian modules over non-commutative rings. In Chapter 9 (Example 9.21) we shall show that it fails for finitely presented modules over serial rings. Recall that in Section 2.12 we saw that Krull-Schmidt holds for artinian modules over rings which are either right noetherian or commutative.

8.1 Krull-Schmidt fails for finitely generated modules

Now we present a class of examples discovered by [Levy]. The examples show that there exist commutative noetherian rings S and finitely generated S-modules for which the Krull-Schmidt Theorem, the cancellation property and the n-th root property do not hold. Throughout the section m is an integer greater than one and R denotes a commutative principal ideal domain with at least $2(m-1)$ maximal ideals.

Let G be the following graph with m vertices $1, 2, \ldots, m$ and $2(m-1)$ edges $P_1, P_{-1}, P_2, P_{-2}, \ldots, P_{m-1}, P_{-(m-1)}$:

$$
\begin{array}{ccccccccc}
1 & \dfrac{P_1}{P_{-1}} & 2 & \dfrac{P_2}{P_{-2}} & 3 & \cdots & m-1 & \dfrac{P_{m-1}}{P_{-(m-1)}} & m \\
\bullet & & \bullet & & \bullet & & \bullet & & \bullet
\end{array}
$$

In the graph G each pair of consecutive vertices is joined by two edges. The edges are labeled by a sequence $P_1, P_{-1}, P_2, P_{-2}, \ldots, P_{m-1}, P_{-(m-1)}$ of

distinct maximal ideals of R. Let p_i be a generator of P_i for every $i = 1, 2, \ldots,$
$m - 2, m - 1, -1, -2, \ldots, -(m - 2), -(m - 1)$.

Define

$$S = \{ (r_1, \ldots, r_m) \in R^m \mid r_i \equiv r_{i+1} \bmod p_i p_{-i} \text{ for every } i = 1, 2, \ldots, m - 1 \}.$$

Note that for every $r, r' \in R$, $r \equiv r' \pmod{p_i p_{-i}}$ if and only if $r \equiv r'$
$(\bmod\, p_i)$ and $r \equiv r' \pmod{p_{-i}}$. The set S is a subring of the ring direct
product R^m of m copies of R. Since R is noetherian, the ring S is a module-
finite R-algebra. In particular, S is a noetherian commutative ring.

Now put $\mathbf{S} = R/P_1 \times \cdots \times R/P_{m-1} \times R/P_{-1} \times \cdots \times R/P_{-(m-1)}$, so that
$\mathbf{S}$ is the direct product of $2(m - 1)$ fields. Elements of $\mathbf{S}$ will be written as
boldface letters $\mathbf{w}$. If $\mathbf{w}$ is an element of $\mathbf{S}$, coordinates of $\mathbf{w}$ will be denoted by
$\overline{w_i}$, so that $\mathbf{w} = (\overline{w_1}, \ldots, \overline{w_{m-1}}, \overline{w_{-1}}, \ldots, \overline{w_{-(m-1)}})$ with $\overline{w_i} \in R/P_i$ for every
$i = 1, \ldots, m - 1, -1, \ldots, -(m - 1)$.

For each $\mathbf{w} \in \mathbf{S}$ let $\text{graph}(\mathbf{w})$ denote the graph obtained from G by deleting
the edge P_i for every i such that $\overline{w_i} = \overline{0}$ in R/P_i. For example, $\text{graph}(\mathbf{0})$ is the
graph with m vertices and zero edges, and $\text{graph}(\mathbf{1}) = G$.

Next, for every element $\mathbf{w} = (\overline{w_1}, \ldots, \overline{w_{m-1}}, \overline{w_{-1}}, \ldots, \overline{w_{-(m-1)}})$ of $\mathbf{S}$ define
an R-submodule $M(\mathbf{w})$ of R^m as follows:

$$M(\mathbf{w}) = \{ (x_1, \ldots, x_m) \in R^m \mid \overline{x_i w_i} = \overline{x_{i+1}} \text{ in } R/P_i$$
$$\text{for every } i = 1, 2, \ldots, m - 1 \text{ such that } \overline{w_i} \neq \overline{0}, \text{ and } \overline{x_i w_{-i}} = \overline{x_{i+1}}$$
$$\text{in } R/P_{-i} \text{ for every } i = 1, 2, \ldots, m - 1 \text{ such that } \overline{w_{-i}} \neq \overline{0} \}.$$

Thus the definition of $M(\mathbf{w})$ consists of one identity for each edge in $\text{graph}(\mathbf{w})$.
Note that $M(\mathbf{w})$ is clearly an R-submodule of R^m. For example, $M(\mathbf{0}) = R^m$
and $M(\mathbf{1}) = S$. The next two lemmas describe the modules $M(\mathbf{w})$, $\mathbf{w} \in \mathbf{S}$.

Lemma 8.1 *Put $I = P_1 \cap \cdots \cap P_{m-1} \cap P_{-1} \cap \cdots \cap P_{-(m-1)}$. Then*

$$\underbrace{I \oplus \cdots \oplus I}_{m \text{ times}} \subseteq M(\mathbf{w})$$

*for every $\mathbf{w} \in \mathbf{S}$. In particular, $M(\mathbf{w})$ is a torsion-free R-module of torsion-free
rank m.*

Proof. $\overline{x_i w_i} = \overline{x_{i+1}} = \overline{0}$ in R/P_i and $\overline{x_i w_{-i}} = \overline{x_{i+1}} = \overline{0}$ in R/P_{-i} for every
$x_1, \ldots, x_m \in I$. $\qquad \square$

Lemma 8.2 *For every $\mathbf{w} \in \mathbf{S}$ the canonical projections*

$$\pi_i \colon M(\mathbf{w}) \to R, \quad i = 1, 2, \ldots, m,$$

are surjective.

Proof. Fix an index $i = 1, 2, \ldots, m$. We must prove that given any $x_i \in R$, there exist $x_1, \ldots, x_{i-1}, x_{i+1}, \ldots, x_m \in R$ such that $(x_1, \ldots, x_m) \in M(\mathbf{w})$. By induction it is sufficient to prove that given any $x_i \in R$, there exist $x_{i-1}, x_{i+1} \in R$ such that $\overline{x_i w_i} = \overline{x_{i+1}}$ in R/P_i if $\overline{w_i} \neq \overline{0}$, $\overline{x_i w_{-i}} = \overline{x_{i+1}}$ in R/P_{-i} if $\overline{w_{-i}} \neq \overline{0}$, $\overline{x_{i-1} w_{i-1}} = \overline{x_i}$ in R/P_{i-1} if $\overline{w_{i-1}} \neq \overline{0}$, and $\overline{x_{i-1} w_{-(i-1)}} = \overline{x_i}$ in $R/P_{-(i-1)}$ if $\overline{w_{-(i-1)}} \neq \overline{0}$. This follows immediately from the Chinese Remainder Theorem. $\square$

Note that if $\mathbf{w}, \mathbf{v} \in \mathbf{S}$, then $M(\mathbf{w}), M(\mathbf{v}) \subseteq R^m$, so that from now on by the product of an element of $M(\mathbf{w})$ and an element of $M(\mathbf{v})$ we mean their product in R^m.

Lemma 8.3 *If* $\mathbf{w}, \mathbf{v} \in \mathbf{S}$*, then* $M(\mathbf{w}) \cdot M(\mathbf{v}) \subseteq M(\mathbf{wv})$*.*

Proof. If $(x_1, \ldots, x_m) \in M(\mathbf{w})$, $(y_1, \ldots, y_m) \in M(\mathbf{v})$ and $i = 1, 2, \ldots, m-1$ is an index such that $\overline{w_i v_i} \neq \overline{0}$ in R/P_i, then $\overline{w_i} \neq \overline{0}$ and $\overline{v_i} \neq \overline{0}$, so that $\overline{x_i w_i} = \overline{x_{i+1}}$ and $\overline{y_i v_i} = \overline{y_{i+1}}$. Hence $\overline{x_i y_i w_i v_i} = \overline{x_{i+1} y_{i+1}}$. $\square$

For $\mathbf{v} = \mathbf{1}$ we see that $M(\mathbf{w})S \subseteq M(\mathbf{w})$, i.e., $M(\mathbf{w})$ is an S-submodule of R^m. Since R is principal ideal domain, $M(\mathbf{w})$ is a finitely generated free R-module. Hence it is a noetherian S-module *a fortiori*. Note that $M(\mathbf{w})$ is not necessarily indecomposable. For instance, $M(\mathbf{0}) = R^m$ is a direct sum of m non-zero modules as an R^m-module, and so *a fortiori* as an S-module.

If $\mathbf{e} \in \mathbf{S}$ is idempotent, Lemma 8.3 tells us that $M(\mathbf{e}) \cdot M(\mathbf{e}) \subseteq M(\mathbf{e})$, so that $M(\mathbf{e})$ is an S-subalgebra of R^m.

Lemma 8.4 *If* $\mathbf{w}, \mathbf{v} \in \mathbf{S}$*, then for every S-homomorphism* $f: M(\mathbf{w}) \to M(\mathbf{v})$ *there exists a unique* $y \in R^m$ *such that* $f(x) = yx$ *for every* $x \in M(\mathbf{w})$*.*

Proof. Let Q be the field of fractions of R, that is, its classical quotient ring. Then $M(\mathbf{w}), M(\mathbf{v}) \subseteq R^m \subseteq Q^m$ are torsion-free R-modules of rank m by Lemma 8.1. Hence f, which is an R-homomorphism *a fortiori*, extends uniquely to a Q-linear mapping $Q^m \to Q^m$. Thus there exists a unique $m \times m$ matrix A with entries in Q such that $f(x) = xA$ for every $x \in M(\mathbf{w})$. Hence it is sufficient to show that the matrix A is diagonal and that its entries are in R. Fix an index $i = 1, 2, \ldots, m$. By Lemma 8.2 there exists an element $X_i \in M(\mathbf{w})$ whose i-th coordinate is equal to 1. By Lemma 8.1 there is an element $s_i \in S$ whose components are all zero except for the i-th coordinate. Then $s_i X_i$ is an element of $M(\mathbf{w})$ for every $i = 1, 2, \ldots, m$ and A is the matrix associated to f with respect to the basis $s_1 X_1, \ldots, s_m X_m$ of Q^m. In order to prove that A is a diagonal matrix with entries in R it suffices to show that $f(s_i X_i) \in R s_i X_i$ for every i. But $f(s_i X_i) \in f(s_i M(\mathbf{w})) \subseteq s_i R^m = R s_i X_i$. $\square$

By Lemma 8.4 the R-module $\mathrm{Hom}_S(M(\mathbf{w}), M(\mathbf{v}))$ is isomorphic to a submodule of R^m. Hence $\mathrm{Hom}_S(M(\mathbf{w}), M(\mathbf{v}))$ is a finitely generated free R-module.

Theorem 8.5 *For every* $\mathbf{w} \in \mathbf{S}$, *if* graph($\mathbf{w}$) *has* t *connected components, then* $M(\mathbf{w})$ *is the direct sum of* t *indecomposable* S-*modules.*

Proof. By Lemma 8.4 the endomorphism ring of $M(\mathbf{w})$ is canonically isomorphic to the subring E of R^m whose elements are the $y \in R^m$ such that $yM(\mathbf{w}) \subseteq M(\mathbf{w})$. Let $y = (y_1, \ldots, y_m)$ be an idempotent element of E, so that y_i is equal to 0 or 1 for every i. Let $i, i+1$ be two arbitrary consecutive vertices of graph($\mathbf{w}$) that are connected by an edge. By Lemma 8.2 there is an element $x = (x_1, \ldots, x_m) \in M(\mathbf{w})$ such that $x_i = 1$. Then

$$yx = (y_1 x_1, \ldots, y_m x_m) \in M(\mathbf{w}).$$

Since the vertices i and $i+1$ of graph($\mathbf{w}$) are connected by an edge, we know that either $\overline{w_i} \neq \overline{0}$ in R/P_i or $\overline{w_{-i}} \neq \overline{0}$ in R/P_{-i}. Suppose that $\overline{w_i} \neq \overline{0}$ in R/P_i. From $x \in M(\mathbf{w})$ we obtain that $\overline{x_i w_i} = \overline{x_{i+1}}$, and from $yx \in M(\mathbf{w})$ we know that $\overline{y_i x_i w_i} = \overline{y_{i+1} x_{i+1}}$. Hence in R/P_i we have $\overline{w_i} = \overline{x_{i+1}}$ and $\overline{y_i w_i} = \overline{y_{i+1} w_i}$, so that $\overline{y_i} = \overline{y_{i+1}}$. But y_i, y_{i+1} are equal to 0 or 1. Hence $y_i \equiv y_{i+1} \pmod{P_i}$ forces $y_i = y_{i+1}$. Similarly for $\overline{w_{-i}} \neq \overline{0}$ in R/P_{-i}. This shows that if $i, i+1$ are two arbitrary consecutive vertices of graph($\mathbf{w}$) that are connected by an edge, then $y_i = y_{i+1}$. Therefore the value of y_i is constant on the connected components of graph($\mathbf{w}$).

This shows that if graph($\mathbf{w}$) has t connected components, then E has at most 2^t idempotents. In order to conclude the proof it is sufficient to prove that if C is a connected component of graph($\mathbf{w}$) and $y = (y_1, \ldots, y_m) \in R^m$ is the idempotent element such that $y_i = 1$ for every vertex $i \in C$ and $y_i = 0$ for every $i \notin C$, then $y \in E$. Hence we must show that $yz \in M(\mathbf{w})$ for every $z = (z_1, \ldots, z_m) \in M(\mathbf{w})$. This is immediately verified from the definition of $M(\mathbf{w})$. $\square$

Let us go back to the homomorphisms between the two S-modules $M(\mathbf{w})$, $M(\mathbf{v})$. We are now interested in the case where $\mathbf{w}, \mathbf{v} \in \mathbf{S}$ are idempotents.

Lemma 8.6 *Let* $\mathbf{e}, \mathbf{f}$ *be idempotent elements of* $\mathbf{S}$ *and* $r \in R$. *Suppose that* $r \in P_i$ *for every index* $i = 1, 2, \ldots, m-1, -1, -2, \ldots, -(m-1)$ *such that* P_i *is an edge in* graph($\mathbf{f}$) *and* P_i *is not an edge in* graph($\mathbf{e}$). *Then multiplication by* r *is an* S-*homomorphism* $M(\mathbf{e}) \to M(\mathbf{f})$.

Proof. It is sufficient to show that $rM(\mathbf{e}) \subseteq M(\mathbf{f})$, and this follows immediately from the definitions of $M(\mathbf{e})$ and $M(\mathbf{f})$. $\square$

In the proof of the next corollary, a homomorphism $M(\mathbf{w}) \to M(\mathbf{v})$ given by multiplication by an element of $r \in R$ will be denoted with the same symbol r.

Corollary 8.7 *If* $\mathbf{e}$ *is any idempotent element in* $\mathbf{S}$, *the* S-*modules*

$$M(\mathbf{e}) \oplus M(\mathbf{1} - \mathbf{e}) \quad and \quad \mathbf{S} \oplus M(\mathbf{0})$$

are isomorphic.

Proof. We have that $\mathbf{e} = (\overline{e_1}, \ldots, \overline{e_{m-1}}, \overline{e_{-1}}, \ldots, \overline{e_{-(m-1)}})$ with $e_i = 0$ or 1 for every $i = 1, 2, \ldots, m - 1, -1, -2, \ldots, -(m - 1)$. Let C be the set of all the indices i such that $e_i = 1$ and C' be the set of all the indices i such that $e_i = 0$. By the Chinese Remainder Theorem there exists a non-zero element $c \in R$ such that $c \equiv e_i \pmod{P_i}$ for every $i = 1, 2, \ldots, m-1, -1, -2, \ldots, -(m-1)$. Then $c \in P_i$ for every $i \in C'$ and $c \notin P_i$ for every $i \in C$. Since $c \neq 0$, there are only a finite number of distinct maximal ideals of R containing c, say P_i ($i \in C'$) and Q_j ($j = 1, \ldots, t$). By the Chinese Remainder Theorem there exists $c' \in R$ such that $c' \equiv 1 - e_i \pmod{P_i}$ for every $i = 1, 2, \ldots, m - 1, -1, -2, \ldots, -(m - 1)$ and $c' \equiv 1 \pmod{Q_j}$ for every $j = 1, \ldots, t$. Then c' does not belong to any maximal ideal containing c. Hence there exist $d, d' \in R$ such that $dc' - d'c = 1$.

Note that $S = M(1)$, $M(0) = R^m$, graph$(1) = G$ and graph(0) is the graph with m vertices and no edges. By Lemma 8.6 multiplication by c is a homomorphism $c \colon M(\mathbf{e}) \to M(1)$, because if P_i is not an edge of graph$(\mathbf{e})$, then $e_i = 0$, so that $c \equiv 0 \pmod{P_i}$, that is, $c \in P_i$. Similar elementary verifications show that

$$c' \colon M(1 - \mathbf{e}) \to M(1),$$
$$d \colon M(\mathbf{e}) \to M(0),$$
$$d' \colon M(1 - \mathbf{e}) \to M(0),$$
$$c' \colon M(0) \to M(\mathbf{e}),$$
$$d \colon M(1) \to M(1 - \mathbf{e}),$$
$$d' \colon M(1) \to M(\mathbf{e}),$$
$$c \colon M(0) \to M(1 - \mathbf{e})$$

are S-module homomorphisms (as we have already said, an element of R and multiplication by that element are denoted with the same symbol). It is immediately verified that the composite of the two homomorphisms

$$\begin{pmatrix} c & c' \\ d & d' \end{pmatrix} \colon M(\mathbf{e}) \oplus M(1 - \mathbf{e}) \to M(1) \oplus M(0)$$

and

$$\begin{pmatrix} -d' & c' \\ d & -c \end{pmatrix} \colon M(1) \oplus M(0) \to M(\mathbf{e}) \oplus M(1 - \mathbf{e})$$

are identities. Thus $M(\mathbf{e}) \oplus M(1 - \mathbf{e}) \cong M(1) \oplus M(0)$. $\square$

We are ready to give our first example. It shows the non-uniqueness of the number of indecomposable summands for finitely generated modules over commutative noetherian rings.

Example 8.8 *Let $m \geq 2$ be an integer and let R be a commutative principal ideal domain with at least $2(m - 1)$ maximal ideals. Then there exist a commutative module-finite R-algebra S and a finitely generated S-module M that is the direct sum of 2 indecomposable S-modules, and also the direct sum of 3 indecomposable S-modules, and $\ldots$, and also the direct sum of $m + 1$ indecomposable S-modules.*

Proof. Let S be the ring constructed in this section and $M = S \oplus M(\mathbf{0})$. For every $i = 0, 1, \ldots, m-1$ put

$$\mathbf{e}_i = (\underbrace{\overline{1}, \overline{1}, \ldots, \overline{1}}_{m-1+i}, \underbrace{\overline{0}, \ldots, \overline{0}}_{m-1-i}) \in \mathbf{S}$$

$$= R/P_1 \times \cdots \times R/P_{m-1} \times R/P_{-1} \times \cdots \times R/P_{-(m-1)}.$$

Then graph($\mathbf{e}$) is connected and graph($\mathbf{1} - \mathbf{e}$) is the union of a connected component and i isolated vertices. Hence graph($\mathbf{1} - \mathbf{e}$) has $i+1$ connected components. By Theorem 8.5 the S-module $M(\mathbf{e})$ is indecomposable and $M(\mathbf{1} - \mathbf{e})$ is the direct sum of $i+1$ indecomposable modules, so that $M(\mathbf{e}) \oplus M(\mathbf{1} - \mathbf{e})$ is the direct sum of $i+2$ indecomposable modules. The conclusion follows from Corollary 8.7. $\qquad\square$

Note that the endomorphism ring $\operatorname{End}(M_S)$ of the module M in Example 8.8 is a module-finite R-algebra. More precisely, it is a finitely generated free R-module.

Lemma 8.9 *Let* $\mathbf{e}, \mathbf{f} \in \mathbf{S}$ *be idempotent elements. Then the S-modules $M(\mathbf{e})$ and $M(\mathbf{f})$ are isomorphic if and only if $\mathbf{e} = \mathbf{f}$.*

Proof. We must prove that if $\mathbf{e} \neq \mathbf{f}$, then $M(\mathbf{e})$ and $M(\mathbf{f})$ are not isomorphic. Suppose the contrary, and let $\mathbf{e}, \mathbf{f}$ be idempotents such that $\overline{e_j} \neq \overline{f_j}$ for some index $j = 1, 2, \ldots, m-1, -1, -2, \ldots, -(m-1)$ and $M(\mathbf{e}) \cong M(\mathbf{f})$. By symmetry we may assume that $\overline{e_j} = \overline{0}$ and $\overline{f_j} = \overline{1}$. Let $g \colon M(\mathbf{e}) \to M(\mathbf{f})$ be an isomorphism. By Lemma 8.4 the mapping g and its inverse mapping g^{-1} are given by multiplication by elements of R^m. Hence there exist invertible elements $u_1, \ldots, u_m \in R$ such that $g(x) = (u_1, \ldots, u_m)x$ for every $x \in M(\mathbf{e})$. By the Chinese Remainder Theorem there exists $c \in R$ such that $c \equiv 0 \pmod{P_j}$ and $c \equiv 1 \pmod{P_{-j}}$. Since $M(\mathbf{e}) = \{ (x_1, \ldots, x_m) \in R^m \mid \overline{x_i} = \overline{x_{i+1}} \text{ in } R/P_i$ for every $i = 1, 2, \ldots, m-1$ with $\overline{e_i} = \overline{1}$ and $\overline{x_i} = \overline{x_{i+1}}$ in R/P_{-i} for every $i = 1, 2, \ldots, m-1$ with $\overline{e_{-i}} = \overline{1} \}$, it is immediately seen that

$$y = (\underbrace{1, \ldots, 1}_{|j|}, \underbrace{c, \ldots, c}_{m-|j|}) \in M(\mathbf{e}).$$

Thus $g(y) = (u_1, \ldots, u_{|j|}, u_{|j|+1}c, \ldots, u_m c) \in M(\mathbf{f})$. In particular, we have $\overline{u_{|j|}} = \overline{u_{|j|+1}c}$ in R/P_j. But $c \equiv 0 \pmod{P_j}$, so that $\overline{u_{|j|}} = \overline{0}$ in R/P_j, a contradiction. $\qquad\square$

We now show that there exists a commutative noetherian ring S with four indecomposable finitely generated modules M_1, M_2, M_3, M_4 such that $M_1 \oplus M_2 \cong M_3 \oplus M_4$ and $M_i \not\cong M_j$ for every $i \neq j$, $i, j = 1, 2, 3, 4$.

Example 8.10 *Let R be a commutative principal ideal domain with at least four maximal ideals. Then there exists a module-finite commutative R-algebra S and indecomposable, pairwise nonisomorphic, finitely generated S-modules M_1, M_2, M_3, M_4 such that $M_1 \oplus M_2 \cong M_3 \oplus M_4$.*

Proof. Let S be the module-finite commutative R-algebra constructed in this section for $m = 3$. Let $\mathbf{e}_1, \mathbf{e}_2, \mathbf{e}_3, \mathbf{e}_4 \in \mathbf{S} = R/P_1 \times R/P_2 \times R/P_{-1} \times R/P_{-2}$ be the following idempotents: $\mathbf{e}_1 = (\overline{1}, \overline{1}, \overline{0}, \overline{0})$, $\mathbf{e}_2 = (\overline{0}, \overline{0}, \overline{1}, \overline{1})$, $\mathbf{e}_3 = (\overline{1}, \overline{0}, \overline{0}, \overline{1})$, $\mathbf{e}_4 = (\overline{0}, \overline{1}, \overline{1}, \overline{0})$. Put $M_i = M(\mathbf{e}_i)$ for $i = 1, 2, 3, 4$. By Corollary 8.7 both $M_1 \oplus M_2$ and $M_3 \oplus M_4$ are isomorphic to $S \oplus M(\mathbf{0})$. The modules M_i are indecomposable because the graphs $\mathrm{graph}(\mathbf{e}_i)$ are connected (Theorem 8.5) and they are pairwise nonisomorphic by Lemma 8.9. $\qquad\square$

Observe that the endomorphism ring $\mathrm{End}_S(M_1 \oplus M_2) \cong \mathrm{End}_S(M_3 \oplus M_4)$ in Example 8.10 is a finitely generated free R-module.

Lemma 8.11 *If $\mathbf{w}, \mathbf{v} \in \mathbf{S}$ and every coordinate of $\mathbf{w}$ is non-zero, then*

$$M(\mathbf{w}) \oplus M(\mathbf{v}) \quad \text{and} \quad S \oplus M(\mathbf{wv})$$

are isomorphic S-modules.

Proof. For every $i = 1, 2, \ldots, m - 1, -1, -2, \ldots, -(m-1)$ let

$$w_i \in R \setminus P_i \quad \text{and} \quad v_i \in R$$

be elements such that

$$\mathbf{w} = (\overline{w_1}, \ldots, \overline{w_{m-1}}, \overline{w_{-1}}, \ldots, \overline{w_{-(m-1)}})$$

and

$$\mathbf{v} = (\overline{v_1}, \ldots, \overline{v_{m-1}}, \overline{v_{-1}}, \ldots, \overline{v_{-(m-1)}}).$$

By the Chinese Remainder Theorem there exists $c \in R$ such that $c \equiv w_i \pmod{p_i}$ for every $i = 1, 2, \ldots, m-1, -1, -2, \ldots, -(m-1)$. In particular $c \notin P_i$ for every i, so that $c \neq 0$. Hence there are only finitely many maximal ideals $Q_1, \ldots, Q_t$ of R containing c, and $Q_j \neq P_i$ for every j and i. By the Chinese Remainder Theorem there exists $c' \in R$ such that $c' \equiv v_i \pmod{P_i}$ for every $i = 1, \ldots, m-1, -1, \ldots, -(m-1)$ and $c' \equiv 1 \pmod{Q_j}$ for every $j = 1, \ldots, t$. Then c and c' are relatively prime. Put $q = p_1 \ldots p_{m-1} p_{-1} \ldots p_{-(m-1)}$. Since p_i does not divide c for every i, it follows that c and q are relatively prime. Hence $(c')^{m+1-k}q$ and c^{m+1-k} are relatively prime for every $k = 1, 2, \ldots, m$, so that for every $k = 1, 2, \ldots, m$ there exist $d_k, d'_k \in R$ such that

$$d_k(c')^{m+1-k}q - d'_k c^{m+1-k} = 1. \tag{8.1}$$

Then

$$(c^m, c^{m-1}, \ldots, c^1) \in M(\mathbf{w}^{-1}),$$

because $c \equiv w_i \pmod{p_i}$ for every $i = 1, 2, \ldots, m-1, -1, -2, \ldots, -(m-1)$, so that

$$\overline{c^{m+1-k} w_k^{-1}} = \overline{c^{m-k}} \text{ in } R/P_k \text{ and } \overline{c^{m+1-k} w_{-k}^{-1}} = \overline{c^{m-k}} \text{ in } R/P_{-k}$$

for every $k = 1, 2, \ldots, m - 1$. As $(c^m, \ldots, c^1) \in M(\mathbf{w}^{-1})$, multiplication by $(c^m, \ldots, c^1)$ is an S-homomorphism $(c^m, \ldots, c^1) \colon M(\mathbf{w}) \to M(\mathbf{1}) = S$ and an S-homomorphism $(c^m, \ldots, c^1) \colon M(\mathbf{wv}) \to M(\mathbf{v})$ (Lemma 8.3), where we have denoted an element of R^m and multiplication by that element with the same symbol. Similarly, from Lemma 8.1 we know that $(qd_1, \ldots, qd_m) \in M(\mathbf{v})$, so that by Lemma 8.3 there are S-homomorphisms $(qd_1, \ldots, qd_m) \colon M(\mathbf{w}) \to M(\mathbf{wv})$ and $(qd_1, \ldots, qd_m) \colon M(\mathbf{1}) \to M(\mathbf{v})$. Elementary verifications analogous to these show that

$$((c')^m, \ldots, (c')^1) \colon M(\mathbf{v}) \to M(\mathbf{1}), \quad ((c')^m, \ldots, (c')^1) \colon M(\mathbf{wv}) \to M(\mathbf{w}),$$
$$(d'_1, \ldots, d'_m) \colon M(\mathbf{v}) \to M(\mathbf{wv}), \qquad (d'_1, \ldots, d'_m) \colon M(\mathbf{1}) \to M(\mathbf{w})$$

are S-module homomorphisms. Since the composite of the two mappings

$$\begin{pmatrix} (c^m, \ldots, c^1) & ((c')^m, \ldots, (c')^1) \\ (qd_1, \ldots, qd_m) & (d'_1, \ldots, d'_m) \end{pmatrix} \colon M(\mathbf{w}) \oplus M(\mathbf{v}) \to M(\mathbf{1}) \oplus M(\mathbf{wv})$$

and

$$\begin{pmatrix} -(d'_1, \ldots, d'_m) & ((c')^m, \ldots, (c')^1) \\ (qd_1, \ldots, qd_m) & -(c^m, \ldots, c^1) \end{pmatrix} \colon M(\mathbf{1}) \oplus M(\mathbf{wv}) \to M(\mathbf{w}) \oplus M(\mathbf{v})$$

are identities, it follows that $M(\mathbf{w}) \oplus M(\mathbf{v}) \cong M(\mathbf{1}) \oplus M(\mathbf{wv})$. $\square$

We conclude the section with three examples that show the failure of the cancellation property and the n-th root property for finitely generated modules over commutative noetherian rings.

Example 8.12 *There exists a commutative noetherian ring S and finitely generated S-modules M, N, P such that $M \oplus P \cong N \oplus P$ and $M \not\cong N$.*

Proof. Let R be the ring of integers, $m \geq 2$ an integer and

$$P_1, P_{-1}, P_2, P_{-2}, \ldots, P_{m-1}, P_{-(m-1)}$$

a sequence of $2(m-1)$ distinct maximal ideals of R generated by prime numbers $p_i \geq 5$. Let S be the ring constructed in this section, $\mathbf{w} = (\bar{2}, \bar{2}, \bar{2}, \ldots, \bar{2})$, $M = M(\mathbf{w})$, $N = S$ and $P = M(\mathbf{0})$. Since $2 \notin P_i$ for every i, every coordinate of $\mathbf{w}$ is non-zero. Hence by Lemma 8.11 we need only show that $M(\mathbf{w})$ and $S = M(\mathbf{1})$ are not isomorphic. Suppose the contrary, and let $f \colon M(\mathbf{1}) \to M(\mathbf{w})$ be an isomorphism. By Lemma 8.4 the mapping f and its inverse mapping f^{-1} are given by multiplication by elements of R^m. Hence there exist invertible elements $u_1, \ldots, u_m \in R$ such that $f(x) = (u_1, \ldots, u_m)x$ for every $x \in M(\mathbf{1}) = S$. In particular $f(1) = (u_1, \ldots, u_m) \in M(\mathbf{w})$. Since $\bar{2} \neq \bar{0}$, from the definition of $M(\mathbf{w})$ we obtain that $\overline{u_1 2} = \overline{u_2}$ in R/P_1. As R is the ring of integers, u_1, u_2 are ± 1, so that $\bar{2} = \pm \bar{1}$. This is a contradiction, because $P_1 = (p_1)$ and $p_1 \geq 5$, so that $2 \not\equiv \pm 1 \pmod{p_1}$. $\square$

Example 8.13 *Let $n \geq 2$ be an integer divisible by $p-1$ for at least two distinct prime numbers $p \geq 5$. Then there exist a commutative noetherian ring S and finitely generated S-modules M and N such that $M \not\cong N$ and $M^n \cong N^n$.*

Proof. Let R be the ring of integers and set $m = 2$. Let p_1, p_{-1} be two distinct prime numbers ≥ 5 such that $p_i - 1$ divides n for $i = 1$ and $i = -1$. Let P_1, P_{-1} be the maximal ideals of R generated by p_1, p_{-1} respectively, let S be the ring constructed in this section, let $M = M(\mathbf{w})$ and let $N = S = M(1)$ be the S-modules of Example 8.12. We have already seen that $M \not\cong N$ and that all the coordinates of $\mathbf{w}$ are non-zero. By Lemma 8.11 the modules $M(\mathbf{w}^j) \oplus M(\mathbf{w})$ and $S \oplus M(\mathbf{w}^{j+1})$ are isomorphic for every $j \geq 0$. Thus

$$M(\mathbf{w})^n \cong S^{n-1} \oplus M(\mathbf{w}^n).$$

Now for $i = \pm 1$ the multiplicative group $(R/P_i) \setminus \{\overline{0}\}$ has order $p_i - 1$, so that $\overline{2}^n = \overline{1}$ in R/P_i. Hence $\mathbf{w}^n = 1$ in $\mathbf{S}$, and thus $M(\mathbf{w})^n \cong S^n$. $\qquad \square$

Example 8.14 *There exists a commutative noetherian ring S and finitely generated S-modules P and Q such that $P \not\cong Q$ and $P^2 \cong Q^2$.*

Proof. Since $n = 16$ is divisible by $5 - 1$ and $17 - 1$, by Example 8.13 there exist S, M and N such that $M \not\cong N$ and $M^{16} \cong N^{16}$. Let t be the smallest non-negative integer such that $M^{2^t} \cong N^{2^t}$. Then $1 \leq t \leq 4$. If $P = M^{2^{t-1}}$ and $Q = N^{2^{t-1}}$, then P and Q have the required properties. $\qquad \square$

8.2 Krull-Schmidt fails for artinian modules

Let R be a noetherian commutative ring. As we have already said in Section 2.13 (page 73), if M is a simple R-module and $E(M)$ is the injective envelope of M, then $E(M)$ is an artinian module [Matlis]. If R is also semilocal, then a set $\{M_1, \ldots, M_n\}$ of representatives of the simple R-modules up to isomorphism is necessarily finite, so that the minimal injective cogenerator

$$E(\oplus_{i=1}^n M_i) = \oplus_{i=1}^n E(M_i)$$

is artinian. This property is used in the next lemma.

Recall that if S is an algebra over a commutative ring R and C is an R-module, then the left S-module structure on $\mathrm{Hom}_R(S, C)$ is defined by $(s\psi)(s') = \psi(s's)$ for every $s, s' \in S$, $\psi \in \mathrm{Hom}_R(S, C)$.

Lemma 8.15 *If S is a module-finite algebra over a semilocal commutative noetherian ring R, ${}_R C$ is the minimal injective cogenerator in R-Mod and ${}_S N_R$ is the S-R-bimodule ${}_S N_R = \mathrm{Hom}_R({}_R S, {}_R C)$, then N_R is artinian and ${}_S N$ is a cogenerator in S-Mod.*

Proof. Since S is a module-finite R-algebra, for some $n \geq 1$ there is an exact sequence of left R-modules $R^n \to {}_R S \to 0$. Applying the functor $\operatorname{Hom}_R(-, {}_R C)$ yields an exact sequence of right R-modules

$$0 \to \operatorname{Hom}_R({}_R S, {}_R C) \to \operatorname{Hom}_R(R^n, {}_R C),$$

so that $N_R = \operatorname{Hom}_R({}_R S, {}_R C)$ is isomorphic to a submodule of

$$C_R^n \cong \operatorname{Hom}_R(R^n, {}_R C).$$

Since R is a semilocal commutative noetherian ring, C_R and N_R are artinian.

To prove that ${}_S N$ is a cogenerator in S-Mod, let ${}_S B$ be an arbitrary left S-module and let x be a non-zero element of ${}_S B$. Since ${}_R C$ is a cogenerator in R-Mod, there is an R-homomorphism $\varphi \colon {}_R B \to {}_R C$ such that $\varphi(x) \neq 0$. If $f \colon {}_S B \to {}_S N$ is the homomorphism of left S-modules defined by

$$(f(b))(s) = \varphi(sb)$$

for every $s \in S$, $b \in B$, then $f(x) \neq 0$. $\qquad\square$

Lemmas 8.15 and 8.16 and Proposition 8.17 are preparatory results to prove that every module-finite algebra over a semilocal commutative noetherian ring is the endomorphism ring of an artinian module (Proposition 8.18).

Lemma 8.16 *Every module-finite algebra over a noetherian commutative ring is isomorphic to a subring of an artinian ring.*

Proof. Suppose the statement is false, and let R be a commutative noetherian ring and S be a module-finite R-algebra such that S is not isomorphic to any subring of any artinian ring. Since S is a noetherian ring, the set of the ideals I of S such that S/I is not isomorphic to any subring of an artinian ring has a maximal element K. Then $T = S/K$ is not isomorphic to any subring of an artinian ring, but for every proper homomorphic image of T such an isomorphism exists. Put $A = R/\operatorname{ann}_R T$, so that T is a module-finite A-algebra and we may suppose that $A \subseteq T$.

Since A is a noetherian commutative ring, A has finitely many minimal prime ideals. Let X be the complement of the union of the set of the minimal prime ideals of A. Then the quotient ring $T[X^{-1}]$ with respect to X is a module-finite $A[X^{-1}]$-algebra and $A[X^{-1}]$ is a commutative noetherian ring in which every prime ideal is maximal. Hence $A[X^{-1}]$ is an artinian ring, so that $T[X^{-1}]$ also is an artinian ring. In particular the canonical mapping $\varphi \colon T \to T[X^{-1}]$ is not injective, and thus its kernel $\ker(\varphi) = \{\, t \in T \mid xt = 0 \text{ for some } x \in X \,\}$ is a non-zero ideal of T. As T is a noetherian A-module, the ideal $\ker(\varphi)$ is a finitely generated A-module, so that there exists $y \in X$ such that $y \ker(\varphi) = 0$.

Since $y \in X \subseteq A \subseteq T$ is non-zero, the two-sided ideal yT of T is non-zero. Now $\ker(\varphi) \cap yT = 0$, because if $k = yt$ with $k \in \ker(\varphi)$ and $t \in T$, then

$yk = 0$, so that $y^2t = 0$. Hence $t \in \ker(\varphi)$, and so $yt = 0$. Therefore there is a canonical injective ring homomorphism $T \to T/\ker(\varphi) \times T/yT$. As there are ring embeddings of $T/\ker(\varphi)$ and T/yT into artinian rings, the same is true for T, a contradiction. □

Lemma 8.16 cannot be generalized to arbitrary noetherian rings, because there exist noetherian rings that are not embeddable into a right artinian ring [Dean and Stafford].

Proposition 8.17 *Let S be a subring of a right artinian ring A. Suppose that there exists a ring R and an S-R-bimodule $_SN_R$ such that N_R is artinian and $_SN$ cogenerates $_S(A/S)$. Then there exists a ring T and an artinian cyclic T-module M_T such that $S \cong \mathrm{End}(M_T)$.*

Proof. In this proof, whenever we consider morphisms of left modules, we write homomorphisms on the right. Let $_AH_R$ be the bimodule

$$_AH_R = \mathrm{Hom}_S(_SA_A, {_SN_R})$$

and $L = \{f \in H \mid (S)f = 0\}$, so that $_SL_R$ is a subbimodule of $_SH_R$. Set

$$T = \begin{pmatrix} A & H \\ 0 & R \end{pmatrix} \text{ and } I = \begin{pmatrix} 0 & L \\ 0 & R \end{pmatrix}.$$

Then T is a ring and I is a right ideal of T. We shall show that $M_T = T/I$ has the required properties.

In order to show that M_T is artinian, note that T contains the direct product $A \times R$ as a subring. Hence it is sufficient to show that M is artinian as a right $A \times R$-module. But the right $A \times R$-module $M = T/I$ is clearly isomorphic to $A \oplus H/L$. Hence it suffices to prove that A_A and $(H/L)_R$ are artinian. Now A is right artinian, and if we apply the functor $\mathrm{Hom}_S(-, {_SN_R})$ to the exact sequence $0 \to {_SS} \to {_SA} \to {_S(A/S)} \to 0$ we get the exact sequence of right R-modules

$$0 \to \mathrm{Hom}_S(_S(A/S), {_SN_R}) \to H_R \to N_R.$$

In this exact sequence the image of the first homomorphism is L, so that H/L is isomorphic to an R-submodule of N_R. Since N_R is artinian, it follows that $(H/L)_R$ is artinian, as desired.

Finally, we must prove that $S \cong \mathrm{End}(M_T)$. Since $M_T = T/I$ is cyclic, it is easily seen that $\mathrm{End}(M_T) \cong I'/I$, where $I' = \{t \in T \mid tI \subseteq I\}$ is the *idealizer* of I in T (that is, the greatest subring of T containing I in which I is a two-sided ideal). Let us prove that $I' = \begin{pmatrix} S & L \\ 0 & R \end{pmatrix}$. The inclusion $\supseteq$ is easily checked. Conversely, let $t = \begin{pmatrix} a & h \\ 0 & r \end{pmatrix}$ be an element of T that belongs to I'. Then $aL + hR \subseteq L$, so that $h \in L$ and $aL \subseteq L$. Suppose $a \notin S$. Since $_SN$ cogenerates $_S(A/S)$, there exists a morphism of left S-modules $g \colon A/S \to N$ such that $(a + S)g \neq 0$. Hence there exists an S-homomorphism $g' \colon A \to N$ such that $(a)g' \neq 0$ and $(S)g' = 0$. Thus $g' \in L$. But $(S)ag' = (Sa)g' \neq 0$, and thus $ag' \notin L$. Hence $aL \not\subseteq L$, a contradiction. This proves that $a \in S$.

Therefore $\mathrm{End}_T(T/I) \cong I'/I \cong S$. $\square$

From Lemmas 8.15 and 8.16 and Proposition 8.17 we obtain

Proposition 8.18 *If S is a module-finite algebra over a semilocal commutative noetherian ring, then there exists a ring T and an artinian cyclic T-module M_T such that $S \cong \mathrm{End}(M_T)$.* $\square$

In particular, every semilocal commutative noetherian ring S is isomorphic to the endomorphism ring of a suitable artinian module. If we take a semilocal commutative noetherian integral domain S that is not local we see that

Example 8.19 *There exists a ring T and an indecomposable artinian T-module M_T whose endomorphism ring $\mathrm{End}(M_T)$ is not local.* $\square$

[Crawley and Jónsson, p. 855] asked whether artinian modules have the exchange property. By Theorem 2.8 and Example 8.19 there exist artinian modules that do not have the exchange property. This result, due to [Camps and Menal], answers Crawley and Jónsson's question.

We are ready to show that Krull-Schmidt fails for artinian modules over noncommutative rings. This will be done in the two examples that follow. Recall that the Krull-Schmidt Theorem holds for right artinian modules over commutative or right noetherian rings (Proposition 2.63).

Example 8.20 *There exists a ring T and indecomposable, pairwise nonisomorphic, artinian right T-modules M_1, M_2, M_3, M_4 such that*

$$M_1 \oplus M_2 \cong M_3 \oplus M_4.$$

Proof. Let R be a semilocal commutative principal ideal domain with at least 4 maximal ideals. As we saw in Example 8.10, there exists a module-finite commutative R-algebra S and indecomposable, pairwise nonisomorphic, finitely generated S-modules N_1, N_2, N_3, N_4 such that $N = N_1 \oplus N_2 \cong N_3 \oplus N_4$. Then $E = \mathrm{End}(N_S)$ is also a module-finite R-algebra. Let Proj-E be the full subcategory of Mod-E whose objects are all projective right E-modules and Add(N_S) denote the full subcategory of Mod-S whose objects are all the modules isomorphic to direct summands of direct sums of copies of N. By Theorem 4.7 the functors $\mathrm{Hom}_S(N, -) \colon \mathrm{Mod}\text{-}S \to \mathrm{Mod}\text{-}E$ and $- \otimes_E N \colon \mathrm{Mod}\text{-}E \to \mathrm{Mod}\text{-}S$ induce an equivalence between Add(N_S) and Proj-E, and if $e \in E$ is idempotent, the direct summand eN of N_S corresponds to the projective E-module eE. Hence there are complete sets $\{e_1, e_2\}$ and $\{e_3, e_4\}$ of orthogonal idempotents of E with $e_1 E, e_2 E, e_3 E, e_4 E$ indecomposable, pairwise nonisomorphic E-modules. By Proposition 8.18 there exists a ring T and an artinian cyclic T-module M_T such that $E \cong \mathrm{End}(M_T)$, and there is a similar equivalence between $\mathrm{Add}(M_T)$ and Proj-E such that, in particular, if e is any idempotent element of E, the direct summand eM of M_T corresponds to the E-module eE. Therefore the T-modules $M_i = e_i M$, $i = 1, 2, 3, 4$, have the required properties. $\qquad\square$

Example 8.21 *Given an integer $n > 2$, there exists a ring T and an artinian right T-module M that is the direct sum of 2 indecomposable modules, and also the direct sum of 3 indecomposable modules, and $\dots$, and also the direct sum of n indecomposable modules.*

Proof. Let R be a semilocal commutative principal ideal domain with at least $2(n-2)$ maximal ideals. In Example 8.8 we saw that there exists a commutative module-finite R-algebra S and a finitely generated S-module N_S that is the direct sum of t indecomposable S-modules for every t with $2 \leq t \leq n$. Then $E = \mathrm{End}(N_S)$ is also a module-finite R-algebra and its identity is the direct sum of t orthogonal primitive idempotents for every t with $2 \leq t \leq n$. By Proposition 8.18 there is a ring T and an artinian T-module M_T such that $E \cong \mathrm{End}(M_T)$. Then M_T is the direct sum of t indecomposable T-modules for every t between 2 and n. $\qquad\square$

8.3 Notes on Chapter 8

The results in Section 8.1 are due to [Levy]. His examples are constructed over rings R more general than principal ideal domains. For instance, in Example 8.8 R can be any commutative integral domain with at least four maximal ideals. Our presentation has followed that given in Rowen's book [Rowen 88].

The Krull-Schmidt Theorem holds for finitely generated modules over complete local noetherian commutative rings [Swan 60, remark on page 566], but it can fail for finitely generated modules over arbitrary local noetherian commutative rings [Evans]. More generally, the Krull-Schmidt Theorem holds for finitely generated modules over Henselian local commutative rings ([Siddoway] and [Vámos 90]), where a local commutative ring R is said to be *Henselian* if for every ideal I of any finitely generated R-algebra Λ, idempotents lift from Λ/I to Λ [Azumaya 51]. For the existence and uniqueness of direct sum decompositions of finitely generated modules and finite rank torsion-free modules over valuation domains, see [Vámos 90] and the survey [Vámos 95].

A complete description of the direct sum decompositions of a finitely generated module over a Dedekind domain was determined by Kaplansky. In this case Krull-Schmidt also fails. For a nice presentation of these results see [Sharpe and Vámos, Section 6.3].

As shown in Example 8.12 the cancellation property does not hold for finitely generated modules over a commutative noetherian ring S. Another example of this, due to [Swan 62], is the following. Let

$$S = \mathbf{R}[X, Y, Z]/(X^2 + Y^2 + Z^2 - 1)$$

be the coordinate ring of the sphere, and let x, y, z denote the images of X, Y, Z in S. Let $f \colon S \oplus S \oplus S \to S$ be the S-module homomorphism defined by $f(s_1, s_2, s_3) = xs_1 + ys_2 + zs_3$ for every $(s_1, s_2, s_3) \in S \oplus S \oplus S$. Then f is an epimorphism because $f(x, y, z) = 1$. Hence $M = \ker(f)$ is a direct summand of $S \oplus S \oplus S$, and $S \oplus S \oplus S \cong M \oplus S$. It is possible to prove that $S \oplus S \not\cong M$ [Swan 62, Theorem 3]. Again, this shows that the cancellation property does not hold for finitely generated modules over commutative noetherian rings. In this setting there is an interesting result proved recently by [Brookfield]: let M, N, P be right modules over an arbitrary ring S, and suppose that $M \oplus P \cong N \oplus P$ and P is noetherian. Then there exists a positive integer n, two chains of submodules $0 = M_0 \subseteq M_1 \subseteq \cdots \subseteq M_n = M$ and $0 = N_0 \subseteq N_1 \subseteq \cdots \subseteq N_n = N$, and a permutation σ of $\{1, 2, \ldots, n\}$ such that $M_i/M_{i-1} \cong N_{\sigma(i)}/N_{\sigma(i)-1}$ for every $i = 1, 2, \ldots, n$.

Other results concerning the cancellation property and the n-th root property for finitely generated modules over suitable module-finite R-algebras, where R is a commutative ring, were proved by [Goodearl and Warfield 76]. In Examples 8.12 and 8.13 we have seen that there exist finitely generated modules over a module-finite $\mathbf{Z}$-algebra S, where $\mathbf{Z}$ is the ring of integers, for which the cancellation property and the n-th root property do not hold. But these two properties hold for finitely generated S-modules if S is a *locally module-finite algebra* over a commutative ring R which is von Neumann regular modulo its Jacobson radical $J(R)$ [Goodearl and Warfield 76] (An R-algebra S is a *locally module-finite algebra* if the localization S_M is a module-finite R_M-algebra for each maximal ideal M of R.) These results were later extended to module-finite algebras over other classes of commutative rings R (*LG rings*) by [Estes and Guralnick].

Further nice examples of rings with particular sets of idempotents were constructed by [Osofsky 70]. She proved that there exists a ring R that can be expressed as a finite direct sum of indecomposable left ideals but that contains an infinite set of primitive orthogonal idempotents. And there exists another ring S that can be expressed as a direct sum of n indecomposable left ideals for each integer $n \geq 0$ but that contains no infinite set of orthogonal idempotents.

Proposition 8.17 is due to Camps and Menal. It is stated as a remark on page 2092 of [Camps and Menal], and its proof is a generalization of the proof of [Camps and Menal, Proposition 3.4]. A complete proof was published in [Camps and Facchini, Proposition 1.3]. The other results presented in Section 8.2 are due to [Facchini, Herbera, Levy and Vámos]. Note that in particular Examples 8.20 and 8.21 answer the question posed by Krull in 1932 (see the

Preface of this book and Section 2.13). Apart from the applications given in this chapter, the construction in the proof of Proposition 8.17 has other applications. For instance, making use of that construction [Herbera and Shamsuddin, Example 10] proved that every ring that can be embedded in a local ring is isomorphic to the endomorphism ring of a local module.

For further examples of artinian modules for which the Krull-Schmidt Theorem fails see [Yakovlev].

Chapter 9

Biuniform Modules

9.1 First properties of biuniform modules

In this section we shall present some elementary results on biuniform modules. We shall study the structure of the endomorphism ring of a biuniform module (Theorem 9.1) and introduce the type of a biuniform module and its monogeny and epigeny classes.

Recall that a module is *biuniform* (p. 109) if it is uniform and couniform. In particular, every biuniform module is non-zero and is indecomposable. For example, if $f: C \to B$ is a non-zero homomorphism of a couniform module C into a uniform module B, then $f(C)$ is biuniform. A non-zero uniserial module over any ring is biuniform. If R is a local ring and M_R is an indecomposable injective module, every cyclic submodule of M_R is biuniform.

Theorem 9.1 *Let A_R be a biuniform module over an arbitrary ring R and let $E = \mathrm{End}(A_R)$ be its endomorphism ring. Let I be the subset of E whose elements are all the endomorphisms of A_R that are not injective, and K be the subset of E whose elements are all the endomorphisms of A_R that are not surjective. Then I and K are two-sided completely prime ideals of E, and every proper right ideal of E and every proper left ideal of E is contained either in I or in K. Moreover exactly one of the following two conditions hold:*

(a) *Either the ideals I and K are comparable, so that E is a local ring and $I \cup K$ is its maximal ideal, or*

(b) *I and K are not comparable, $J(E) = I \cap K$, and $E/J(E)$ is canonically isomorphic to the direct product of the two division rings E/I and E/K.*

Proof. The subset I of E is additively closed because A_R is uniform. Similarly K is additively closed because A_R is couniform. By Lemma 6.26 the subsets I and K of E are two-sided completely prime ideals.

Let L be any proper right or left ideal of E. The set $I \cup K$ is exactly the set of non-invertible elements of E, so that $L \subseteq I \cup K$. If there exist $x \in L \setminus I$ and $y \in L \setminus K$, then $x + y \in L$, $x \in K$, and $y \in I$. Hence $x + y \notin I$ and $x + y \notin K$. Thus $x + y \notin I \cup K$. But $x + y \in L$, a contradiction. This shows that L is contained either in I or in K. In particular, the unique maximal right ideals of E are at most I and K. Similarly, the unique maximal left ideals of E are at most I and K.

If I and K are comparable, then $I \cup K$ is the unique maximal right (and left) ideal of E and case (a) holds. If I and K are not comparable, then E has exactly two maximal right ideals I and K, so that $J(E) = I \cap K$, and there is a canonical injective ring homomorphism $E/J(E) \to E/I \times E/K$. But $I + K = E$, hence this ring homomorphism is surjective by the Chinese Remainder Theorem.   □

Note that we already knew that the endomorphism ring of a biuniform module is a semilocal ring of dual Goldie dimension ≤ 2 (Corollary 4.16). In particular, *biuniform modules cancel from direct sums* (Corollary 4.6).

Let A be a biuniform module. Then we have by Theorem 9.1 that either $\operatorname{End}(A)/J(\operatorname{End}(A))$ is a division ring (i.e., $\operatorname{End}(A)$ is a local ring) or $\operatorname{End}(A)/J(\operatorname{End}(A))$ is the direct product of exactly two division rings. A biuniform module is said to be *of type* 1 if its endomorphism ring is local, and *of type* 2 otherwise. For instance, if R is a right chain ring, the uniserial module R_R is of type 1. Further examples of uniserial modules of type 1 and 2 will be given in Sections 9.6 and 9.7.

Lemma 9.2 *Let A be a uniform module and let B be a couniform module over a ring R.*

(a) *If $f, g \colon A \to B$ are two homomorphisms, f is injective and not surjective, and g is surjective and not injective, then $f + g$ is an isomorphism.*

(b) *If $f_1, \ldots, f_n \colon A \to B$ are n homomorphisms and $f_1 + \cdots + f_n$ is an isomorphism, then either one of the f_i is an isomorphism or there exist two distinct indices $i, j = 1, 2, \ldots, n$ such that f_i is injective and not surjective, and f_j is surjective and not injective.*

Proof. (a) Since $\ker(f) \supseteq \ker(g) \cap \ker(f + g)$, if f is injective and g is not injective, then $f + g$ must be injective. Dually for surjectivity.

(b) Since $\ker(f_1 + \cdots + f_n) \supseteq \bigcap_i \ker(f_i)$ and $(f_1 + \cdots + f_n)(A) \subseteq \sum_i f_i(A)$, if $f_1 + \cdots + f_n$ is an isomorphism, then there exist i, j such that $\ker(f_i) = 0$ and $f_j(A) = A$. If f_i or f_j is an isomorphism, we are done. Otherwise $i \neq j$, and the conclusion follows easily.   □

If A and B are two modules, we say that A and B *belong to the same monogeny class*, and write $[A]_m = [B]_m$, if there exist a monomorphism $A \to B$ and a monomorphism $B \to A$. Similarly, we say that A and B *belong to the same epigeny class*, and write $[A]_e = [B]_e$, if there exist an epimorphism $A \to B$ and an epimorphism $B \to A$. Clearly, this defines two equivalence relations in the class of all right modules over a ring.

Proposition 9.3 *Let A be a uniform module and let B be a couniform module over a ring R. Then $A \cong B$ if and only if $[A]_m = [B]_m$ and $[A]_e = [B]_e$.*

Proof. Suppose that $[A]_m = [B]_m$ and $[A]_e = [B]_e$. Then there is a monomorphism $f: A \to B$ and an epimorphism $g: A \to B$. If one of these two homomorphisms is an isomorphism, then $A \cong B$. If neither f nor g is an isomorphism, then $f + g$ is an isomorphism by Lemma 9.2(a). Hence $A \cong B$ in any case. The converse is obvious. $\square$

From Proposition 9.3 it follows that to check that two biuniform modules A and B are isomorphic, we may separately check that they belong to the same monogeny and epigeny classes.

Lemma 9.4 *Let A be a module over a ring R and let B, C be biuniform R-modules such that $[A]_m = [B]_m$ and $[A]_e = [C]_e$. Then:*

(a) *A is biuniform;*

(b) *$A \oplus D \cong B \oplus C$ for some R-module D;*

(c) *the module D in (b) is unique up to isomorphism and is biuniform;*

(d) *if B and C are uniserial, then A and D are uniserial.*

Proof. (a) By hypothesis there exist two monomorphisms $\alpha_1: A \to B$ and $\alpha_2: B \to A$ and two epimorphisms $\beta_1: A \to C$ and $\beta_2: C \to A$. Since $B \neq 0$, it follows that $A \neq 0$, and $\dim(A) \leq \dim(B) = 1$. Hence A is uniform. Similarly, A is couniform, and so biuniform.

(b) By Lemma 9.2(a) at least one of the three mappings $\alpha_2\alpha_1$, $\beta_2\beta_1$ or $\gamma = \alpha_2\alpha_1 + \beta_2\beta_1$ is an isomorphism. If $\alpha_2\alpha_1: A \to B \to A$ is an isomorphism, then $\alpha_1: A \to B$ is an isomorphism by Lemma 6.26(b). Hence the module $D = C$ has the property required in (b). Similarly, if $\beta_2\beta_1$ is an isomorphism, then $\beta_1: A \to C$ is an isomorphism and $D = B$ has the required property.

Suppose γ is an isomorphism. The composite of the homomorphisms

$$\begin{pmatrix} \alpha_1 \\ \beta_1 \end{pmatrix} : A \to B \oplus C \quad \text{and} \quad \begin{pmatrix} \alpha_2 & \beta_2 \end{pmatrix} : B \oplus C \to A$$

is γ. Hence the composite

$$\left(\gamma^{-1} \begin{pmatrix} \alpha_2 & \beta_2 \end{pmatrix} \right) \circ \begin{pmatrix} \alpha_1 \\ \beta_1 \end{pmatrix} : A \to B \oplus C \to A$$

is the identity mapping of A, so that A is isomorphic to a direct summand of $B \oplus C$, and a module D with the required property exists.

(c) If D, D' are modules such that $A \oplus D \cong B \oplus C$ and $A \oplus D' \cong B \oplus C$, then $D \cong D'$ because biuniform modules cancel from direct sums. As A, B, C are biuniform, from $A \oplus D \cong B \oplus C$ it follows that D is biuniform as well.

(d) Suppose that B and C are uniserial. In order to prove that A and D are uniserial, it is sufficient to prove that every uniform submodule U of $B \oplus C$ is uniserial. Let $\pi_1 \colon B \oplus C \to B$ and $\pi_2 \colon B \oplus C \to C$ be the canonical projections. If U is a uniform submodule of $B \oplus C$, then from $U \cap \ker(\pi_1) \cap \ker(\pi_2) = 0$ it follows that either $U \cap \ker(\pi_1) = 0$ or $U \cap \ker(\pi_2) = 0$. Thus either the restriction of π_1 to U or the restriction of π_2 to U is a monomorphism. Hence U is isomorphic to a submodule of B or C. In both cases U is uniserial. □

9.2 Some technical lemmas

In this section we prove a series of technical results that will be used in the subsequent sections. The first proposition is the "two-dimensional analogue" of Lemma 2.7.

Proposition 9.5 *Let* $A, B, C_1, \ldots, C_n$ *($n \geq 2$) be modules. Suppose A is biuniform and $A \oplus B = C_1 \oplus \cdots \oplus C_n$. Then there are two distinct indices $i, j = 1, \ldots, n$ and a direct sum decomposition $A' \oplus B' = C_i \oplus C_j$ of $C_i \oplus C_j$ such that $A \cong A'$ and $B \cong B' \oplus (\oplus_{k \neq i,j} C_k)$.*

Proof. Let $\varepsilon_A, \pi_A, \varepsilon_B, \pi_B$ and ε_i, π_i ($i = 1, 2, \ldots, n$) be the embeddings and the canonical projections with respect to the two direct sum decompositions $A \oplus B$ and $C_1 \oplus \cdots \oplus C_n$. In the ring $E = \mathrm{End}(A)$ we have that

$$1_E = \pi_A \varepsilon_A = \pi_A \left(\sum_{i=1}^{n} \varepsilon_i \pi_i \right) \varepsilon_A = \sum_{i=1}^{n} \pi_A \varepsilon_i \pi_i \varepsilon_A.$$

By Lemma 9.2(b) either one of summands $\pi_A \varepsilon_i \pi_i \varepsilon_A$ is an isomorphism or there exist two distinct indices $i, j = 1, 2, \ldots, n$ such that $\pi_A \varepsilon_i \pi_i \varepsilon_A$ is injective and not surjective, and $\pi_A \varepsilon_j \pi_j \varepsilon_A$ is surjective and not injective.

If there exists an i such that $\pi_A \varepsilon_i \pi_i \varepsilon_A$ is an isomorphism, then the composite of $\pi_i \varepsilon_A \colon A \to C_i$ and $(\pi_A \varepsilon_i \pi_i \varepsilon_A)^{-1} \pi_A \varepsilon_i \colon C_i \to A$ is the identity mapping of A. Thus A is isomorphic to a direct summand A' of C_i, and the statement holds for every index $j \neq i$ because biuniform modules cancel from direct sums.

Thus we may suppose that there exist two distinct indices $i, j = 1, 2, \ldots, n$ such that $\pi_A \varepsilon_i \pi_i \varepsilon_A$ is injective and not surjective and $\pi_A \varepsilon_j \pi_j \varepsilon_A$ is surjective and not injective. Then $\gamma = \pi_A \varepsilon_i \pi_i \varepsilon_A + \pi_A \varepsilon_j \pi_j \varepsilon_A$ is an automorphism of A by Lemma 9.2(a). The composite of the homomorphisms

$$\begin{pmatrix} \pi_i \varepsilon_A \\ \pi_j \varepsilon_A \end{pmatrix} \colon A \to C_i \oplus C_j \quad \text{and} \quad \begin{pmatrix} \gamma^{-1} \pi_A \varepsilon_i & \gamma^{-1} \pi_A \varepsilon_j \end{pmatrix} \colon C_i \oplus C_j \to A$$

is the identity of A, so that A is isomorphic to a direct summand of $C_i \oplus C_j$. Let A', B' be such that $A \cong A'$ and $A' \oplus B' = C_i \oplus C_j$. Then $B' \oplus (\oplus_{k \neq i,j} C_k) \cong B$ because A cancels from direct sums. $\qquad \square$

We now consider the direct sum of an arbitrary (possibly infinite) family $\{ C_j \mid j \in J \}$ of uniform modules.

Proposition 9.6 *Suppose that $M = \oplus_{j \in J} C_j = A \oplus B$, where A and C_j are uniform modules for every $j \in J$. Let $\varepsilon_j : C_j \to M$, $\varepsilon_A : A \to M$, $\pi_j : M \to C_j$ and $\pi_A : M \to A$ be the embeddings and the canonical projections relative to these direct sum decompositions of M. Then there exists $k \in J$ such that $\pi_A \varepsilon_k \pi_k \varepsilon_A$ is a monomorphism. In particular, $[A]_m = [C_k]_m$.*

Proof. Fix a non-zero element $x \in A$. There exists a finite subset $F \subseteq J$ such that $x \in \oplus_{j \in F} C_j$, and if $D = \oplus_{j \in J \setminus F} C_j$, then $M = D \oplus (\oplus_{j \in F} C_j)$. Let $\varepsilon_D : D \to M$ and $\pi_D : M \to D$ denote the embedding and the canonical projection associated to this direct sum decomposition. Then

$$1_A = \pi_A \varepsilon_A = \pi_A \left(\varepsilon_D \pi_D + \left(\sum_{j \in F} \varepsilon_j \pi_j \right) \right) \varepsilon_A$$
$$= \pi_A \varepsilon_D \pi_D \varepsilon_A + \sum_{j \in F} \pi_A \varepsilon_j \pi_j \varepsilon_A,$$

so that

$$\ker(\pi_A \varepsilon_D \pi_D \varepsilon_A) \cap \left(\bigcap_{j \in F} \ker(\pi_A \varepsilon_j \pi_j \varepsilon_A) \right) = 0.$$

Since A is uniform, one of the kernels in this intersection must be zero, and $\ker(\pi_A \varepsilon_D \pi_D \varepsilon_A) \neq 0$ because it contains x. Hence there exists $k \in F$ such that $\ker(\pi_A \varepsilon_k \pi_k \varepsilon_A) = 0$, i.e., $\pi_A \varepsilon_k \pi_k \varepsilon_A$ is a monomorphism. By Lemma 6.26(a) both $\pi_A \varepsilon_k : C_k \to A$ and $\pi_k \varepsilon_A : A \to C_k$ are monomorphisms. Thus $[A]_m = [C_k]_m$. $\qquad \square$

Proposition 9.7 *Let $\{ C_1, C_2, \ldots, C_n \}$ be a finite family of couniform modules over an arbitrary ring R. Then for every couniform direct summand A of $\oplus_{j=1}^n C_j$ there exists $k \in J$ such that $[A]_e = [C_k]_e$.*

Proof. Put $M = \oplus_{j=1}^n C_j = A \oplus B$. Let $\varepsilon_j : C_j \to M$, $\varepsilon_A : A \to M$, $\pi_j : M \to C_j$ and $\pi_A : M \to A$ denote the embeddings and the projections relative to these direct sum decompositions of M. Then

$$1_A = \pi_A \varepsilon_A = \pi_A \left(\sum_{j=1}^n \varepsilon_j \pi_j \right) \varepsilon_A = \sum_{j-1}^n \pi_A \varepsilon_j \pi_j \varepsilon_A,$$

so that

$$\sum_{j=1}^n \pi_A \varepsilon_j \pi_j \varepsilon_A(A) = A.$$

But A is couniform, so there exists $k = 1, 2, \ldots, n$ such that $\pi_A \varepsilon_k \pi_k \varepsilon_A(A) = A$. Since $\pi_A \varepsilon_k \pi_k \varepsilon_A$ is an epimorphism, by Lemma 6.26(b) both $\pi_A \varepsilon_k : C_k \to A$ and $\pi_k \varepsilon_A : A \to C_k$ are epimorphisms. Thus $[A]_e = [C_k]_e$. $\qquad \square$

The next lemma shows that if A, B, C, D are biuniform modules with $A \oplus D \cong B \oplus C$, then the monogeny (respectively, epigeny) classes of A and D coincide with those of B and C up to the order.

Lemma 9.8 *Let A, B, C, D be biuniform modules such that $A \oplus D \cong B \oplus C$. Then $\{[A]_m, [D]_m\} = \{[B]_m, [C]_m\}$ and $\{[A]_e, [D]_e\} = \{[B]_e, [C]_e\}$.*

Proof. By symmetry, it is sufficient to prove that

$$\{[A]_m, [D]_m\} \subseteq \{[B]_m, [C]_m\} \quad \text{and} \quad \{[A]_e, [D]_e\} \subseteq \{[B]_e, [C]_e\}.$$

Hence it suffices to show that if X is a biuniform direct summand of $B \oplus C$, then $[X]_m \in \{[B]_m, [C]_m\}$ and $[X]_e \in \{[B]_e, [C]_e\}$. This is proved in Propositions 9.6 and 9.7. $\qquad\square$

We conclude the section with two more technical results.

Proposition 9.9 *Let R be an arbitrary ring, let $\{C_j \mid j \in J\}$ be a family of biuniform R-modules and let $A_1, A_2, \ldots, A_n$ be uniform R-modules. If*

$$A_1 \oplus \cdots \oplus A_n$$

is a direct summand of $\oplus_{j \in J} C_j$, there exist n distinct elements $k_1, \ldots, k_n \in J$ such that $[A_i]_m = [C_{k_i}]_m$ for every $i = 1, 2, \ldots, n$.

Proof. We may suppose $[A_1]_m = \cdots = [A_n]_m$, because the general case, with the $[A_i]_m$ possibly different, follows immediately from this. We shall argue by induction on n. The case $n = 1$ is given by Proposition 9.6. Fix $n > 1$ and suppose the statement is true for $n - 1$ uniform modules in the same monogeny class. Let $\{C_j \mid j \in J\}$ be a family of biuniform modules and let $A_1, \ldots, A_n$ be n uniform modules such that $[A_1]_m = \cdots = [A_n]_m$ and $A_1 \oplus \cdots \oplus A_n$ is a direct summand of $\oplus_{j \in J} C_j$. By Proposition 9.6 there exists $k_1 \in J$ such that $[A_1]_m = [C_{k_1}]_m$. Put $M = \oplus_{j \in J} C_j = A_1 \oplus \cdots \oplus A_n \oplus D$. Apply Proposition 9.5 to the biuniform direct summand C_{k_1} of the finite direct sum $M = A_1 \oplus \cdots \oplus A_n \oplus D$. There are two possible cases.

In the first case there are two distinct indices $i, j = 1, \ldots, n$ and a direct sum decomposition $C' \oplus B' = A_i \oplus A_j$ of $A_i \oplus A_j$ such that

$$C_{k_1} \cong C' \quad \text{and} \quad \oplus_{\ell \neq k_1} C_\ell \cong B' \oplus (\oplus_{t \neq i,j} A_t) \oplus D. \tag{9.1}$$

Since A_i, A_j and C' are uniform, B' must also be uniform. From Proposition 9.6 we obtain that $[B']_m = [A_i]_m = [A_j]_m$. By the induction hypothesis applied to the decompositions (9.1) and the $n - 1$ uniform modules in the same monogeny class B' and A_t with $t \neq i, j$, there exist $n - 1$ distinct elements

$$k_t \in J \setminus \{k_1\}, \quad t = 2, \ldots, n,$$

with $[C_{k_t}]_m = [A_1]_m$ for every $t = 2, \ldots, n$. Thus the monogeny classes of the n biuniform modules C_{k_t}, $t = 1, \ldots, n$, coincide with the monogeny class of A_1.

In the second case there is an index $i = 1, \ldots, n$ and a direct sum decomposition $C' \oplus B' = A_i \oplus D$ of $A_i \oplus D$ such that

$$C_{k_1} \cong C' \quad \text{and} \quad \oplus_{\ell \neq k_1} C_\ell \cong B' \oplus (\oplus_{t \neq i} A_t).$$

Now the conclusion follows immediately from the induction hypothesis. $\qquad \square$

Proposition 9.10 *Let $A_1, \ldots, A_n, C_1, \ldots, C_m$ be biuniform right modules over an arbitrary ring R. If $A_1 \oplus \cdots \oplus A_n$ is isomorphic to a direct summand of $C_1 \oplus \cdots \oplus C_m$, then there exist n distinct indices $k_1, \ldots, k_n \in \{1, 2, \ldots, m\}$ such that $[A_i]_e = [C_{k_i}]_e$ for every $i = 1, \ldots, n$.*

Proof. Suppose that $A_1 \oplus \cdots \oplus A_n \oplus B \cong C_1 \oplus \cdots \oplus C_m$. The proof is by induction on m. The cases $m = 0$ and $m = 1$ are trivial. Suppose that $m \geq 2$. By Propositions 9.6 and 9.7, there exist $i, j \in \{1, \ldots, m\}$ such that $[A_1]_m = [C_i]_m$ and $[A_1]_e = [C_j]_e$.

If $i = j$, then C_i is isomorphic to A_1 (Proposition 9.3). Then we can renumber the C_k's so that $A_1 \cong C_1$. Since biuniform modules cancel from direct sums, $A_2 \oplus \cdots \oplus A_n \oplus B \cong C_2 \oplus \cdots \oplus C_m$, and we conclude by the induction hypothesis.

If $i \neq j$, we can renumber the C_k's so that

$$[A_1]_m = [C_1]_m \quad \text{and} \quad [A_1]_e = [C_2]_e.$$

By Lemma 9.4 there is a biuniform module D such that $A_1 \oplus D \cong C_1 \oplus C_2$, and from Lemma 9.8 it follows that $[D]_e = [C_1]_e$. Then

$$A_1 \oplus \cdots \oplus A_n \oplus B \cong C_1 \oplus \cdots \oplus C_m \cong A_1 \oplus D \oplus C_3 \oplus \cdots \oplus C_m.$$

If we cancel the biuniform module A_1 we obtain that

$$A_2 \oplus \cdots \oplus A_n \oplus B \cong D \oplus C_3 \oplus \cdots \oplus C_m.$$

Since $[D]_e = [C_1]_e$, by the induction hypothesis there exist $n-1$ distinct indices $k_2, \ldots, k_n \in \{1, 3, 4, \ldots, m\}$ such that $[A_i]_e = [C_{k_i}]_e$ for every $i = 2, 3, \ldots, n$. Set $k_1 = 2$. Then $[A_i]_e = [C_{k_i}]_e$ for every $i = 1, \ldots, n$. $\qquad \square$

9.3   A sufficient condition

The aim of this section is to prove the following theorem.

Theorem 9.11 *Let $\{A_i \mid i \in I\}$ be an arbitrary family of modules over a ring R and let $\{B_j \mid j \in J\}$ be a family of biuniform R-modules. Assume that there exist two bijections $\sigma, \tau\colon I \to J$ such that $[A_i]_m = [B_{\sigma(i)}]_m$ and $[A_i]_e = [B_{\tau(i)}]_e$ for every $i \in I$. Then all the modules A_i are biuniform and*

$$\oplus_{i \in I} A_i \cong \oplus_{j \in J} B_j.$$

Proof. From $[A_i]_m = [B_{\sigma(i)}]_m$ and $[A_i]_e = [B_{\tau(i)}]_e$, it follows that the module A_i is non-zero, isomorphic to a submodule of $B_{\sigma(i)}$ and a homomorphic image of $B_{\tau(i)}$. Hence every A_i is biuniform. Let S_I be the symmetric group on the set I, that is, the group whose elements are all bijections $\alpha\colon I \to I$. The mapping $S_I \times I \to I$, $(\alpha, i) \mapsto \alpha(i)$, defines a natural action of S_I on the set I. Let $C = \{(\tau^{-1}\sigma)^z \mid z \in \mathbf{Z}\}$ be the cyclic subgroup of S_I generated by $\tau^{-1}\sigma$. Then the action of S_I on I restricts to an action of C on I. For every element $i \in I$ let $Ci = \{(\tau^{-1}\sigma)^z(i) \mid z \in \mathbf{Z}\}$ be the C-orbit of i and let $\sigma(Ci) \subseteq J$ be the image of Ci via the bijection $\sigma\colon I \to J$.

We claim that

$$\oplus_{k \in Ci} A_k \cong \oplus_{\ell \in \sigma(Ci)} B_\ell \tag{9.2}$$

for every $i \in I$. If we prove the claim, we are done, because the set

$$\mathcal{F} = \{Ci \mid i \in I\}$$

is a partition of I, so that its image $\mathcal{G} = \{\sigma(Ci) \mid i \in I\}$ via the bijection $\sigma\colon I \to J$ is a partition of J. Hence the conclusion follows immediately from (9.2).

In order to prove the claim, fix an index $i \in I$. For simplicity of notation, for every $z \in \mathbf{Z}$ define $i_z = (\tau^{-1}\sigma)^z(i)$, $j_z = \sigma(i_z)$, $A_z = A_{i_z}$ and $B_z = B_{j_z}$. Thus if the orbit $Ci = \{i_z \mid z \in \mathbf{Z}\}$ is infinite, then $\sigma(Ci) = \{j_z \mid z \in \mathbf{Z}\}$ is infinite, and $A_z = A_w$ if and only if $z = w$. Whereas if the orbit $Ci = \{i_z \mid z \in \mathbf{Z}\}$ is a finite set with q elements, then $A_z = A_w$ if and only if $z \equiv w \pmod{q}$. For every $z \in \mathbf{Z}$ we have

$$\tau(i_z) = \tau(\tau^{-1}\sigma)^z(i) = \sigma(\tau^{-1}\sigma)^{z-1}(i) = \sigma(i_{z-1}) = j_{z-1}.$$

Hence from the hypothesis $[A_k]_m = [B_{\sigma(k)}]_m$ and $[A_k]_e = [B_{\tau(k)}]_e$ for every $k \in I$ we have

$$[A_z]_m = [B_z]_m \quad \text{and} \quad [A_z]_e = [B_{z-1}]_e \tag{9.3}$$

for every $z \in \mathbf{Z}$.

We shall argue by induction on the integer $n \geq 0$ and show that for every $n \geq 0$ there exist biuniform modules C_n, D_n satisfying the following properties:

(a) $[C_n]_m = [A_{-n-1}]_m$ and $[C_n]_e = [A_{n+1}]_e$ for every $n \geq 0$;

(b) $C_n \oplus D_n \cong A_{n+1} \oplus A_{-n-1}$ for every $n \geq 0$;

(c) $B_0 \oplus B_{-1} \cong A_0 \oplus C_0$ and $B_n \oplus B_{-n-1} \cong C_n \oplus D_{n-1}$ for every $n \geq 1$.

Since $[A_0]_m = [B_0]_m$ and $[A_0]_e = [B_{-1}]_e$, by Lemma 9.4 there is a biuniform module C_0 such that $A_0 \oplus C_0 \cong B_0 \oplus B_{-1}$. Thus C_0 satisfies property (c), and from Lemma 9.8 we have that $[C_0]_m = [B_{-1}]_m$ and $[C_0]_e = [B_0]_e$. Hence $[C_0]_m = [A_{-1}]_m$ and $[C_0]_e = [A_1]_e$ because of (9.3), that is, property (a) is satisfied. By Lemma 9.4 there exists a biuniform module D_0 such that $C_0 \oplus D_0 \cong A_1 \oplus A_{-1}$, i.e., D_0 satisfies property (b) as well.

Fix an integer $n \geq 1$ and suppose that there exist C_{n-1}, D_{n-1} satisfying properties (a) and (b), i.e., such that $[C_{n-1}]_m = [A_{-n}]_m$, $[C_{n-1}]_e = [A_n]_e$ and $C_{n-1} \oplus D_{n-1} \cong A_n \oplus A_{-n}$. From Lemma 9.8 we obtain $[D_{n-1}]_m = [A_n]_m$ and $[D_{n-1}]_e = [A_{-n}]_e$. From (9.3) we get that

$$[D_{n-1}]_m = [B_n]_m \quad \text{and} \quad [D_{n-1}]_e = [B_{-n-1}]_e.$$

Hence by Lemma 9.4 there exists a biuniform module C_n such that

$$D_{n-1} \oplus C_n \cong B_n \oplus B_{-n-1},$$

that is, (c) holds. From Lemma 9.8 it follows that $[C_n]_m = [B_{-n-1}]_m$ and $[C_n]_e = [B_n]_e$. Thus $[C_n]_m = [A_{-n-1}]_m$ and $[C_n]_e = [A_{n+1}]_e$ by (9.3), i.e., property (a) holds. By Lemma 9.4 there exists a biuniform module D_n such that $C_n \oplus D_n \cong A_{n+1} \oplus A_{-n-1}$. This shows that (b) holds and completes the construction of the modules C_n and D_n.

Now we shall prove the claim (9.2) distinguishing the following cases: the orbit Ci is infinite, or finite with an even number of elements, or finite with one element, or finite with an odd number $q \geq 3$ of elements.

If the orbit Ci is infinite, then

$$
\begin{aligned}
\oplus_{k \in Ci} A_k = \oplus_{z \in \mathbf{Z}} A_z &= A_0 \oplus (\oplus_{n \geq 0}(A_{n+1} \oplus A_{-n-1})) \\
&\cong A_0 \oplus (\oplus_{n \geq 0}(C_n \oplus D_n)) = A_0 \oplus C_0 \oplus (\oplus_{n \geq 1}(C_n \oplus D_{n-1})) \\
&\cong B_0 \oplus B_{-1} \oplus (\oplus_{n \geq 1}(B_n \oplus B_{-n-1})) = \oplus_{z \in \mathbf{Z}} B_z = \oplus_{\ell \in \sigma(Ci)} B_\ell.
\end{aligned}
$$

If the orbit Ci is a finite set with an even number $q = 2r$ of elements, where $r \geq 1$, then $-r \equiv r \pmod q$, so that $A_{-r} = A_r$. From (a) we have that $[C_{r-1}]_m = [A_{-r}]_m = [A_r]_m$ and $[C_{r-1}]_e = [A_r]_e$. Hence $C_{r-1} \cong A_r$ by Proposition 9.3. Thus

$$
\begin{aligned}
\oplus_{k \in Ci} A_k &= A_0 \oplus (\oplus_{n=1}^{r-1}(A_n \oplus A_{-n})) \oplus A_r \\
&\cong A_0 \oplus (\oplus_{n=1}^{r-1}(C_{n-1} \oplus D_{n-1})) \oplus C_{r-1} \\
&= A_0 \oplus C_0 \oplus (\oplus_{n=1}^{r-1}(C_n \oplus D_{n-1})) \\
&\cong B_0 \oplus B_{-1} \oplus (\oplus_{n=1}^{r-1}(B_n \oplus B_{-n-1})) = \oplus_{\ell \in \sigma(Ci)} B_\ell.
\end{aligned}
$$

If the orbit Ci has exactly one element, then $0 \equiv -1 \pmod{1}$ forces $B_0 = B_{-1}$, so that $[A_0]_m = [B_0]_m$ and $[A_0]_e = [B_{-1}]_e = [B_0]_e$ by (9.3). Hence $A_0 \cong B_0$ by Proposition 9.3 as desired.

If the orbit Ci is a finite set with an odd number $q = 2r + 1$ of elements ($r \geq 1$), then $-r - 1 \equiv r \pmod{q}$, so that $B_{-r-1} = B_r$. Hence

$$C_r \oplus D_{r-1} \cong B_r \oplus B_{-r-1} = B_r \oplus B_r.$$

It follows that $[D_{r-1}]_m = [B_r]_m$ and $[D_{r-1}]_e = [B_r]_e$ (Lemma 9.8). Hence $D_{r-1} \cong B_r$ (Proposition 9.3). Therefore

$$
\begin{aligned}
\oplus_{k \in Ci} A_k &= A_0 \oplus (\oplus_{n=1}^{r}(A_n \oplus A_{-n})) \\
&\cong A_0 \oplus (\oplus_{n=1}^{r}(C_{n-1} \oplus D_{n-1})) \\
&= A_0 \oplus C_0 \oplus (\oplus_{n=1}^{r-1}(C_n \oplus D_{n-1})) \oplus D_{r-1} \\
&\cong B_0 \oplus B_{-1} \oplus (\oplus_{n=1}^{r-1}(B_n \oplus B_{-n-1})) \oplus B_r = \oplus_{\ell \in \sigma(Ci)} B_\ell.
\end{aligned}
$$

This concludes the proof. $\square$

9.4 Weak Krull-Schmidt Theorem for biuniform modules

In this section we shall try to invert the implication proved in Theorem 9.11. Half of that implication can be inverted in general, as the next theorem shows.

Theorem 9.12 Let $\{\, U_i \mid i \in I \,\}$, $\{\, V_j \mid j \in J \,\}$ be two families of biuniform right modules over an arbitrary ring R such that $\oplus_{i \in I} U_i \cong \oplus_{j \in J} V_j$. Then there exists a bijection $\sigma \colon I \to J$ such that $[U_i]_m = [V_{\sigma(i)}]_m$ for every $i \in I$.

Proof. Without loss of generality we may suppose

$$M = \oplus_{i \in I} U_i = \oplus_{j \in J} V_j,$$

where the U_i, V_j are biuniform.

For every uniform direct summand Z of M put

$$I_Z = \{\, i \in I \mid [U_i]_m = [Z]_m \,\} \quad \text{and} \quad J_Z = \{\, j \in J \mid [V_j]_m = [Z]_m \,\}.$$

By Proposition 9.6 the sets I_Z and J_Z are non-empty. It is obvious that the I_Z form a partition of I and the J_Z form a partition of J. In order to prove the theorem it suffices to show that the cardinalities $|I_Z|$ of I_Z and $|J_Z|$ of J_Z are equal for every uniform direct summand Z of M.

Suppose that either I_Z or J_Z is a finite set and that $|I_Z| \neq |J_Z|$. By symmetry we may assume $|I_Z| < |J_Z|$. Put $n = |I_Z|$. Then in J_Z there are $n + 1$ indices $j_1, \ldots, j_{n+1}$ with $[C_{j_t}]_m = [Z]_m$, so that the same holds in I_Z by Proposition 9.9, contradiction. Hence $|I_Z| = |J_Z|$ if either I_Z or J_Z is finite.

Suppose that I_Z and J_Z are both infinite. By symmetry, we only have to prove that $|J_Z| \le |I_Z|$. Let $\varepsilon_k \colon U_k \to M$, $e_\ell \colon V_\ell \to M$, $\pi_k \colon M \to U_k$ and $p_\ell \colon M \to V_\ell$ denote the embeddings and the canonical projections. For every $t \in I$ put $A(t) = \{\, \ell \in J \mid \pi_t e_\ell p_\ell \varepsilon_t \text{ is a monomorphism} \,\}$.

For each $t \in I$ the subset $A(t)$ of J is finite, because if $x \in U_t$ is non-zero, there is a finite subset F of J such that $x \in \oplus_{j \in F} V_j$, and then $\pi_t e_\ell p_\ell \varepsilon_t(x) = 0$ for every $\ell \in J \setminus F$.

Now

$$J_Z \subseteq \bigcup_{t \in I_Z} A(t), \qquad (9.4)$$

because if $j \in J_Z$, there is an index $t \in I$ such that $p_j \varepsilon_t \pi_t e_j$ is a monomorphism and $[U_t]_m = [V_j]_m = [Z]_m$ (Proposition 9.6). Hence both $p_j \varepsilon_t$ and $\pi_t e_j$ are monomorphisms (Lemma 6.26(a)), so that $\pi_t e_j p_j \varepsilon_t$ is a monomorphism, that is, $j \in A(t)$. As $[U_t]_m = [Z]_m$, we have that $t \in I_Z$. This proves (9.4).

From (9.4) it follows that $|J_Z| \le \aleph_0 |I_Z| = |I_Z|$. Hence $|I_Z| = |J_Z|$ also in the case where I_Z and J_Z are both infinite. $\qquad \square$

We shall now prove a weak form of the Krull-Schmidt Theorem for finite direct sums of biuniform modules.

Theorem 9.13 (Weak Krull-Schmidt Theorem for biuniform modules) *Let U_1, ..., U_n, V_1, ..., V_t be biuniform modules over an arbitrary ring. Then the direct sums $U_1 \oplus \cdots \oplus U_n$ and $V_1 \oplus \cdots \oplus V_t$ are isomorphic if and only if $n = t$ and there are two permutations σ, τ of $\{1, 2, \ldots, n\}$ such that $[U_i]_m = [V_{\sigma(i)}]_m$ and $[U_i]_e = [V_{\tau(i)}]_e$ for every $i = 1, 2, \ldots, n$.*

Proof. If $n = t$ and there are two permutations σ, τ of $\{1, 2, \ldots, n\}$ that preserve the monogeny classes and the epigeny classes, then

$$U_1 \oplus \cdots \oplus U_n \cong V_1 \oplus \cdots \oplus V_n$$

by Theorem 9.11.

Conversely, suppose that $U_1 \oplus \cdots \oplus U_n \cong V_1 \oplus \cdots \oplus V_t$. By Theorem 9.12 there exists a bijection $\sigma \colon \{1, 2, \ldots, n\} \to \{1, 2, \ldots, t\}$ such that $[U_i]_m = [V_{\sigma(i)}]_m$ for every $i = 1, 2, \ldots, n$. In particular, $n = t$. By Proposition 9.10 there is an injective mapping $\tau \colon \{1, 2, \ldots, n\} \to \{1, 2, \ldots, t\}$ such that $[U_i]_e = [V_{\tau(i)}]_e$ for every $i = 1, \ldots, n$. But $n = t$ and τ injective imply that τ is a permutation. $\qquad \square$

It will follow from Example 9.20 that for any two permutations σ, τ of $\{1, 2, \ldots, n\}$, there is a suitable serial ring R and $2n$ finitely presented uniserial R-modules $U_1, \ldots, U_n, V_1, \ldots, V_n$ such that $[U_i]_m = [V_{\sigma(i)}]_m$ and $[U_i]_e = [V_{\tau(i)}]_e$ for every $i = 1, 2, \ldots, n$. Thus if a module M is a finite direct sum of n biuniform modules, then the isomorphism classes of the biuniform direct summands may depend on the decomposition. This proves that Theorem 9.13 cannot be improved even if the base ring R is serial and the modules in question are finitely presented and uniserial.

Definition 9.14 *A biuniform R-module A is called a* Krull-Schmidt module *if either*

(a) *for every biuniform R-module B, $[A]_m = [B]_m$ implies $A \cong B$, or*

(b) *for every biuniform R-module B, $[A]_e = [B]_e$ implies $A \cong B$.*

Thus a biuniform module is a Krull-Schmidt module if and only if it is the only biuniform module up to isomorphism in its monogeny class or the only biuniform module up to isomorphism in its epigeny class.

Corollary 9.15 *A biuniform module A is a Krull-Schmidt module if and only if whenever $B_1, \ldots, B_n$ are biuniform modules and A is isomorphic to a direct summand of $B_1 \oplus \cdots \oplus B_n$, then A is isomorphic to B_i for some $i = 1, 2, \ldots, n$.*

Proof. If A is a biuniform Krull-Schmidt module, $B_1, \ldots, B_n$ are biuniform modules and A is isomorphic to a direct summand of $U_1 \oplus \cdots \oplus U_n$, then there exist $i, j = 1, 2, \ldots, n$ such that $[A]_m = [B_i]_m$ and $[A]_e = [B_j]_e$ (Propositions 9.6 and 9.7). But A is Krull-Schmidt, so that $A \cong B_i$ or $A \cong B_j$.

Conversely, if a biuniform module A is not a Krull-Schmidt module, then there is a biuniform module $B \not\cong A$ with $[B]_m = [A]_m$ and a biuniform module $C \not\cong A$ with $[C]_e = [A]_e$. By Lemma 9.4 the module A is isomorphic to a direct summand of $B \oplus C$. $\qquad\qquad\square$

From Lemma 2.7, Theorem 2.8 and Corollary 9.15 we obtain

Proposition 9.16 *Every biuniform module of type 1 is a Krull-Schmidt module.*
$$\square$$

In Example 9.26 we shall show that the uniserial cyclic module constructed in Example 1.16 is a Krull-Schmidt module of type 2.

9.5   Krull-Schmidt holds for finitely presented modules over chain rings

In Theorem 9.13 we proved a weak form of the Krull-Schmidt Theorem for finite direct sums of biuniform modules. In particular, Theorem 9.13 can be applied to finitely presented modules over serial rings, because every finitely presented module over a serial ring is a finite direct sum of uniserial modules (Corollary 3.30). The aim of this section is to prove that the Krull-Schmidt Theorem holds for finitely presented modules over chain rings. In the next section we shall see that Krull-Schmidt fails for finitely presented modules over serial rings.

The results in this section are due to G. Puninski.

Lemma 9.17 *Let R be a right chain ring and let $r, s \in R$. Then $R/rR \cong R/sR$ if and only if there are invertible elements $u, v \in R$ such that $urv = s$.*

Proof. If $urv = s$ with u, v invertible elements, then left multiplication by u induces an isomorphism of R/rR onto R/urR. But $urR = sv^{-1}R = sR$.

Conversely, let $f : R/rR \to R/sR$ be an isomorphism. Then there exists a homomorphism $g : R_R \to R_R$ that induces f, so that $g(R) + sR = R$ and $g(rR) = sR$. Hence there is an element $u \in R$ such that $uR + sR = R$ and $urR = sR$. If u is not invertible, then $sR = R$, so that $rR = R$, i.e., both s and r are invertible, and the result follows easily. Hence we may suppose that u is invertible. Then $urR = sR$ forces the existence of two elements $a, b \in R$ such that $ur = sa$ and $s = urb$. If a is invertible, then u and $v = a^{-1}$ are the required elements. If a is not invertible, then a is in the Jacobson radical, so that $1 - ab$ is invertible. But $s = urb = sab$, so $s(1 - ab) = 0$, and thus $s = 0$. Therefore $r = 0$, and in this case the result holds for any invertible u, v. $\square$

Lemma 9.18 *Let r, s be two elements of a chain ring R such that R/rR and R/rsR are not isomorphic. Then $Rrsv \subset Rr$ for every invertible element $v \in R$.*

Proof. Let r, s, v be elements of a chain ring R such that v is invertible and $Rrsv$ is not properly contained in Rr. Since R is a chain ring, $Rrsv \supseteq Rr$, so that $r = ursv$ for some $u \in R$. If u is invertible, then $R/rR \cong R/rsR$ by Lemma 9.17. If s is invertible, then $rsR = rR$, so that $R/rR = R/rsR$. If neither u nor s are invertible, then both $u + 1$ and $1 + sv$ are invertible, and $r(1 + sv) = r + rsv = ursv + rsv = (u + 1)rsv$. Hence $r = (u + 1)rsv(1 + sv)^{-1}$, so that $R/rR \cong R/rsR$ by Lemma 9.17. Therefore $R/rR \cong R/rsR$ in any case. $\square$

We are ready to prove the main result of this section, that is, the Krull-Schmidt Theorem for finitely presented modules over a chain ring.

Theorem 9.19 (Puninski) *Let M_R be a finitely presented module over a chain ring R. Then M_R is the direct sum of cyclic modules, and any two decompositions of M_R as direct sums of cyclic modules are isomorphic.*

Proof. The finitely presented module M_R is a direct sum of uniserial modules by Corollary 3.30, and every finitely presented uniserial module is cyclic. Suppose that $M_R = U_1 \oplus \cdots \oplus U_n = V_1 \oplus \cdots \oplus V_t$ for suitable cyclic modules $U_1, \ldots, U_n, V_1, \ldots, V_t$. From Theorem 9.13 it follows that $n = t$ and there is a permutation τ of $\{1, 2, \ldots, n\}$ such that $[U_i]_e = [V_{\tau(i)}]_e$ for every $i = 1, 2, \ldots, n$. Hence it is sufficient to prove that if U, V are cyclic finitely presented modules over a chain ring R and $[U]_o = [V]_e$, then $U \cong V$. Suppose the contrary, and let U, V be cyclic finitely presented R-modules such that $[U]_e = [V]_e$ and $U \not\cong V$. Then $U \cong R/rR$ for some $r \in R$. Since $[U]_e = [V]_e$, we know that U is isomorphic to a quotient of V and V is isomorphic to a quotient of U. Hence we may assume that $V \cong R/rsR$ and $U \cong R/rstR$ for some $s, t \in R$. By Lemma 9.17 there are invertible elements $u, v \in R$ such that $urstv = r$. If tv is invertible, then $U \cong V$ by Lemma 9.17, a contradiction. Hence tv is not invertible, so that $1 + tv$ is invertible. From Lemma 9.18 we have that $Rrs \subset Rr$ and $Rrs(1 + tv) \subset Rr$, so that

$$Rr = R(urs + r - urs) \subseteq R(urs + r) + Rurs = R(urs + urstv) + Rurs$$
$$= Rurs(1 + tv) + Rurs = Rrs(1 + tv) + Rrs \subset Rr,$$

contradiction.   □

9.6   Krull-Schmidt fails for finitely presented modules over serial rings

We know that every finitely presented module over a serial ring is a direct sum of uniserial modules (Corollary 3.30). In the example we consider in this section we show that there exists a finitely presented module over a suitable serial ring with non-isomorphic direct sum decompositions. More precisely, we construct an example that proves that the permutations σ and τ in the statement of the weak Krull-Schmidt Theorem can be completely arbitrary. Hence the Krull-Schmidt Theorem fails for finitely presented modules over serial rings. Observe that in Theorem 9.19 we saw that the Krull-Schmidt Theorem holds for finitely presented modules over chain rings.

Example 9.20 *If $n \geq 2$ is an integer, then there exist n^2 pairwise non-isomorphic finitely presented uniserial modules $U_{i,j}$ ($i, j = 1, 2, \ldots, n$) over a suitable serial ring R satisfying the following properties:*

(a) *for every $i, j, k, \ell = 1, 2, \ldots, n$, $[U_{i,j}]_m = [U_{k,\ell}]_m$ if and only if $i = k$;*

(b) *for every $i, j, k, \ell = 1, 2, \ldots, n$, $[U_{i,j}]_e = [U_{k,\ell}]_e$ if and only if $j = \ell$.*

Hence

$$U_{1,1} \oplus U_{2,2} \oplus \cdots \oplus U_{n,n} \cong U_{\sigma(1),\tau(1)} \oplus U_{\sigma(2),\tau(2)} \oplus \cdots \oplus U_{\sigma(n),\tau(n)}$$

for every pair of permutations σ, τ of $\{1, 2, \ldots, n\}$.

Proof. Let $\mathbf{M}_n(\mathbf{Q})$ be the ring of all $n \times n$-matrices over the field $\mathbf{Q}$ of rational numbers. Let $\mathbf{Z}$ be the ring of integers and let $\mathbf{Z}_p, \mathbf{Z}_q$ be the localizations of $\mathbf{Z}$ at two distinct maximal ideals (p) and (q) of $\mathbf{Z}$ (here $p, q \in \mathbf{Z}$ are distinct prime numbers). Let Λ_p denote the subring of $\mathbf{M}_n(\mathbf{Q})$ whose elements are $n \times n$-matrices with entries in $\mathbf{Z}_p$ on and above the diagonal and entries in $p\mathbf{Z}_p$ under the diagonal, that is,

$$\Lambda_p = \begin{pmatrix} \mathbf{Z}_p & \mathbf{Z}_p & \cdots & \mathbf{Z}_p \\ p\mathbf{Z}_p & \mathbf{Z}_p & \cdots & \mathbf{Z}_p \\ \vdots & & \ddots & \\ p\mathbf{Z}_p & p\mathbf{Z}_p & \cdots & \mathbf{Z}_p \end{pmatrix} \subseteq \mathbf{M}_n(\mathbf{Q}).$$

Similarly, put

$$\Lambda_q = \begin{pmatrix} \mathbf{Z}_q & \mathbf{Z}_q & \cdots & \mathbf{Z}_q \\ q\mathbf{Z}_q & \mathbf{Z}_q & \cdots & \mathbf{Z}_q \\ \vdots & & \ddots & \\ q\mathbf{Z}_q & q\mathbf{Z}_q & \cdots & \mathbf{Z}_q \end{pmatrix} \subseteq \mathbf{M}_n(\mathbf{Q}).$$

If

$$R = \begin{pmatrix} \Lambda_p & 0 \\ \mathbf{M}_n(\mathbf{Q}) & \Lambda_q \end{pmatrix},$$

then R is a subring of the ring $\mathbf{M}_{2n}(\mathbf{Q})$ of $2n \times 2n$-matrices with rational entries.

It is easily seen that the Jacobson radicals of these rings are

$$J(\Lambda_p) = \begin{pmatrix} p\mathbf{Z}_p & \mathbf{Z}_p & \cdots & \mathbf{Z}_p \\ p\mathbf{Z}_p & p\mathbf{Z}_p & \cdots & \mathbf{Z}_p \\ \vdots & & \ddots & \\ p\mathbf{Z}_p & p\mathbf{Z}_p & \cdots & p\mathbf{Z}_p \end{pmatrix} \quad \text{and} \quad J(R) = \begin{pmatrix} J(\Lambda_p) & 0 \\ \mathbf{M}_n(\mathbf{Q}) & J(\Lambda_q) \end{pmatrix}.$$

For every $i = 1, 2, \ldots, 2n$ let e_i be the matrix with 1 in (i, i)-position and 0's elsewhere. Then $\{e_1, \ldots, e_{2n}\}$ is a complete set of orthogonal idempotents for R. We leave the reader to check that the left R-modules Re_i and the right R-modules $e_i R$ are uniserial. Hence R is a serial ring. A complete set of representatives of the simple right R-modules is given by the $2n$ modules $e_i R / e_i J(R)$, $i = 1, 2, \ldots, 2n$.

The modules $U_{i,j}$ we are constructing are submodules of homomorphic images of the right ideal

$$e_{n+1}R = \begin{pmatrix} 0 & \cdots & 0 & 0 & \cdots & 0 \\ \vdots & & \vdots & \vdots & & \vdots \\ 0 & \cdots & 0 & 0 & \cdots & 0 \\ \mathbf{Q} & \cdots & \mathbf{Q} & \mathbf{Z}_q & \cdots & \mathbf{Z}_q \\ 0 & \cdots & 0 & 0 & \cdots & 0 \\ \vdots & & \vdots & \vdots & & \vdots \\ 0 & \cdots & 0 & 0 & \cdots & 0 \end{pmatrix}$$

of R. In order to simplify the notation, it is convenient to remark that if

$$V = (\underbrace{\mathbf{Q}, \ldots, \mathbf{Q}}_{n}, \underbrace{\mathbf{Z}_q, \ldots, \mathbf{Z}_q}_{n})$$

(V is a set of $1 \times 2n$-matrices), and V is endowed with the right R-module structure given by matrix multiplication (observe that elements of V are $1 \times 2n$-matrices and elements of the ring R are $2n \times 2n$-matrices), then the right ideal $e_{n+1}R$ and the right R-module V are isomorphic. In particular, V is uniserial.

For every $i = 1, 2, \ldots, n$ put

$$V_i = (\underbrace{\mathbf{Q}, \ldots, \mathbf{Q}}_{n}, \underbrace{q\mathbf{Z}_q, \ldots, q\mathbf{Z}_q}_{i-1}, \underbrace{\mathbf{Z}_q, \ldots, \mathbf{Z}_q}_{n-i+1})$$

and

$$X_i = (\underbrace{p\mathbf{Z}_p, \ldots, p\mathbf{Z}_p}_{i-1}, \underbrace{\mathbf{Z}_p, \ldots, \mathbf{Z}_p}_{n-i+1}, \underbrace{0, \ldots, 0}_{n}).$$

It is easily seen that V_i and X_i are R-submodules of V and that

(a) $V = V_1 \supset V_2 \supset \cdots \supset V_n \supset qV_1 \supset X_1 \supset X_2 \supset \cdots \supset X_n \supset pX_1$;

(b) $X_i J(R) = X_{i+1}$ for every $i = 1, 2, \ldots, n-1$, $X_n J(R) = pX_1$;

(c) $V_i J(R) = V_{i+1}$ for every $i = 1, 2, \ldots, n-1$, $V_n J(R) = qV_1$;

(d) $X_i/X_{i+1} \cong e_i R/e_i J(R)$ for every $i = 1, 2, \ldots, n-1$;

(e) $X_n/pX_1 \cong e_n R/e_n J(R)$;

(f) $V_j/V_{j+1} \cong e_{n+j} R/e_{n+j} J(R)$ for every $j = 1, 2, \ldots, n-1$;

(g) $V_n/qV_1 \cong e_{2n} R/e_{2n} J(R)$.

The n^2 uniserial R-modules $U_{i,j} = V_j/X_i$, $i, j = 1, 2, \ldots, n$, are the modules with the required properties. In fact, if $[U_{i,j}]_m = [U_{k,\ell}]_m$, then $U_{i,j}$ and $U_{k,\ell}$ have isomorphic socles. By (d) and (e) the socle of $U_{i,j}$ is isomorphic to $e_{i-1} R/e_{i-1} J(R)$ for $i = 2, 3, \ldots, n$ and is isomorphic to $e_n R/e_n J(R)$ for $i = 1$. Similarly for $U_{k,\ell}$. Thus $[U_{i,j}]_m = [U_{k,\ell}]_m$ implies $i = k$. Conversely, in order to show that $[U_{i,j}]_m = [U_{i,\ell}]_m$ for every i, j, ℓ, it is necessary to find a monomorphism $U_{i,j} \to U_{i,\ell}$ and a monomorphism $U_{i,\ell} \to U_{i,j}$. Suppose that $j \leq \ell$. Then $U_{i,\ell} \subseteq U_{i,j}$, so that the embedding $U_{i,\ell} \to U_{i,j}$ is one of the required monomorphisms. The other monomorphism $U_{i,j} \to U_{i,\ell}$ is given by multiplication by q. We leave the details as an exercise for the reader. $\square$

In particular, if in the previous example σ and τ are two permutations of $\{1, 2, \ldots, n\}$ such that $\sigma(i) \neq \tau(i)$ for every i, put $U_i = U_{i,i}$ and $V_i = U_{\sigma(i), \tau(i)}$ for every $i = 1, 2, \ldots, n$. Then

Example 9.21 *Let $n \geq 2$ be an integer. Then there exists a serial ring R and $2n$ pairwise non-isomorphic finitely presented uniserial right R-modules $U_1, U_2, \ldots, U_n, V_1, V_2, \ldots, V_n$ such that*

$$U_1 \oplus U_2 \oplus \cdots \oplus U_n \cong V_1 \oplus V_2 \oplus \cdots \oplus V_n.$$

$\square$

9.7 Further examples of biuniform modules of type 1

In this section we have collected some classes of examples of biuniform modules.

Proposition 9.22 *A biuniform module M_R over an arbitrary ring R is of type 1 if at least one of the following conditions hold:*
  (a) M_R *is projective;*
  (b) M_R *is injective;*
  (c) M_R *is artinian;*
  (d) M_R *is noetherian.*

Proof. (a) If M_R is projective and K is the ideal of $\mathrm{End}(M_R)$ whose elements are all the endomorphisms of M_R that are not surjective, then K is the set of all non-invertible elements of $\mathrm{End}(M_R)$, because if $f \in \mathrm{End}(M_R) \setminus K$, then $f \colon M_R \to M_R$ is surjective, and so splits. But M_R is indecomposable, so that f is an automorphism of M_R. This proves that $\mathrm{End}(M_R)$ is local.

  (b) By Lemma 2.25 the endomorphism ring of M_R is local.

  (c) By Lemma 2.16(b) every injective endomorphism of an artinian module M_R is an automorphism. Hence the endomorphism ring of an artinian uniform module M_R is local.

  (d) Lemma 2.17(b). □

  Observe that the proofs of (c) and (d) show that the endomorphism ring of an artinian uniform module is local and, dually, the endomorphism ring of a noetherian couniform module is local. These facts, generalizations of (c) and (d), were already used in the proof of Theorems 2.18 and 2.19.

  The proof of the following proposition is due to Dolors Herbera and is modelled on the proof of [Faith and Herbera, Lemma 4.10].

Proposition 9.23 *Every biuniform module over a commutative noetherian ring is of type 1.*

Proof. Let M be a biuniform module over a commutative noetherian ring R. Let

$$P_1 = \{\, r \in R \mid \text{multiplication by } r \text{ is a non-injective endomorphism of } M \,\}$$

and

$$P_2 = \{\, r \in R \mid \text{multiplication by } r \text{ is a non-surjective endomorphism of } M \,\}.$$

Then P_1 and P_2 are prime ideals of R. Let R_S be the quotient ring of R with respect to $S = R \setminus (P_1 \cup P_2)$. Then R_S is a commutative noetherian ring and M has a canonical structure as a biuniform R_S-module.

  Suppose that $P_1 \not\subseteq P_2$. Then $P_1 R_S$ is a maximal ideal of R_S. Let I be an ideal of R_S that is maximal in the set $\{\, \mathrm{ann}_{R_S}(x) \mid x \in M, \ x \neq 0 \,\}$, let $x_0 \neq 0$

be an element of M with $\mathrm{ann}_{R_S}(x_0) = I$, and let C be the cyclic R_S-submodule of M generated by x_0. For every $r \in P_1$ let $\mu_r\colon C \to C$ be the endomorphism of C given by multiplication by r. Since C is essential in M and $r \in P_1$, we have that $\ker\mu_r \neq 0$. Hence there exists $y \in C$, $y \neq 0$, with $\mu_r y = 0$. Then $\mathrm{ann}_{R_S}(y) \supseteq \mathrm{ann}_{R_S}(C) = I$ and $r/1 \in \mathrm{ann}_{R_S}(y)$. From the choice of I it follows that $\mathrm{ann}_{R_S}(y) = I$, so that $r/1 \in I$, that is, $\mu_r = 0$. This shows that $P_1 C = 0$, so that the uniform R_S-module C is annihilated by the maximal ideal $P_1 R_S$ of R_S, that is, C is a simple R_S-module. Since C is essential in M, the injective envelope of the R_S-module M is the injective envelope of the simple R_S-module C, so that it is artinian ([Matlis], see the Notes on Chapter 2). Thus the R_S-module M is artinian, so that $\mathrm{End}_{R_S}(M)$ is local by Proposition 9.22(c). But $\mathrm{End}_{R_S}(M) = \mathrm{End}_R(M)$, so that M is a biuniform R-module of type 1.

Suppose that $P_1 \subseteq P_2$, so that R_S is a local ring with maximal ideal $P_2 R_S$. Let $r_1, \ldots, r_n$ be a finite set of generators of P_2. Then for every $i = 1, \ldots, n$ and every $q \in R_S$ we know that $r_i q M \subseteq r_i M$, so that $r_i R_S M \subseteq r_i M \subset M$. Hence $P_2 R_S M \subset M$ because M is couniform. Then $M/P_2 R_S M$ is a couniform $R_S/P_2 R_S$-module, and so is a simple R_S-module. Thus M has a maximal R_S-submodule, so it is colocal, in particular M is a cyclic R_S-module. Thus $\mathrm{End}_{R_S}(M)$ is a homomorphic image of R_S, and so $\mathrm{End}_{R_S}(M) = \mathrm{End}_R(M)$ is a local ring. This proves that M is a biuniform R-module of type 1 in any case. $\qquad\square$

Recall that a uniserial module is *shrinkable* if it is isomorphic to a proper submodule of a proper homomorphic image of itself (Section 1.3). Otherwise a uniserial module is said to be *unshrinkable*.

Proposition 9.24 *Every unshrinkable uniserial module is of type 1. In particular, if R is a ring that is either commutative or right noetherian, then every uniserial right R-module is of type 1.*

Proof. Let U_R be a uniserial module of type 2 over an arbitrary ring R. There exist $f, g \in \mathrm{End}(U_R)$ with f surjective and non-injective and g injective and non-surjective. Then g induces an isomorphism $\widetilde{g}\colon U \to g(U)$, so that $g(U)/g(\ker(f)) \cong U/\ker(f) \cong U$. Thus U is a shrinkable module. This proves the first part of the statement. The second part follows from Proposition 1.18. $\qquad\square$

From Proposition 9.24 and Corollary 2.54 we obtain:

Corollary 9.25 *Let M be a serial right module over a ring that is either commutative or right noetherian. Then any two direct sum decompositions of M into indecomposable direct summands are isomorphic, and any indecomposable direct summand of M is uniserial.* $\qquad\square$

We conclude the section with an example of a Krull-Schmidt uniserial module of type 2.

Example 9.26 *The uniserial cyclic module of Example* 1.16 *is a Krull-Schmidt biuniform module of type* 2.

Proof. In the notation of Example 1.16 it is easily seen that the idealizer $H' = \{ r \in R \mid rH \subseteq H \}$ of H in R is

$$ H' = \begin{pmatrix} \mathbf{Z}_p & 0 \\ \mathbf{Z}_p & \mathbf{Z}_p \cap \mathbf{Z}_q \end{pmatrix}. $$

Hence $\operatorname{End}(R/H) \cong H'/H \cong \mathbf{Z}_p \cap \mathbf{Z}_q$ is a ring with two maximal ideals. Thus R/H is a uniserial module of type 2.

It is easily seen that the submodules of R/H isomorphic to R/H are exactly the modules

$$ \begin{pmatrix} \mathbf{Z}_p & 0 \\ \mathbf{Q} & q^n \mathbf{Z}_q \end{pmatrix} \Big/ H, \qquad n \geq 0. $$

It follows that every submodule of R/H that contains a submodule isomorphic to R/H is isomorphic to R/H. Hence R/H is a Krull-Schmidt uniserial right R-module. $\qquad\square$

9.8 Quasi-small uniserial modules

Let $M = \oplus_{i \in I} A_i = \oplus_{j \in J} B_j$ be a direct sum of two families of biuniform modules $\{ A_i \mid i \in I \}$ and $\{ B_j \mid j \in J \}$. By Theorem 9.12 the sets I and J have the same cardinality. In view of Theorems 9.11 and 9.12 it is natural to ask whether there exists a bijection $\tau \colon I \to J$ such that $[A_i]_e = [B_{\tau(i)}]_e$ for every $i \in I$. If I is finite (which happens if and only if J is finite by Theorem 9.12), such a bijection τ exists (Theorem 9.13).

Little is known when I and J are infinite. The next theorem considers the case of cyclic biuniform modules. Note that a biuniform module is finitely generated if and only if it is cyclic, if and only if it is local.

Theorem 9.27 *Let R be an arbitrary ring and let $\{ U_i \mid i \in I \}$, $\{ V_j \mid j \in J \}$ be two families of biuniform R-modules such that $\oplus_{i \in I} U_i \cong \oplus_{j \in J} V_j$. If*

$$ I' = \{ i \in I \mid U_i \text{ is cyclic} \} \quad and \quad J' = \{ j \in J \mid V_j \text{ is cyclic} \}, $$

then there exists a bijection $\tau' \colon I' \to J'$ such that $[U_i]_e = [V_{\tau'(i)}]_e$ for every $i \in I'$.

Proof. The proof is similar to that of Theorem 9.12. Without loss of generality we may assume that $M = \oplus_{i \in I} U_i = \oplus_{j \in J} V_j$.

For every cyclic biuniform direct summand Z of M put

$$ I_Z = \{ i \in I \mid [U_i]_e = [Z]_e \} \quad and \quad J_Z = \{ j \in J \mid [V_j]_e = [Z]_e \}. $$

Since Z is cyclic, there is a finite subset $F \subseteq I$ such that Z is isomorphic to a direct summand of $\oplus_{i \in F} U_i$. By Proposition 9.7 there exists $k \in F$ with $[Z]_e = [U_k]_e$. Hence the set I_Z is not empty. Similarly, J_Z is not empty. Note that Z cyclic and $[U_i]_e = [Z]_e$ imply U_i cyclic, so that $I_Z \subseteq I'$ for every cyclic biuniform direct summand Z of M. Similarly, $J_Z \subseteq J'$. Obviously, the I_Z, where Z ranges over the set of cyclic biuniform direct summands of M, form a partition of I', and the J_Z form a partition of J'.

In order to prove the theorem it suffices to show that $|I_Z| = |J_Z|$ for every cyclic biuniform direct summand Z of M.

Suppose that either I_Z or J_Z is a finite set and $|I_Z| \neq |J_Z|$. By symmetry we may assume that $|I_Z| < |J_Z|$. Put $n = |I_Z|$. Then in J_Z there are $n+1$ indices $j_1, \ldots, j_{n+1}$ with $[V_{j_t}]_e = [Z]_e$. Now $V_{j_1} \oplus \cdots \oplus V_{j_{n+1}}$ is finitely generated, so that there exists a finite subset $G \subseteq I$ such that $V_{j_1} \oplus \cdots \oplus V_{j_{n+1}} \subseteq \oplus_{i \in G} U_i$. Thus $V_{j_1} \oplus \cdots \oplus V_{j_{n+1}}$ is a direct summand of $\oplus_{i \in G} U_i$, whence by Proposition 9.10 there exist $n + 1$ distinct indices $i \in G$ such that $[U_i]_e = [Z]_e$. In particular $|I_Z| \geq n + 1$, a contradiction. Thus $|I_Z| = |J_Z|$ if either I_Z or J_Z is finite.

Suppose that I_Z and J_Z are both infinite. By symmetry we only have to prove that $|J_Z| \leq |I_Z|$. Let $\varepsilon_k \colon U_k \to M$, $e_\ell \colon V_\ell \to M$, $\pi_k \colon M \to U_k$ and $p_\ell \colon M \to V_\ell$ denote the embeddings and the canonical projections. If $t \in I'$ put
$$A(t) = \{ \ell \in J \mid \pi_t e_\ell p_\ell \varepsilon_t \text{ is an epimorphism} \}.$$
The sets $A(t)$ are finite for every $t \in I'$, because U_t is cyclic. Hence there is a finite subset F of J such that $U_t \subseteq \oplus_{j \in F} V_j$, and then $\pi_t e_\ell p_\ell \varepsilon_t(U_t) = 0$ for every $\ell \in J \setminus F$. In particular, $\pi_t e_\ell p_\ell \varepsilon_t$ is not surjective for every $\ell \in J \setminus F$, so that $A(t) \subseteq F$ is finite. Moreover, $A(t) \subseteq J'$, because if $\ell \in A(t)$, then $\pi_t e_\ell p_\ell \varepsilon_t$ is an epimorphism, so that $p_\ell \varepsilon_t \colon U_t \to V_\ell$ is an epimorphism by Lemma 6.26(b). In particular, V_ℓ is cyclic, that is, $\ell \in J'$.

We now show that $J_Z \subseteq \bigcup_{t \in I_Z} A(t)$. If $j \in J_Z$, then V_j is cyclic, so that there is a finite subset H of I such that $V_j \subseteq \oplus_{i \in H} U_i$. Then $\sum_{i \in H} p_j \varepsilon_i \pi_i e_j$ is the identity of V_j. In particular, there exists $i \in H$ such that $p_j \varepsilon_i \pi_i e_j$ is surjective. By Lemma 6.26(b) both $p_j \varepsilon_i$ and $\pi_i e_j$ are surjective, so that $[U_i]_e = [V_j]_e = [Z]_e$ and $\pi_i e_j p_j \varepsilon_i$ is surjective. Thus $j \in A(i)$ and $i \in I_Z$. This proves that $J_Z \subseteq \bigcup_{t \in I_Z} A(t)$, so that $|J_Z| \leq \aleph_0 |I_Z| = |I_Z|$. Hence $|I_Z| = |J_Z|$. $\square$

In the remaining part of this section we will restrict our attention to the case in which all the modules A_i, B_j ($i \in I, j \in J$) are uniserial, and show that in this case there exists a bijection $\tau'' \colon I'' \to J''$ such that $[A_i]_e = [B_{\tau''(i)}]_e$ for every $i \in I''$, where $I'' \subseteq I$ and $J'' \subseteq J$ are in some sense the largest possible subsets for which a weak Krull-Schmidt Theorem can hold.

An R-module N_R is said to be *quasi-small* if for every family $\{ M_i \mid i \in I \}$ of R-modules such that N_R is isomorphic to a direct summand of $\oplus_{i \in I} M_i$, there is a finite subset $F \subseteq I$ such that N_R is isomorphic to a direct summand of $\oplus_{i \in F} M_i$. For instance, a semisimple module is quasi-small if and only if its Goldie dimension is finite.

Example 9.28 *Every small module is quasi-small.* $\square$

In particular, every finitely generated module is quasi-small.

Example 9.29 *Every module with local endomorphism ring is quasi-small.*

Proof. Suppose that N is a module with a local endomorphism ring and

$$\{ M_i \mid i \in I \}$$

is a family of modules with N isomorphic to a direct summand of $\oplus_{i \in I} M_i$, that is, suppose that $G = N' \oplus P = \oplus_{i \in I} M_i$ with $N' \cong N$. By Theorem 2.8 the module N has the exchange property, so that for every $i \in I$ there is a direct sum decomposition $M_i = B_i \oplus C_i$ such that $G = N' \oplus (\oplus_{i \in I} B_i)$. Then $N' \cong \oplus_{i \in I} C_i$. But $N' \cong N$ is indecomposable. Hence there exists an index $j \in I$ with $N' \cong C_j$. Thus N is isomorphic to a direct summand of M_j. In particular, M is quasi-small. $\square$

Proposition 9.30(a) shows that the class of quasi-small uniserial modules is the largest class of uniserial modules for which we can possibly find a bijection preserving the epigeny classes.

Proposition 9.30 *Let N be a uniserial R-module that is not quasi-small. Then the following statements hold true:*

(a) *There exists a countable family $\{ A_n \mid n \geq 1 \}$ of uniserial R-modules such that $N \oplus (\oplus_{n \geq 1} A_n) \cong \oplus_{n \geq 1} A_n$ and $[A_n]_e \neq [N]_e$ for every $n \geq 1$.*

(b) *Every non-zero homomorphic image of N is not quasi-small.*

Proof. If N is a uniserial module that is not quasi-small, then there exists a family $\{ M_i \mid i \in I \}$ of modules such that N is isomorphic to a direct summand of $\oplus_{i \in I} M_i$ and M_R is not isomorphic to a direct summand of $\oplus_{i \in F} M_i$ for every finite subset $F \subseteq I$. Hence there exist homomorphisms $f \colon N \to \oplus_{i \in I} M_i$ and $g \colon \oplus_{i \in I} M_i \to N$ such that $gf = 1_N$. For every $j \in I$ let $\varepsilon_j \colon M_j \to \oplus_{i \in I} M_i$ and $\pi_j \colon \oplus_{i \in I} M_i \to M_j$ denote the embedding and the canonical projection.

By Proposition 2.45 the module N has a countable set $\{ x_n \mid n \geq 0 \}$ of generators. Since N is not cyclic (Example 9.28), without loss of generality we may assume that $0 = x_0 R \subset x_1 R \subset x_2 R \subset \dots$.

We claim that there exists a chain $0 = A_0 \subseteq A_1 \subseteq A_2 \subseteq \dots$ of cyclic submodules of N and endomorphisms f_n, $n \geq 0$, of N satisfying the following conditions:

(a) $x_n \in A_n$ for every $n \geq 0$;

(b) $f_n(x) = x$ for every $x \in A_{n-1}$ and every $n \geq 1$;

(c) $f_n(N) \subseteq A_n$ for every $n \geq 0$.

The proof of the claim is by induction on n. For $n = 0$ put $A_0 = 0$ and $f_0 = 0$.

Suppose that $n \geq 1$ and that $A_0, \ldots, A_{n-1}, f_0, \ldots, f_{n-1}$ with the required properties exist. Since A_{n-1} is cyclic and N is not cyclic, there exists $y \in N$, $y \notin A_{n-1}$. As $f(y) \in \oplus_{i \in I} M_i$, there exists a finite subset F of I such that

$$f(y) = \sum_{i \in F} \varepsilon_i \pi_i f(y).$$

Then $y = gf(y) = \sum_{i \in F} g\varepsilon_i \pi_i f(y)$. Put $f_n = \sum_{i \in F} g\varepsilon_i \pi_i f$, so that from $f_n(y) = y$ it follows that $f_n \in \mathrm{End}(N)$ is a monomorphism and satisfies property (b) (because $A_{n-1} \subseteq yR$). If $f_n(N) = N$, then f_n is an automorphism of N, so that the composite of

$$\oplus_{i \in F} \pi_i f \colon N \to \oplus_{i \in F} M_i \quad \text{and} \quad \oplus_{i \in F} f_n^{-1} g\varepsilon_i \colon \oplus_{i \in F} M_i \to N$$

is the identity mapping of N. Hence N is isomorphic to a direct summand of $\oplus_{i \in F} M_i$, contradiction. This shows that $f_n(N) \subset N$, whence $f_n(N) + x_n R \subset N$. Therefore there exists a cyclic submodule A_n of N with $f_n(N) + x_n R \subseteq A_n$. This concludes the proof of the claim.

Since N is not cyclic and the A_n are cyclic, we have $[A_n]_e \neq [N]_e$ for every $n \geq 1$.

Put $g_n = f_n - f_{n-1}$ for every $n \geq 1$ and let $g \colon N \to \oplus_{n \geq 1} A_n$ be the mapping defined by $g(x) = (g_n(x))_{n \geq 1}$. This mapping is well-defined, because $g_n(N) \subseteq A_n$ for $n \geq 1$ and $g_n(x) = 0$ for every $x \in A_{n-2}$, $n \geq 2$, so that for every element $x \in N$ one has that $g_n(x) = 0$ for almost all n. Let

$$h \colon \oplus_{n \geq 1} A_n \to N$$

be the homomorphism defined by $h((x_n)_{n \geq 1}) = \sum_{n \geq 1} x_n$. If $x \in A_n$, then $hg(x) = \sum_{i=1}^{n+1} g_i(x) = f_{n+1}(x) - f_0(x) = x$. Since $N = \bigcup_{n \geq 1} A_n$, we obtain that $hg = 1_N$. Hence if $K = \ker h$, then $N \oplus K \cong \oplus_{n \geq 1} A_n$.

Let $\varepsilon \colon \oplus_{n \geq 1} A_n \to \oplus_{n \geq 1} A_n$ be the homomorphism defined by

$$\varepsilon(y_1, y_2, y_3, \ldots) = (y_1, y_2 - y_1, y_3 - y_2, y_4 - y_3, \ldots).$$

Then $h\varepsilon = 0$, so that the image of ε is contained in $\ker h = K$. Conversely, if $z = (z_1, z_2, z_3, \ldots) \in \oplus_{n \geq 1} A_n$ and $z \in K$, then there exists n such that $z_i = 0$ for every $i \geq n$, and $\sum_{i=1}^{n-1} z_i = 0$. If $y_j = \sum_{i=1}^{j} z_i$ for every $j \geq 1$, then $y = (y_1, y_2, y_3, \ldots) \in \oplus_{n \geq 1} A_n$ and $\varepsilon(y) = z$. Hence the image of ε is equal to K. Finally, it is easily verified that ε is a monomorphism. Hence $K \cong \oplus_{n \geq 1} A_n$. This proves (a).

Now let P be a proper submodule of N. Then there exists an $n \geq 1$ such that $P \subset A_n$. In particular, $f_i(x) = x$ for every $x \in P$ and every $i > n$, so that f_i induces an endomorphism $\widetilde{f}_i$ of N/P such that $\widetilde{f}_i(\overline{x}) = \overline{x}$ for every $\overline{x} \in A_{i-1}/P$ and $\widetilde{f}_i(N/P) \subseteq A_i/P$ for every $i > n$. Put $\widetilde{g}_{n+1} = \widetilde{f}_{n+1}$ and $\widetilde{g}_i = \widetilde{f}_i - \widetilde{f}_{i-1}$ for every $i > n + 1$. Let $\widetilde{g} \colon N/P \to \oplus_{i \geq n+1} A_i/P$ be the mapping defined by $\widetilde{g}(\overline{x}) = (\widetilde{g}_i(\overline{x}))_{i \geq n+1}$, and let $\widetilde{h} \colon \oplus_{i \geq n+1} A_i/P \to N/P$ be defined by $\widetilde{h}((\overline{x}_i)_{i \geq n+1}) = \sum_{i \geq n+1} \overline{x}_i$. Then $\widetilde{h}\widetilde{g} = 1_{N/P}$, so that N/P is isomorphic to a direct summand of $\oplus_{i \geq n+1} A_i/P$. If N/P were quasi-small, then there would exist a finite set F with N/P isomorphic to a direct summand of $\oplus_{i \in F} A_i/P$. But each A_i is cyclic, so that N/P would be finitely generated. Hence N would be cyclic, a contradiction. Therefore N/P is not quasi-small. □

We do not know an example of a uniserial module that is not quasi-small.

9.9 A necessary condition for families of uniserial modules

Let $\{ U_i \mid i \in I \}$, $\{ V_j \mid j \in J \}$ be two families of non-zero uniserial modules over an arbitrary ring R such that $\oplus_{i \in I} U_i \cong \oplus_{j \in J} V_j$. It is not known whether there exists a one-to-one correspondence $\tau \colon I \to J$ that preserves the epigeny classes. The aim of this section is to prove that if $\oplus_{i \in I} U_i \cong \oplus_{j \in J} V_j$, then, necessarily, there exists a one-to-one correspondence that preserves the epigeny classes of quasi-small uniserial modules.

Our first lemma is an extension of Proposition 9.10 to the case of an infinite family of C_j's. It holds when the modules A_i's are quasi-small.

Lemma 9.31 *Let R be an arbitrary ring, let $A_1, \ldots, A_n$ be biuniform quasi-small R-modules and let $\{ C_j \mid j \in J \}$ be a family of biuniform modules. If $A_1 \oplus \cdots \oplus A_n$ is isomorphic to a direct summand of $\oplus_{j \in J} C_j$, then there exist n distinct indices $j_1, \ldots, j_n \in J$ such that $[A_i]_e = [C_{j_t}]_e$ for every $t = 1, \ldots, n$.*

Proof. Suppose that $A_1 \oplus \cdots \oplus A_n \oplus B \cong \oplus_{j \in J} C_j$. The proof is by induction on n. The case $n = 0$ is trivial. Suppose that $n \geq 1$. By Proposition 9.6 there exists $k \in J$ such that $[A_1]_m = [C_k]_m$. Since A_1 is quasi-small, there exists a finite subset $F \subseteq J$ such that A_1 is isomorphic to a direct summand of $\oplus_{j \in F} C_j$. Hence by Proposition 9.7 there exists $j_1 \in F$ such that $[A_1]_e = [C_{j_1}]_e$. If $k = j_1$, then $A_1 \cong C_{j_1}$, so that $A_2 \oplus \cdots \oplus A_n \oplus B \cong \oplus_{j \in J, \, j \neq j_1} C_j$ and we obtain the result by the induction hypothesis. If $k \neq j_1$, then

$$A_1 \oplus D \cong C_k \oplus C_{j_1}$$

for a biuniform module D with $[D]_m = [C_{j_1}]_m$ and $[D]_e = [C_k]_e$. Then

$$A_1 \oplus \cdots \oplus A_n \oplus B \cong \oplus_{j \in J} C_j \cong A_1 \oplus D \oplus (\oplus_{j \in J, \, j \neq k, j_1} C_j),$$

so that $A_2 \oplus \cdots \oplus A_n \oplus B \cong D \oplus (\oplus_{j \in J, \, j \neq k, j_1} C_j)$. Since $[D]_e = [C_k]_e$, by the induction hypothesis there exist $n - 1$ distinct indices $j_2, \ldots, j_n \in J \setminus \{j_1\}$ such that $[A_i]_e = [C_{j_i}]_e$ for every $i = 2, \ldots, n$. Thus $j_1, \ldots, j_n$ have the required property. □

The next theorem shows that for the class of quasi-small uniserial modules a bijection preserving the epigeny classes exists.

Theorem 9.32 *Let R be an arbitrary ring and let $\{U_i \mid i \in I\}$, $\{V_j \mid j \in J\}$ be two families of non-zero uniserial R-modules such that $\oplus_{i \in I} U_i \cong \oplus_{j \in J} V_j$. If $I'' = \{i \in I \mid U_i \text{ is quasi-small}\}$ and $J'' = \{j \in J \mid V_j \text{ is quasi-small}\}$, then there exists a bijection $\tau'': I'' \to J''$ such that $[U_i]_e = [V_{\tau''(i)}]_e$ for every $i \in I''$.*

Proof. The proof is similar to that of Theorem 9.27. We may assume $M = \oplus_{i \in I} U_i = \oplus_{j \in J} V_j$.

For every non-zero uniserial quasi-small direct summand Z of M set

$$I_Z = \{i \in I'' \mid [U_i]_e = [Z]_e\} \quad \text{and} \quad J_Z = \{j \in J'' \mid [V_j]_e = [Z]_e\}.$$

Since Z is quasi-small, there is a finite subset $F \subseteq I$ such that Z is isomorphic to a direct summand of $\oplus_{i \in F} U_i$. By Proposition 9.7 there exists $k \in F$ with $[Z]_e = [U_k]_e$. In particular, U_k is quasi-small by Proposition 9.30(b). Thus $k \in I_Z$, so that the set I_Z is non-empty. Similarly, J_Z is non-empty. It is obvious that if Z ranges in the set of all non-zero uniserial quasi-small direct summands of M, then the I_Z form a partition of I'' and the J_Z form a partition of J''. As in the proof of Theorem 9.27, in order to prove the statement it suffices to show that $|I_Z| = |J_Z|$ for every non-zero uniserial quasi-small direct summand Z of M.

If either I_Z or J_Z is a finite set and $|I_Z| \neq |J_Z|$ we may assume $|I_Z| < |J_Z|$ by symmetry. Set $n = |I_Z|$. Then in J_Z there are $n+1$ indices $j_1, \ldots, j_{n+1}$ with $[V_{j_t}]_e = [Z]_e$. By Lemma 9.31 $[U_i]_e = [Z]_e$ for at least $n+1$ distinct indices $i \in I$. By Proposition 9.30(b) these $n+1$ modules U_i are quasi-small, so that $|I_Z| \geq n+1$, a contradiction. Hence $|I_Z| = |J_Z|$ if either I_Z or J_Z is a finite set.

If I_Z and J_Z are both infinite, it is sufficient to prove that $|J_Z| \leq |I_Z|$. Let $\varepsilon_k: U_k \to \oplus_{i \in I} U_i$ and $e_\ell: V_\ell \to \oplus_{j \in J} V_j$ be the embeddings, and $\pi_k: \oplus_{i \in I} U_i \to U_k$ and $p_\ell: \oplus_{j \in J} V_j \to V_\ell$ be the canonical projections. For every $t \in I$ set $A(t) = \{j \in J \mid \pi_t e_j p_j \varepsilon_t: U_t \to U_t \text{ is an epimorphism}\}$.

Each set $A(t)$, $t \in I$, is countable, because by Proposition 2.45 the uniserial module U_t is either small or countably generated, so that there is a countable subset C of J such that $U_t \subseteq \oplus_{j \in C} V_j$, and then $\pi_t e_\ell p_\ell \varepsilon_t(U_t) = 0$ for every $\ell \in J \setminus C$. Hence $A(t) \subseteq C$ is countable.

We claim that $J'' \subseteq \bigcup_{t \in I} A(t)$. To prove this, suppose the contrary, so that there exists $j \in J''$ such that $j \notin A(t)$ for every $t \in I$. Then $\pi_t e_j p_j \varepsilon_t: U_t \to U_t$ is not an epimorphism for every $t \in I$, so that

$$p_j \varepsilon_t \pi_t e_j: V_j \to V_j$$

is not an epimorphism for every $t \in I$ by Lemma 6.26(b). Hence for every $t \in I$ there is a cyclic proper submodule $C_t \subset V_j$ such that $p_j \varepsilon_t \pi_t e_j(V_j) \subseteq C_t$. For every $x \in V_j$ there are only a finite number of $t \in I$ such that $\pi_t e_j(x) \neq 0$, so that it is possible to define a homomorphism $\psi : V_j \to \oplus_{t \in I} C_t$ by

$$\psi(x) = (p_j \varepsilon_t \pi_t e_j(x))_{t \in I}.$$

Let $\omega : \oplus_{t \in I} C_t \to V_j$ be defined by $\omega((c_t)_{t \in I}) = \sum_{t \in I} c_t$. Then $\omega \psi = 1_{V_j}$, so that V_j is isomorphic to a direct summand of $\oplus_{t \in I} C_t$. As $j \in J''$, the module V_j is quasi-small. Hence there is a finite subset $F \subseteq I$ such that V_j is isomorphic to a direct summand of $\oplus_{t \in F} C_t$. In particular, V_j is finitely generated, and so cyclic. Let v be a generator of V_j. There exists a finite subset $G \subseteq I$ such that $v \in \oplus_{t \in G} U_t$. Then $\sum_{t \in G} p_j \varepsilon_t \pi_t e_j(v) = v$ forces

$$v \in \sum_{t \in G} p_j \varepsilon_t \pi_t e_j(V_j) \subseteq \sum_{t \in G} C_t \subset V_j,$$

a contradiction. This proves the claim.

Now $J_Z \subseteq \bigcup_{t \in I_z} A(t)$, because if $j \in J_Z$, then by the claim there exists $t \in I$ such that $j \in A(t)$. The mapping $\pi_t e_j p_j \varepsilon_t$ is an epimorphism, so that $[U_t]_e = [V_j]_e = [Z]_e$ by Lemma 6.26(b). Since Z is quasi-small, U_t must be quasi-small by Proposition 9.30(b), that is, $t \in I_Z$. Therefore

$$|J_Z| \leq \aleph_0 |I_Z| = |I_Z|. \qquad \square$$

In particular,

Corollary 9.33 *Let R be a ring and let $\{U_i \mid i \in I\}$ $\{V_j \mid j \in J\}$ be two families of non-zero quasi-small uniserial R-modules. Then $\oplus_{i \in I} U_i \cong \oplus_{j \in J} V_j$ if and only if there exist two bijections $\sigma, \tau : I \to J$ such that $[U_i]_m = [V_{\sigma(i)}]_m$ and $[U_i]_e = [V_{\tau(i)}]_e$ for every $i \in I$.* $\qquad \square$

9.10 Notes on Chapter 9

For an arbitrary module A_R, the analogues of the ideals I and K described in Theorem 9.1 are the sets of all the endomorphisms of A_R with essential kernel and all the endomorphisms of A_R with superfluous image. These are always two two-sided ideals of $\mathrm{End}(A_R)$.

A left or right artinian ring is a *QF-2 ring* if each of its indecomposable projective left and right modules has a simple socle (see [Anderson and Fuller]). If R is a left or right artinian ring, then R is semiperfect, so that it has a basic set of primitive idempotents $\{e_1, \ldots, e_m\}$, and $\{e_1 R, \ldots, e_m R\}$ and $\{R e_1, \ldots, R e_m\}$ are complete sets of representatives of the indecomposable projective left and right R-modules. In particular, every indecomposable projective left and right module is local. It follows that a left or right artinian ring is a *QF-2 ring* if and only if every indecomposable projective left and right module is biuniform, if and only if R_R and $_RR$ are direct sums of biuniform modules, if and only if $\dim(R_R) = \dim(_RR) = \mathrm{codim}(R)$.

Almost all the results in this chapter are due to [Facchini 96] and [Dung and Facchini a]. The only exceptions are Lemmas 9.17, 9.18 and Theorem 9.19, due to Puninski, and Proposition 9.23, due to Herbera. The results in this chapter were proved by [Facchini 96] and [Dung and Facchini a] for uniserial modules, but the results hold for biuniform modules as well, as was observed independently by Dung and Herbera. The papers [Facchini 96] and [Dung and Facchini 97] contain other results concerning direct sums of uniserial modules. For instance, in [Facchini 96, Corollary 1.13] it is proved that if $U_1, \ldots, U_n$ are Krull-Schmidt uniserial modules, $V_1, \ldots, V_r$ are non-Krull-Schmidt uniserial modules, $t_1, \ldots, t_n, u_1, \ldots, u_r$ are non-negative integers, and

$$M = U_1^{t_1} \oplus \cdots \oplus U_n^{t_n} \oplus V_1^{u_1} \oplus \cdots \oplus V_r^{u_r},$$

then M has at most $n + r^2$ non-isomorphic uniserial direct summands $\neq 0$.

Note that Theorem 9.13 and Example 9.20 answer Warfield's question about the uniqueness for decompositions of a finitely presented module over a serial ring into uniserial summands (see the Preface).

For an axiomatic approach to the results in this chapter and for an application to a class of modules that are torsionfree with respect to a suitable hereditary torsion theory, see [Bican].

In Proposition 9.24 we have seen that every unshrinkable uniserial module is of type 1, that is, it has a local endomorphism ring, so that the Krull-Schmidt Theorem holds for serial modules that are direct sums of unshrinkable uniserial modules (Corollary 9.25). [Facchini and Salce, Theorem 4] have proved the following extension of this result: Let $V_1, \ldots, V_n, U_1, \ldots, U_m$ be uniserial right modules over an arbitrary ring R. Let M be an R-module that is both a homomorphic image of $V_1 \oplus \cdots \oplus V_n$ and a submodule of $U_1 \oplus \cdots \oplus U_m$. Suppose that either

(a) for every $i = 1, 2, \ldots, m$ every homomorphic image of V_i is unshrinkable, or that

(b) for every $j = 1, 2, \ldots, n$ every submodule of U_j is unshrinkable.

Then M is serial.

A different kind of condition on the families of uniserials is considered in [Dung and Facchini 97, Section 5]. Recall that a family of modules

$$\{ M_i \mid i \in I \}$$

is called *locally semi-T-nilpotent* if, for any countably infinite set of non-isomorphisms $\{ f_n \colon M_{i_n} \to M_{i_{n+1}} \mid n \geq 0 \}$ with all the i_n distinct in I, and for any $x \in M_{i_1}$, there exists a positive integer k (depending on x) such that $f_k \ldots f_1(x) = 0$. [Zimmermann-Huisgen and Zimmermann 84] have shown that there is a connection between the locally semi-T-nilpotency of a family of modules with local endomorphism rings and the exchange property of their direct sum, because they proved that if M_i, $i \in I$, are modules with $\text{End}(M_i)$ local for all $i \in I$, then $M = \oplus_{i \in I} M_i$ has the (finite) exchange property if and only if the family $\{ M_i \mid i \in I \}$ is locally semi-T-nilpotent.

If $\{ U_i \mid i \in I \}$ and $\{ V_j \mid j \in J \}$ are two locally semi-T-nilpotent families of non-zero uniserial modules over an arbitrary ring R, then $\oplus_{i \in I} U_i \cong \oplus_{j \in J} V_j$ if and only if there are two bijections $\sigma, \tau : I \to J$ such that $[U_i]_m = [V_{\sigma(i)}]_m$ and $[U_i]_e = [V_{\tau(i)}]_e$ for every $i \in I$ [Dung and Facchini 97, Theorem 5.4].

Chapter 10

Σ-pure-injective Modules and Artinian Modules

The aim of this chapter is to study the endomorphism rings of artinian modules (Section 10.2) and, in particular, to characterize the endomorphism rings of artinian modules that are serial rings (Theorem 10.23). If M_S is an artinian module over an arbitrary ring S and $R = \mathrm{End}(M_S)$, then $_RM_S$ is a bimodule and $_RM$ is a faithful Σ-pure-injective module (Example 1.41). This leads us to study rings with a faithful Σ-pure-injective module (Section 10.1). In Section 10.3 we proceed to consider distributive modules, because every Σ-pure-injective module over a serial ring is a direct sum of distributive modules as a module over its endomorphism ring. In Sections 10.4 and 10.5 we consider Σ-pure-injective modules over chain rings and describe them making use of localization at a completely prime ideal. The main result of Section 10.5 is Theorem 10.20, which says that a module over a serial ring is Σ-pure-injective if and only if it is artinian as a module over its endomorphism ring. Sections 10.6 and 10.7 are devoted to characterizing the endomorphism rings of artinian modules that are serial rings via Krull dimension. In Section 10.8 we extend the results about Σ-pure-injective modules over chain rings found in the previous sections to Σ-pure-injective modules over serial rings.

10.1 Rings with a faithful Σ-pure-injective module

In this section we shall prove that a ring with a faithful Σ-pure-injective module satisfies the a.c.c. on right annihilators and its nil subrings are nilpotent. Proposition 10.1 is somewhat more precise.

Proposition 10.1 *Let R be an arbitrary ring and let M_R be a faithful R-module.*

(a) *If M satisfies the d.c.c. on additive subgroups of the form $(0 :_M F)$, where F is a finite subset of R, then R satisfies the a.c.c. on right annihilators.*

(b) *If M satisfies the d.c.c. on additive subgroups of the form $(0 :_M F)$, where F is a finite subset of R, and the d.c.c. on additive subgroups of the form Mr, $r \in R$, then every nil subring of R is nilpotent.*

Proof. (a) Suppose that M satisfies the d.c.c. on additive subgroups of the form $(0 :_M F)$, where F is a finite subset of R. Let $\mathcal{L}$ be the set of all right annihilators of R and let $\mathcal{L}' = \{ (0 :_M F) \mid F$ a finite subset of $R \}$. If $I \in \mathcal{L}$, then $(0 :_M I)$ is the intersection of all the subgroups $(0 :_M F)$, when F ranges in the set of finite subsets of I. By the d.c.c. on these subgroups, $(0 :_M I) \in \mathcal{L}'$. Hence there is an inclusion reversing mapping $\varphi \colon \mathcal{L} \to \mathcal{L}'$ defined by $\varphi(I) = (0 :_M I)$ for every $I \in \mathcal{L}$. Since M_R is a faithful module, for every subset $A \subseteq R$ one has that the right annihilator $I = \mathrm{r.ann}_R(A)$ of A in R is equal to $(0 :_R N)$, where $N = \{ ma \mid m \in M, \ a \in A \}$. Then $N \subseteq (0 :_M (0 :_R N)) = (0 :_M I)$, so that $(0 :_R (0 :_M I)) \subseteq (0 :_R N) = I$. Conversely, $I \subseteq (0 :_R (0 :_M I))$. Hence $I = (0 :_R (0 :_M I))$. It follows that φ is injective. Therefore the d.c.c. in $\mathcal{L}'$ yields the a.c.c. in $\mathcal{L}$.

(b) Let N be a nil subring of R that is not nilpotent. By (a) the ring R has the a.c.c. on right annihilators, so that by Proposition 6.8 the subring N is not right T-nilpotent. Hence by Proposition 6.7 there exists a sequence $y_0, y_1, y_2, \ldots$ of elements of N such that $y_n y_{n-1} \cdots y_1 y_0 \neq 0$ for every $n \geq 0$ and $y_i y_n y_{n-1} \cdots y_1 y_0 = 0$ for every $n \geq i \geq 0$.

Set $a_n = y_n \cdot \ldots \cdot y_0$ for every $n \geq 0$. Since the chain $Ma_0 \supseteq Ma_1 \supseteq \ldots$ is stationary, there exists a positive integer k such that $Ma_n = Ma_k$ for every $n \geq k$. Then $My_{n+1}a_n = Ma_{n+1} = Ma_n$, from which $M = (0 :_M a_n) + My_{n+1}$ for every $n \geq k$. From $y_i a_n = 0$ it follows that $My_i \subseteq (0 :_M a_n)$ for every $n \geq i \geq 0$. By induction we shall prove that

$$M = \left(\bigcap_{m=k}^{n} (0 :_M a_m) \right) + \sum_{i=k+1}^{n+1} My_i \tag{10.1}$$

for every $n \geq k$. We already know that the case $n = k$ holds. Suppose that equality (10.1) holds for $n - 1$. Then

$$(0 :_M a_n) = (0 :_M a_n) \cap M = (0 :_M a_n) \cap \left(\left(\bigcap_{m=k}^{n-1} (0 :_M a_m) \right) + \sum_{i=k+1}^{n} My_i \right)$$

by the induction hypothesis, so that

$$(0 :_M a_n) = \left(\bigcap_{m=k}^{n} (0 :_M a_m) \right) + \sum_{i=k+1}^{n} My_i$$

by the modular identity. Hence

$$M = (0 :_M a_n) + My_{n+1} = \left(\bigcap_{m=k}^{n} (0 :_M a_m) \right) + \sum_{i=k+1}^{n+1} My_i,$$

which proves equality (10.1).

As M satisfies the d.c.c. on subgroups of the form $(0 :_M F)$, where F is a finite subset of R, there exists a positive integer $t \geq k$ such that

$$\bigcap_{m=k}^{t} (0 :_M a_m) = \bigcap_{m=k}^{t+1} (0 :_M a_m).$$

Then both

$$\bigcap_{m=k}^{t} (0 :_M a_m) \quad \text{and} \quad \sum_{i=k+1}^{t+1} My_i$$

are contained in $(0 :_M a_{t+1})$. Hence $M = (0 :_M a_{t+1})$. But M_R is faithful, contradiction. $\qquad\square$

Corollary 10.2 *Let R be a ring with a faithful Σ-pure-injective right module. Then R satisfies the a.c.c. on right annihilators and every nil subring of R is nilpotent.*

Proof. By Theorem 1.40 every Σ-pure-injective module M_R satisfies the d.c.c. on finitely definable subgroups. Since the additive subgroups of the form $(0 :_M F)$, where F is a finite subset of R, and of the form Mr, $r \in R$, are finitely definable, the corollary follows from Proposition 10.1. $\qquad\square$

Corollary 10.2 allows us to characterize the commutative semiprime rings with a faithful Σ-pure-injective module:

Example 10.3 *Let R be a commutative semiprime ring. The following conditions are equivalent:*

(a) *R has a faithful Σ-pure-injective module;*

(b) *R has the a.c.c. on annihilators;*

(c) *R has only a finite number of minimal prime ideals.*

Proof. (a) $\Rightarrow$ (b). Corollary 10.2.

(b) $\Rightarrow$ (c). Let R be a commutative ring with the a.c.c. on annihilators. Suppose that there exist proper annihilators that do not contain the product of a finite number of prime ideals. Let I be maximal among these proper annihilators. In particular, I is not prime, so that there exist $r, s \in R$, $r, s \notin I$ such that $rs \in I$. Set $J = \text{ann}(\text{ann}(I)r)$ and $K = \text{ann}(\text{ann}(I)J)$. Then $J, K \supseteq I$. From

$\operatorname{ann}(I)rs = 0$ we obtain that $s \in \operatorname{ann}(\operatorname{ann}(I)r) = J$. Hence $J \supset I$. Similarly, $\operatorname{ann}(I)rJ = 0$ implies that $r \in \operatorname{ann}(\operatorname{ann}(I)J) = K$. Hence $K \supset I$. Now J is a proper ideal, because otherwise $r \in \operatorname{ann}(\operatorname{ann}(I)) = I$, a contradiction. And K is a proper ideal, because otherwise $\operatorname{ann}(I)J = 0$. Hence $J \subseteq \operatorname{ann}(\operatorname{ann}(I)) = I$, contradiction. By the maximality of I, both J and K contain finite products of primes, so that JK contains a finite product of primes. But $\operatorname{ann}(I)JK = 0$, so that $JK \subseteq \operatorname{ann}(\operatorname{ann}(I)) = I$, contradiction. This proves that every proper annihilator contains a finite product of prime ideals. In particular, there exist prime ideals $P_1, \ldots, P_n$ such that $P_1 \ldots P_n = 0$. Then every prime ideal of R contains one of the ideals $P_1, \ldots, P_n$, that is, R has at most n minimal prime ideals.

(c) $\Rightarrow$ (a). If a commutative semiprime ring R has finitely many minimal prime ideals, then R embeds in a finite direct product of commutative integral domains. Hence R embeds in a finite direct product of fields, that is, in a semisimple artinian ring S. Obviously, the S-module S is a faithful Σ-pure-injective S-module. Thus it is a faithful Σ-pure-injective R-module by Corollary 1.36. $\qquad\square$

10.2 Rings isomorphic to endomorphism rings of artinian modules

Let $\mathcal{A}$ be the class of rings that are isomorphic to endomorphism rings of artinian modules:

$$\mathcal{A} = \{\, R \mid R \cong \operatorname{End}(M_S) \text{ for some artinian module } M_S \text{ over some ring } S \,\}.$$

In this section we shall concentrate our attention on the class $\mathcal{A}$.

We have already seen that $\mathcal{A}$ contains all module-finite algebras over a semilocal commutative noetherian ring (Proposition 8.18) and that every ring in $\mathcal{A}$ is semilocal (Theorem 4.12). In particular, every ring in $\mathcal{A}$ has a complete set of orthogonal primitive idempotents, so that every ring in $\mathcal{A}$ has a direct product decomposition into indecomposable rings (Theorem 5.13). It is easily seen that if R and T are rings and $R \times T$ is their direct product, then $R \times T \in \mathcal{A}$ if and only if both R and T belong to $\mathcal{A}$.

We have proved that Krull-Schmidt fails for artinian modules (Examples 8.20 and 8.21). Hence the identity of a ring belonging to $\mathcal{A}$ is not the sum of orthogonal primitive idempotents in a unique way up to isomorphism. More precisely, there is a ring $R \in \mathcal{A}$ and indecomposable, pairwise nonisomorphic, direct summands P_1, P_2, P_3, P_4 of R_R such that $R_R = P_1 \oplus P_2 = P_3 \oplus P_4$ (Example 8.20 and Theorem 4.7). And for every integer $n > 2$ there is a ring $R_n \in \mathcal{A}$ such that the right R_n-module R_n is the direct sum of i indecomposable modules for each i where $2 \leq i \leq n$ (Example 8.21 and Theorem 4.7).

Recall that two rings R and T are *Morita equivalent* if there is an additive category equivalence Mod-$R \to$ Mod-T. Equivalently, R and T are Morita equivalent if and only if $T \cong \text{End}(P_R)$ for some progenerator P_R (a right R-module P_R is a *progenerator* if it is finitely generated, projective, and every right R-module is a homomorphic image of a direct sum of copies of P_R). If R and T are Morita equivalent rings, then there is a integer $m > 0$ and an idempotent $m \times m$ matrix $A \in \mathbf{M}_m(R)$ such that $T \cong A\mathbf{M}_m(R)A$ [Anderson and Fuller, §§21 and 22].

The next proposition shows that the class $\mathcal{A}$ is closed under Morita equivalence.

Proposition 10.4 *If R and T are Morita equivalent rings and $R \in \mathcal{A}$, then $T \in \mathcal{A}$.*

Proof. Let R and T be Morita equivalent rings and suppose that $R \in \mathcal{A}$, so that $R \cong \text{End}(M_S)$ for an artinian right S-module N_S. There is an integer $m > 0$ and an idempotent matrix $A \in \mathbf{M}_m(R)$ such that $T \cong A\mathbf{M}_m(R)A$. But $\mathbf{M}_m(R) \cong \text{End}(N_S^m)$, so that AN^m is a direct summand of N^m. Hence it is an artinian S-module, and $\text{End}(AN_S^m) \cong A\text{End}(N_S^m)A \cong A\mathbf{M}_m(R)A \cong T$. Thus $T \in \mathcal{A}$. $\qquad\square$

If M_S is an artinian right module over a ring S, then M is a Σ-pure-injective left module over its endomorphism ring $\text{End}(M_S)$ (Example 1.41). Hence every ring $R \in \mathcal{A}$ has a faithful Σ-pure-injective left module. From the left-right dual of Corollary 10.2 we immediately obtain that

Corollary 10.5 *Every ring $R \in \mathcal{A}$ satisfies the a.c.c. on left annihilators, and every nil subring of R is nilpotent.* $\qquad\square$

We shall now prove that there is a bound on the nilpotency indices of nil subrings of a ring belonging to $\mathcal{A}$. Recall that every artinian module has finite dual Goldie dimension by Corollary 2.41.

Proposition 10.6 *Let M_S be an artinian module over a ring S and suppose that $\text{codim}(M_S) = n$. Set $R = \text{End}(M_S)$ and*

$$H = \{ f \in R \mid f(M) \text{ is superfluous in } M \}.$$

Then:

(a) *H is a nilpotent two-sided ideal of R.*

(b) *Every strictly descending chain of left annihilators of the ring R/H has at most $n + 1$ elements.*

Proof. (a) It is easily seen that H is a two-sided ideal of R (Lemma 5.8(b)). By Corollary 10.5 the ring R satisfies the a.c.c. on left annihilators, so that by the left-right dual of Proposition 6.8 it suffices to show that H is left T-nilpotent. Let $h_0, h_1, h_2, \ldots$ be a sequence of elements of H. Set

$$f_n = h_0 \ldots h_n$$

for every $n \geq 0$. Since M_S is artinian, there exists an m such that

$$f_m(M) = f_{m+1}(M).$$

Then $f_m(M) = f_{m+1}(M) = f_m h_{m+1}(M)$ forces $M = h_{m+1}(M) + \ker(f_m)$. But $h_{m+1}(M)$ is superfluous in M, so that $M = \ker(f_m)$, i.e., $f_m = 0$. This shows that H is left T-nilpotent.

(b) Let $A_1 \supset A_2 \supset \cdots \supset A_{m+1}$ be a strictly descending chain of left annihilators of the ring R/H. We must show that $m \leq n$. Since

$$A_i = \mathrm{l.ann}_{R/H}(\mathrm{r.ann}_{R/H}(A_i)),$$

the chain $\mathrm{r.ann}_{R/H}(A_1) \subset \mathrm{r.ann}_{R/H}(A_2) \subset \cdots \subset \mathrm{r.ann}_{R/H}(A_{m+1})$ is strictly ascending. Let V_i be the right ideal of R containing H such that

$$V_i/H = \mathrm{r.ann}_{R/H}(A_i)$$

for every $i = 1, \ldots, m+1$, and set $(H : V_i) = \{r \in R \mid rV_i \subseteq H\}$, so that $(H : V_i)/H = \mathrm{l.ann}_{R/H}(V_i/H) = A_i$. The chain

$$(H : V_1) \supset (H : V_2) \supset \cdots \supset (H : V_{m+1})$$

is obviously strictly descending, so that for every $i = 1, 2, \ldots, m$ there is an element $r_i \in (H : V_i)$, $r_i \notin (H : V_{i+1})$. Hence $r_i V_i \subseteq H$ and there exists an element $v_{i+1} \in V_{i+1}$ such that $r_i v_{i+1} \notin H$. Thus $r_i v_j \in H$ for every $j = 1, 2, \ldots, i$ and $r_i v_{i+1}(M)$ is not superfluous in M. For every $i = 1, 2, \ldots, m$ let N_i be a proper submodule of M such that $N_i + r_i v_{i+1}(M) = M$. Then

$$M = r_i^{-1}(N_i + r_i v_{i+1}(M)) = r_i^{-1}(N_i) + v_{i+1}(M) \qquad (10.2)$$

and

$$N_i + r_i(M) = M. \qquad (10.3)$$

Observe that $r_i^{-1}(N_i)$ is a proper submodule of M, otherwise $N_i \supseteq r_i(M)$, so that $N_i + r_i(M) = M$ forces $N_i = M$, contradiction.

For every i, j with $1 \leq i \leq j \leq m$ define a module $M_{i,j}$ as follows: $M_{i,i} = r_i^{-1}(N_i)$ for $i = 1, 2, \ldots, m$, and $M_{i,j} = v_{i+1}(M) + \cdots + v_j(M) + r_j^{-1}(N_j)$ for $1 \leq i < j \leq m$. We prove that the $M_{i,j}$ are proper submodules of M for every $1 \leq i \leq j \leq m$. We already know that $M_{i,i}$ is proper. If $i < j$ and $M_{i,j} = M$, then $v_{i+1}(M) + \cdots + v_j(M) + r_j^{-1}(N_j) = M$ implies

$$r_j v_{i+1}(M) + \cdots + r_j v_j(M) + N_j \supseteq r_j(M).$$

From (10.3) it follows that

$$r_j v_{i+1}(M) + \cdots + r_j v_j(M) + N_j \supseteq N_j + r_j(M) = M.$$

Since $r_j v_{i+1}, \ldots, r_j v_j \in H$, the submodules $r_j v_{i+1}(M), \ldots, r_j v_j(M)$ are superfluous in M, so that $N_j = M$, a contradiction. Hence all the submodules $M_{i,j}$ of M are proper.

We claim that the family $\{M_{i,i}, M_{i,i+1}, \ldots, M_{i,m}\}$ is coindependent for every $i = 1, 2, \ldots, m$. If we prove the claim, then, in particular, the family $\{M_{1,1}, M_{1,2}, \ldots, M_{1,m}\}$ is coindependent, so that $m \le \mathrm{codim}(M_S) = n$, which concludes the proof of the proposition.

The proof of the claim is by induction on $k = m - i$. The case $k = 0$ is trivial, because a family consisting of one proper submodule is always coindependent.

Suppose that the family $\{M_{i+1,i+1}, M_{i+1,i+2}, \ldots, M_{i+1,m}\}$ is coindependent. We must prove that

$$\{M_{i,i}, M_{i,i+1}, \ldots, M_{i,m}\}$$

is coindependent. As $M_{i,j} \supseteq v_{i+1}(M)$ for every $j = i+1, \ldots, m$, we have that

$$M_{i,i} + \bigcap_{j=i+1}^{m} M_{i,j} \supseteq r_i^{-1}(N_i) + v_{i+1}(M) = M$$

by (10.2). Hence we only have to show that if $\ell = i+1, i+2, \ldots, m$, then $M_{i,\ell} + \bigcap_{i \le j \le m,\, j \ne \ell} M_{i,j} = M$. Now

$$
\begin{aligned}
&M_{i,\ell} + \bigcap_{i \le j \le m,\, j \ne \ell} M_{i,j} && \text{(since } M_{i,\ell} \supseteq v_{i+1}(M)) \\
&= M_{i,\ell} + \Big(\bigcap_{i \le j \le m,\, j \ne \ell} M_{i,j}\Big) + v_{i+1}(M) \\
&= M_{i,\ell} + \Big(\big(\bigcap_{i+1 \le j \le m,\, j \ne \ell} M_{i,j}\big) \cap r_i^{-1}(N_i)\Big) + v_{i+1}(M) && \text{(by modularity)} \\
&= M_{i,\ell} + \Big(\bigcap_{i+1 \le j \le m,\, j \ne \ell} M_{i,j} \cap \big(r_i^{-1}(N_i) + v_{i+1}(M)\big)\Big) && \text{(by (10.2))} \\
&= M_{i,\ell} + \Big(\bigcap_{i+1 \le j \le m,\, j \ne \ell} M_{i,j} \cap M\Big) \\
&= M_{i,\ell} + \bigcap_{i+1 \le j \le m,\, j \ne \ell} M_{i,j} && \text{(as } M_{i,j} \supseteq M_{i+1,j}) \\
&\supseteq M_{i+1,\ell} + \bigcap_{i+1 \le j \le m,\, j \ne \ell} M_{i+1,j} = M && \text{(because the family } \{M_{i+1,j} \mid j = \\
& && i+1, \ldots, m\} \text{ is coindependent)}
\end{aligned}
$$

$$(10.4)$$

This concludes the proof of the claim. □

Proposition 10.7 *Let M_S be an artinian module over a ring S and suppose* $\mathrm{codim}(M_S) = n$. *Set $R = \mathrm{End}(M_S)$ and let k be the nilpotency index of the nilpotent ideal $H = \{\, f \in R \mid f(M) \text{ is superfluous in } M \,\}$ of R. Then*

$$L^{(n+1)k} = 0$$

for every nil subring L of R.

Proof. It is easily seen that if T is a ring with the property that every strictly descending chain of left annihilators of T has at most $n + 1$ elements, then $A^{n+1} = 0$ for every nilpotent subring A of T.

Now the subring L of R is nilpotent by Corollary 10.5, so that the subring $L + H/H$ of R/H is nilpotent. By Proposition 10.6(b) $(L/H)^{n+1} = 0$. Hence $L^{n+1} \subseteq H$, so that $L^{(n+1)k} \subseteq H^k = 0$. $\qquad\qquad\square$

10.3   Distributive modules

A lattice $(L, \vee, \wedge)$ is called *distributive* if $x \wedge (y \vee z) = (x \wedge y) \vee (x \wedge z)$ for all $x, y, z \in L$, or, equivalently, $x \vee (y \wedge z) = (x \vee y) \wedge (x \vee z)$ for all $x, y, z \in L$. A right module M_R is said to be a *distributive* module if the lattice $\mathcal{L}(M_R)$ of its submodules is distributive. For instance, every uniserial module is distributive.

Theorem 10.8 *The following are equivalent for a right R-module M_R:*

(a) *The module M_R is distributive.*

(b) *For all $a, b, c \in M_R$, $aR \cap (bR + cR) = (aR \cap bR) + (aR \cap cR)$.*

(c) *For all $a, b \in M_R$, $aR + bR = (a + b)R + (aR \cap bR)$.*

(d) *For all $a, b \in M_R$,*
$$(aR : b) + (bR : a) = R,$$
where $(aR : b) = \{r \in R \mid br \in aR\}$ and $(bR : a) = \{r \in R \mid ar \in bR\}$.

(e) *For all $a, b \in M_R$, $(a + b)R = (aR \cap (a + b)R) + (bR \cap (a + b)R)$.*

Proof. . (a) $\Rightarrow$ (b). This is obvious.

(b) $\Rightarrow$ (c). Suppose that (b) holds. We show that
$$aR + bR = (a + b)R + (aR \cap bR).$$

The inclusion $\supseteq$ is obvious. In order to prove the inclusion $\subseteq$, it is sufficient to prove that $a \in (a + b)R + (aR \cap bR)$ by symmetry. Now
$$a \in aR \cap (bR + (a + b)R) = (aR \cap bR) + (aR \cap (a + b)R)$$

by (b). And $(aR \cap bR) + (aR \cap (a + b)R) \subseteq (aR \cap bR) + (a + b)R$.

(c) $\Rightarrow$ (d). Suppose that (c) holds. It suffices to show that $1 = r + s$ with $r \in (aR : b)$ and $s \in (bR : a)$. By (c) we have that $a \in (a + b)R + (aR \cap bR)$, so that $a = (a + b)r + y$, where $r \in R$ and $y \in aR \cap bR$. Hence $br = a - ar - y \in aR$ forces $r \in (aR : b)$. Then
$$a(1 - r) = a - ar = br + y \in bR,$$

so that $s = 1 - r \in (bR : a)$.

(d) $\Rightarrow$ (e). The inclusion $(a + b)R \supseteq (aR \cap (a + b)R) + (bR \cap (a + b)R)$ is obvious. Conversely, by (d) there exist $r \in (aR : b)$ and $s \in (bR : a)$ with $r + s = 1$. Then $(a+b)r \in aR$, so that $(a+b)r \in aR \cap (a+b)R$. Similarly, $(a+b)s \in bR \cap (a+b)R$. Hence $a+b = (a+b)r + (a+b)s \in (aR \cap (a+b)R) + (bR \cap (a+b)R)$.

(e) $\Rightarrow$ (a). Let A, B, C be submodules of M_R. The inclusion

$$(A \cap C) + (B \cap C) \subseteq (A + B) \cap C$$

always holds. Conversely, suppose that $a + b = c \in (A + B) \cap C$, with $a \in A$ and $b \in B$. Then from (e) it follows that

$$c \in (a + b)R = (aR \cap (a + b)R) + (bR \cap (a + b)R) \subseteq (A \cap C) + (B \cap C). \quad \Box$$

From characterization (d) of Theorem 10.8 we obtain

Corollary 10.9 *Every distributive module over a local ring is uniserial.* $\quad \Box$

In the proof of Theorem 10.11 we need the following technical result of lattice theory.

Theorem 10.10 (Birkhoff, 1940) *Let L be a modular lattice. If C_0, C_1 are two linearly ordered subsets of L, then the sublattice of L generated by $C_0 \cup C_1$ is distributive.* $\quad \Box$

Since Theorem 10.10 is a result of lattice theory, we shall not prove it here. For a proof, see [Grätzer, Theorem IV.1.13].

Theorem 10.11 *If R is a serial ring, $\{e_1, \ldots, e_n\}$ is a complete set of orthogonal primitive idempotents in R, M_R is a Σ-pure-injective R-module and $S = \text{End}(M_R)$ is its endomorphism ring, then the left S-module $_S M$ has a direct sum decomposition $_S M = \oplus_{i=1}^n Me_i$ where each $_S Me_i$ is a distributive S-module with the d.c.c. on cyclic submodules.*

Proof. Since $\{e_1, \ldots, e_n\}$ is a complete set of orthogonal idempotents in $\text{End}(_S M)$, we know that $_S M = \oplus_{i=1}^n Me_i$. Hence it suffices to show that for every $i = 1, 2, \ldots, n$ the left S-module Me_i is distributive and has the d.c.c. on cyclic submodules.

Let $\mathcal{L}(_\mathbf{Z} Me_i)$ be the modular lattice of all additive subgroups of Me_i. Since R is a serial ring, the right R-module $e_i R$ is uniserial, hence the subgroups $(0 :_{Me_i} e_i r)$, $r \in R$, form a linearly ordered subset C_0 of $\mathcal{L}(_\mathbf{Z} Me_i)$. Similarly, the left R-module Re_i is uniserial, so that the subgroups Mre_i, $r \in R$, form a linearly ordered subset C_1 of $\mathcal{L}(_\mathbf{Z} Me_i)$. By Theorem 10.10 the subset $C_0 \cup C_1$ generates a distributive sublattice $\mathcal{D}$ of $\mathcal{L}(_\mathbf{Z} Me_i)$. As the elements of $C_0 \cup C_1$ are finitely definable subgroups of M_R, the groups in $\mathcal{D}$ are finitely definable subgroups of M_R (Lemma 1.33). But M_R is Σ-pure-injective, so that the distributive lattice $\mathcal{D}$ is an artinian poset (Theorem 1.40(b)). Let $\mathcal{L}_{\text{cyc}}(_S Me_i)$

be the sublattice of $\mathcal{L}(_{\mathbf{Z}}Me_i)$ generated by all the cyclic submodules Sme_i of $_SMe_i$, $m \in M$. If we prove that $Sme_i \in \mathcal{D}$ for every $m \in M$, then $\mathcal{L}_{\text{cyc}}(_SMe_i)$ is a sublattice of $\mathcal{D}$. Hence $\mathcal{L}_{\text{cyc}}(_SMe_i)$ is a distributive lattice, so that $_SMe_i$ is a distributive module by Theorem 10.8 (a)$\Leftrightarrow$(b). Moreover $\mathcal{D}$ is an artinian poset, hence $\mathcal{L}_{\text{cyc}}(_SMe_i)$ is an artinian poset, that is, $_SMe_i$ has the d.c.c. on cyclic submodules.

Thus it suffices to prove that $Sme_i \in \mathcal{D}$ for every $m \in M$. We claim that for every $m, n \in M$, we have $n \in Sme_i$ if and only if $n \in G$ for every $G \in \mathcal{D}$ with $me_i \in G$. If we prove the claim, then $Sme_i = \bigcap G$, where G ranges in the set of the subgroups $G \in \mathcal{D}$ with $me_i \in G$. Since the distributive lattice $\mathcal{D}$ is an artinian poset, the set of the subgroups $G \in \mathcal{D}$ with $me_i \in G$ has a smallest element H. Then $Sme_i = H \in \mathcal{D}$, and this concludes the proof of the theorem.

Let us prove the claim. If $m, n \in M$ and $n \in Sme_i$, then n belongs to every submodule of $_SM$ that contains me_i. Now $(0 :_{Me_i} e_i r)$ and $Me_i r e_i$ are submodules of $_SM$, so that all the elements of $\mathcal{D}$ are submodules of $_SM$. Hence $n \in G$ for every $G \in \mathcal{D}$ with $me_i \in G$.

Conversely, let m, n be elements of M and suppose that $n \in G$ for every $G \in \mathcal{D}$ with $me_i \in G$. In order to prove that $n \in Sme_i$ we must show that there exists an endomorphism g of M_R such that $g(me_i) = n$. By Proposition 1.37 it suffices to prove that for every finitely presented module F_R, every element $a \in Fe_i$ and every homomorphism $f : F_R \to M_R$ such that $f(a) = me_i$ there exists a homomorphism $f' : F_R \to M_R$ such that $f'(a) = n$. By Corollary 3.30 we may assume

$$F_R = e_{i_1} R / e_{i_1} r_1 R \oplus \cdots \oplus e_{i_k} R / e_{i_k} r_k R,$$

where $i_1, \ldots, i_k \in \{1, 2, \ldots, n\}$ and $r_1, \ldots, r_k \in R$. Fix an element $a \in Fe_i$ and a homomorphism $f : F_R \to M_R$ such that $f(a) = me_i$. Let $s_1, \ldots, s_k$ be elements of R such that $a = (e_{i_1} s_1 + e_{i_1} r_1 R, \ldots, e_{i_k} s_k + e_{i_k} r_k R)$, so that

$$a = ae_i = (e_{i_1} s_1 e_i + e_{i_1} r_1 R, \ldots, e_{i_k} s_k e_i + e_{i_k} r_k R).$$

For every $t = 1, \ldots, k$ set

$$m_t = f(e_{i_1} r_1 R, \ldots, e_{i_t} + e_{i_t} r_t R, \ldots, e_{i_k} r_k R).$$

Then $m_1, \ldots, m_k \in M_R$ are such that $m_t e_{i_t} = m_t$ and $m_t e_{i_t} r_t = 0$ for every $t = 1, \ldots, k$, and $m_1 e_{i_1} s_1 e_i + \cdots + m_k e_{i_k} s_k e_i = me_i$. Thus

$$me_i \in \sum_{t=1}^{k} (0 :_{Me_{i_t}} e_{i_t} r_t) e_{i_t} s_t e_i.$$

If $e_{i_t} s_t e_i R \subseteq e_{i_t} r_t R$, then $(0 :_{Me_{i_t}} e_{i_t} r_t) e_{i_t} s_t e_i = 0$. If $e_{i_t} s_t e_i R \supseteq e_{i_t} r_t R$, then there exists $u_t \in R$ such that $e_{i_t} r_t = e_{i_t} s_t e_i u_t$, and it is easily seen that

$$(0 :_{Me_{i_t}} e_{i_t} r_t) e_{i_t} s_t e_i = (0 :_{Me_i} e_i u_t) \cap Me_{i_t} s_t e_i.$$

Hence $G = \sum_{t=1}^{k} (0 :_{Me_{i_t}} e_{i_t} r_t) e_{i_t} s_t e_i \in \mathcal{D}$ and $me_i \in G$. By hypothesis $n \in G$, so that there are elements $n_t \in M_R$ such that $n_t e_{i_t} r_t = 0$ for every $t = 1, \ldots, k$ and $n = \sum_{t=1}^{k} n_t e_{i_t} s_t e_i$. Let $f' \colon F_R \to M_R$ be the homomorphism defined by $f'(e_{i_1} v_1 + e_{i_1} r_1 R, \ldots, e_{i_k} v_k + e_{i_k} r_k R) = \sum_{t=1}^{k} n_t e_{i_t} v_t$ for every $v_1, \ldots, v_k \in R$. Then

$$f'(a) = f'(e_{i_1} s_1 e_i + e_{i_1} r_1 R, \ldots, e_{i_k} s_k e_i + e_{i_k} r_k R) = \sum_{t=1}^{k} n_t e_{i_t} s_t e_i = n.$$

Therefore there exists an endomorphism g of M_R such that $g(me_i) = n$ by Proposition 1.37. $\qquad \square$

Corollary 10.12 *If R is a serial ring, $\{e_1, \ldots, e_n\}$ is a complete set of orthogonal primitive idempotents in R, M_R is an indecomposable Σ-pure-injective right R-module and $S = \mathrm{End}(M_R)$ is its endomorphism ring, then the left S-module ${}_S M$ has a direct sum decomposition ${}_S M = \oplus_{i=1}^{n} Me_i$ in which every Me_i is a uniserial artinian left S-module.*

Proof. By the previous theorem every ${}_S Me_i$ is distributive and has the d.c.c. on cyclic submodules. The ring S is local by Corollary 2.27. Hence ${}_S Me_i$ is uniserial by Corollary 10.9. And every uniserial module with the d.c.c. on cyclic submodules is artinian. $\qquad \square$

Corollary 10.12 shows that indecomposable Σ-pure-injective modules over a serial ring satisfy the hypotheses of Proposition 6.27. When the serial ring R is a chain ring we obtain the following corollary.

Corollary 10.13 *Let M_R be an indecomposable Σ-pure-injective right module over a chain ring R and let $S = \mathrm{End}(M_R)$ be its endomorphism ring. Then ${}_S M$ is a uniserial artinian left S-module.* $\qquad \square$

In the next proposition we show that every indecomposable Σ-pure-injective module over a serial ring is injective modulo its annihilator.

Proposition 10.14 *Let M_R be an indecomposable Σ-pure-injective right module over a serial ring R. Then M is an injective right $R/\mathrm{ann}_R(M)$-module.*

Proof. Since the indecomposable Σ-pure-injective module M_R is a faithful indecomposable Σ-pure-injective module over $R/\mathrm{ann}_R(M)$, we may assume that M_R is faithful without loss of generality. Let $\{e_1, \ldots, e_n\}$ be a complete set of orthogonal primitive idempotents in R and let $S = \mathrm{End}(M_R)$. By Corollary 10.12 the left S-module Me_1 is a uniserial artinian submodule of ${}_S M$ and $Me_1 \neq 0$ because M_R is faithful. In particular ${}_S Me_1$ has a non-zero socle. Let m be a non-zero element in the socle of ${}_S Me_1$ and let E_R be the injective envelope of the cyclic R-module mR.

We shall show that there exists a homomorphism $g\colon E_R \to M_R$ such that $g(m) = m$ applying Proposition 1.37. Let F_R be a finitely presented R-module, let $a \in Fe_1$ and let $f\colon F_R \to E_R$ be a homomorphism such that $f(a) = m$. We must prove that there exists a homomorphism

$$f'\colon F_R \to M_R$$

such that $f'(a) = m$. By Corollary 3.30 we may assume

$$F_R = e_{i_1}R/e_{i_1}r_1R \oplus \cdots \oplus e_{i_k}R/e_{i_k}r_kR,$$

where $i_1, \ldots, i_k \in \{1, 2, \ldots, n\}$ and $r_1, \ldots, r_k \in R$. Let $s_1, \ldots, s_k$ be elements of R such that $a = (e_{i_1}s_1 + e_{i_1}r_1R, \ldots, e_{i_k}s_k + e_{i_k}r_kR)$, so that

$$a = ae_1 = (e_{i_1}s_1e_1 + e_{i_1}r_1R, \ldots, e_{i_k}s_ke_1 + e_{i_k}r_kR).$$

Set

$$I = \{\, t \mid t = 1, \ldots, k, \ e_{i_t}s_te_1R \supset e_{i_t}r_tR \,\}$$

and

$$J = \{\, t \mid t = 1, \ldots, k, \ e_{i_t}s_te_1R \subseteq e_{i_t}r_tR \,\},$$

so that $\{I, J\}$ is a partition of $\{1, 2, \ldots, k\}$. For every $t \in I$ let $u_t \in R$ be such that $e_{i_t}r_t = e_{i_t}s_te_1u_t$. Since the right ideals e_1u_tR are contained in the uniserial module e_1R, there exists an index $\ell \in I$ such that $e_1u_\ell R \subseteq e_1u_tR$ for every $t \in I$. Then $e_{i_t}s_te_1u_\ell \in e_{i_t}s_te_1u_tR = e_{i_t}r_tR$ for every $t \in I$. And obviously $e_{i_t}s_te_1u_\ell \in e_{i_t}r_tR$ for every $t \in J$. This proves that $e_{i_t}s_te_1u_\ell \in e_{i_t}r_tR$ for every $t = 1, 2, \ldots, k$. Hence $ae_1u_\ell = 0$. Then $me_1u_\ell = f(ae_1u_\ell) = 0$. Now $\ell \in I$, that is, $e_{i_\ell}s_\ell e_1R \supset e_{i_\ell}r_\ell R$, and from this it follows that $e_{i_\ell}s_\ell e_1 \neq 0$. Hence $Me_{i_\ell}s_\ell e_1 \neq 0$ because M_R is faithful. In particular the S-submodule $Me_{i_\ell}s_\ell e_1$ of Me_1 contains m. Thus there exists $y \in M$ with $ye_{i_\ell}s_\ell e_1 = m$. Let $f'\colon F_R \to M_R$ be defined by

$$f'(e_{i_1}v_1 + e_{i_1}r_1R, \ldots, e_{i_k}v_k + e_{i_k}r_kR) = ye_{i_\ell}v_\ell$$

for every $v_1, \ldots, v_k \in R$. This f' is a well-defined homomorphism, because if $e_{i_\ell}v_\ell \in e_{i_\ell}r_\ell R$, then $ye_{i_\ell}v_\ell \in ye_{i_\ell}r_\ell R = ye_{i_\ell}s_\ell e_1u_\ell R = me_1u_\ell R = 0$. Moreover $f'(a) = f'(e_{i_1}s_1e_1 + e_{i_1}r_1R, \ldots, e_{i_k}s_ke_1 + e_{i_k}r_kR)) = ye_{i_\ell}s_\ell e_1 = m$, as desired. Hence we can apply Proposition 1.37. This shows that there exists a homomorphism $g\colon E_R \to M_R$ such that $g(m) = m$.

In particular, g is the identity on the essential submodule mR of E_R, and so $g\colon E_R \to M_R$ is a monomorphism. But E_R is injective, so that the image of g is a direct summand of M_R. As M_R is indecomposable, we obtain that g is an isomorphism. Therefore $M_R \cong E_R$ is injective. $\qquad\square$

10.4 Σ-pure-injective modules over chain rings

We begin this section with an elementary result.

Lemma 10.15 *Let S be a right denominator set in a ring R and*

$$\varphi: R \to R[S^{-1}]$$

the canonical homomorphism of R in the right quotient ring.
A right $R[S^{-1}]$-module is a Σ-pure-injective $R[S^{-1}]$-module if and only if it is a Σ-pure-injective R-module.

Proof. Let A be a right $R[S^{-1}]$-module. If $M = (a_{ij})$ is a $n \times m$ matrix with entries in $R[S^{-1}]$, then the elements of M can be written with a common denominator $\varphi(s)$, that is, $M = M'\varphi(s)^{-1}$ for an $n \times m$ matrix M' with entries in R. Thus $XM = 0$ if and only if $XM' = 0$ for every $X \in A^n$. Hence the finitely definable subgroups of $A_{R[S^{-1}]}$ coincide with the finitely definable subgroups of A_R. The conclusion now follows from Theorem 1.40. $\qquad\square$

In the following theorem we show that every indecomposable Σ-pure-injective module over a chain ring R can be viewed as an injective module over a suitable noetherian chain ring.

Theorem 10.16 *Let M_R be an indecomposable Σ-pure-injective module over a chain ring R. Set $P = \{r \in R \mid (0 :_M r) \neq 0\}$. Then the following statements hold:*

(a) *P is a completely prime two-sided ideal of R;*

(b) *M has a unique structure as a right $R_P/\mathrm{ann}_{R_P}(M)$-module that extends its structure as a right R-module, and is a Σ-pure-injective indecomposable faithful $R_P/\mathrm{ann}_{R_P}(M)$-module;*

(c) *the ring $R_P/\mathrm{ann}_{R_P}(M)$ is a noetherian chain ring;*

(d) *the module M is, as a right $R_P/\mathrm{ann}_{R_P}(M)$-module, the injective envelope of the unique simple $R_P/\mathrm{ann}_{R_P}(M)$-module.*

Proof. (a) Set $S = \mathrm{End}(M_R)$, so that $_SM$ is a uniserial artinian module by Corollary 10.13. There is a canonical ring homomorphism $\varphi: R \to \mathrm{End}(_SM)$, and $I = \{f \in \mathrm{End}(_SM) \mid f \text{ is not injective}\}$ is a two-sided completely prime ideal of $\mathrm{End}(_SM)$ by Theorem 9.1. Hence $P = \varphi^{-1}(I)$ is a two-sided completely prime ideal of R.

(b) The right R-module structure on M_R is defined by a ring homomorphism $\varphi: R \to \mathrm{End}(_ZM)^{\mathrm{op}}$. If $r \in R \setminus P$, then $\varphi(f)$ is an automorphism of $_ZM$ by Lemma 1.43. Hence φ induces a unique ring homomorphism $R_P \to \mathrm{End}(_ZM)^{\mathrm{op}}$ by Proposition 6.1. This shows that M has a unique structure as a right

R_P-module that extends its structure as a right R-module. Hence M has a unique structure as a right $R_P/\mathrm{ann}_{R_P}(M)$-module that extends its structure as a right R-module. Since M_R is indecomposable, $M_{R_P/\mathrm{ann}_{R_P}(M)}$ is indecomposable a fortiori. And $M_{R_P/\mathrm{ann}_{R_P}(M)}$ is Σ-pure-injective by Lemma 10.15.

(c) Note that if $q = rs^{-1} \in R_P$ and $(0 :_M q) = 0$, then $(0 :_M r) = 0$, so that $r \notin P$, and q is invertible in R_P. Hence in the ring $R_P/\mathrm{ann}_{R_P}(M)$ an element q is invertible if and only if $(0 :_M q) = 0$. By (b) we may assume that M_R is a Σ-pure-injective indecomposable faithful module over the chain ring R, that the maximal ideal P of R is the set of the elements $r \in R$ with $(0 :_M r) \neq 0$, and we must prove that R is noetherian.

Let us prove that R has the a.c.c. on principal right ideals. Let

$$r_0 R \subseteq r_1 R \subseteq r_2 R \subseteq \dots$$

be an ascending chain of principal right ideals. Then

$$(0 :_M r_0) \supseteq (0 :_M r_1) \supseteq (0 :_M r_2) \supseteq \dots$$

is a descending chain of finitely definable subgroups of M_R. Hence it is stationary. In order to prove that R has the a.c.c. on principal right ideals it is sufficient to show that $(0 :_M r_n) = (0 :_M r_{n+1})$ implies $r_n R = r_{n+1} R$. Since $r_n R \subseteq r_{n+1} R$, there exists an element $r' \in R$ such that $r_n = r_{n+1} r'$. Set $S = \mathrm{End}(M_R)$, so that $_S M$ is a uniserial artinian module (Corollary 10.13). Let $\rho_{r_n}, \rho_{r_{n+1}}, \rho_{r'}$ be the endomorphisms of $_S M$ given by right multiplication by r_n, r_{n+1}, r' respectively. Then ρ_{r_n} and $\rho_{r_{n+1}}$ induce injective homomorphisms $\rho'_{r_n}, \rho'_{r_{n+1}} : M/(0 :_M r_n) \to M$, and $\rho'_{r_n} = \rho_{r'} \rho'_{r_{n+1}}$. Since ρ'_{r_n} is injective, by Lemma 6.26(a) either $M/(0 :_M r_n) = 0$ or the homomorphism $\rho_{r'}$ is injective. Hence either $M = (0 :_M r_{n+1})$ or $(0 :_M r') = 0$, so that either $r_{n+1} = 0$ or r' is invertible in R. In both cases $r_n R = r_{n+1} R$.

This proves that R has the a.c.c. on principal right ideals, and clearly a chain ring with the a.c.c. on principal right ideals is right noetherian, and so noetherian by Proposition 5.3(b).

(d) As in the proof of (c) we may assume that R is equal to $R_P/\mathrm{ann}_{R_P}(M)$. In other words, we may assume that M_R is a faithful R-module, R is a noetherian chain ring with maximal ideal $P = \{r \in R \mid (0 :_M r) \neq 0\}$, and we must prove that M_R is the injective envelope of the unique simple R-module. By Proposition 10.14 the module M_R is injective. Hence it is sufficient to show that it has a non-zero socle. If $P = P^2$, then $P = 0$ by Nakayama's Lemma, so M is simple. If $P \neq P^2$, fix an element p of $P \setminus P^2$, so that $P = pR$. Since $p \in P$ we have that $(0 :_M p) \neq 0$. If $x \in (0 :_M p)$ and $x \neq 0$, then $xp = 0$ implies $xP = 0$. Hence $\mathrm{ann}_R(x) = P$ and $xR \cong R/P$ is simple. $\square$

10.5   Homogeneous Σ-pure-injective modules

Now that we have a description of the indecomposable Σ-pure-injective modules over a chain ring R we may begin to consider the case of an arbitrary Σ-pure-injective R-module. Such a module is a direct sum of indecomposable modules by Theorem 2.29.

Lemma 10.17 *If* $\{\, M_i \mid i \in I \,\}$ *is a non-empty family of right modules over a chain ring R and $M = \oplus_{i \in I} M_i$ is Σ-pure-injective, then the set*

$$\mathcal{N} = \{\, \mathrm{ann}_R(M_i) \mid i \in I \,\},$$

partially ordered by set inclusion, is well-ordered.

Proof. Since R is a chain ring, the set $\mathcal{N}$ is linearly ordered by set inclusion. If it is not well-ordered, it has an infinite strictly descending chain, that is, there is a sequence $i_0, i_1, i_2, \ldots$ of elements of I such that

$$\mathrm{ann}_R(M_{i_0}) \supset \mathrm{ann}_R(M_{i_1}) \supset \mathrm{ann}_R(M_{i_2}) \supset \ldots .$$

For every $n = 1, 2, 3, \ldots$ let r_n be an element in $\mathrm{ann}_R(M_{i_{n-1}}) \setminus \mathrm{ann}_R(M_{i_n})$. It is easily seen that the chain $Mr_1 \supset Mr_2 \supset Mr_3 \supset \ldots$ is strictly descending. Hence M_R has a strictly descending chain of finitely definable subgroups, and this contradicts Theorem 1.40. □

If M_R is a Σ-pure-injective module over a chain ring R, then $M = \oplus_{i \in I} M_i$, where each M_i is indecomposable, and this direct sum decomposition is essentially unique. Let P be a completely prime ideal in R. The Σ-pure-injective module $M = \oplus_{i \in I} M_i$ over the chain ring R is said to be *P-homogeneous* if $P = \{r \in R \mid (0 :_{M_i} r) \neq 0\}$ for every $i \in I$.

Proposition 10.18 *Let P be a completely prime ideal in a chain ring R and let $M_R \neq 0$ be a P-homogeneous Σ-pure-injective R-module. Then the following assertions hold:*

(a) *the module M has a unique structure as a right $R_P / \mathrm{ann}_{R_P}(M)$-module that extends its structure as a right R-module, and M is a Σ-pure-injective faithful $R_P / \mathrm{ann}_{R_P}(M)$-module;*

(b) *the ring $R_P / \mathrm{ann}_{R_P}(M)$ is a noetherian chain ring;*

(c) *for every $x \in M$, there exists an integer $n \geq 0$ such that $xP^n = 0$;*

(d) *$MQ = 0$ for every completely prime ideal of R properly contained in P.*

Proof. The module M_R is a direct sum of indecomposable modules M_i, $i \in I$ (Theorem 2.29). Since M_R is P-homogeneous, by Theorem 10.16(b) each M_i has a unique structure as a right R_P-module that extends its structure as a right R-module. Hence M has a unique structure as a right R_P-module that extends its R-module structure and $\operatorname{ann}_{R_P}(M) = \bigcap_{i \in I} \operatorname{ann}_{R_P}(M_i)$. By Lemma 10.17 the set $\{\operatorname{ann}_R(M_i) \mid i \in I\}$ has a least element $\operatorname{ann}_R(M_j)$, so that $\operatorname{ann}_{R_P}(M) = \operatorname{ann}_R(M_j)$. Therefore (a) and (b) follow from Lemma 10.15 and Theorem 10.16(c).

In order to prove (c) we may assume that M_R is indecomposable without loss of generality. Since M is a module over the noetherian chain ring $R_P/\operatorname{ann}_{R_P}(M)$, we may also assume that R is a noetherian chain ring with maximal ideal P and M_R is faithful. In this case M_R is the injective envelope of the simple module (Theorem 10.16(d)). If $x \in M$ and $x \neq 0$, there is an element $r \in R$ such that xr is a non-zero element in the socle of M_R. Thus $xrP = 0$. The non-zero right ideal rR of R is a power of P (remark after the proof of Proposition 5.3). Hence $rP = P^{n-1}$ for some integer $n \geq 0$. This proves (c).

For the proof of (d) we may also assume that M_R is indecomposable. If Q is a completely prime ideal of R that is properly contained in P, then Q is contained in every power of P. Therefore Q annihilates every element of M by (c). □

Interest in the notion of homogeneous Σ-pure-injective module is aroused by the fact that every Σ-pure-injective module is a finite direct sum of homogeneous components, as the next theorem shows.

Theorem 10.19 *Let M_R be a Σ-pure-injective module over a chain ring R. Then there is a direct sum decomposition $M_R = M_0 \oplus M_1 \oplus \cdots \oplus M_n$, in which for each $i = 0, 1, \ldots, n$ the module M_i is a P_i-homogeneous Σ-pure-injective R-module for some completely prime ideal P_i of R.*

Proof. We know that the module M_R is a direct sum of indecomposable modules M_i, $i \in I$. For every $i \in I$ let P_i be the completely prime ideal of R such that M_i is P_i-homogeneous. We must prove that the set $\mathcal{P} = \{ P_i \mid i \in I \}$ is finite. Since $\mathcal{P}$ is linearly ordered by inclusion, if it is not finite it has either an infinite strictly ascending chain or an infinite strictly descending chain.

If $\mathcal{P}$ has a strictly ascending chain $P_{i_0} \subset P_{i_1} \subset P_{i_2} \subset \ldots$, let r_n be an element of $P_{i_{n+1}} \setminus P_{i_n}$ for every $n \geq 0$. Then $r_0 R \subset r_1 R \subset r_2 R \subset \ldots$, so that

$$(0 :_M r_0) \supseteq (0 :_M r_1) \supseteq (0 :_M r_2) \supseteq \ldots \tag{10.5}$$

is a descending chain of finitely definable subgroups of M_R. Now M_{i_n} is the injective envelope of the simple $R_{P_{i_n}}/\operatorname{ann}_{R_{P_{i_n}}}(M_{i_n})$-module (Theorem 10.16(d)). Let x_n be a non-zero element in the socle of the $R_{P_{i_n}}/\operatorname{ann}_{R_{P_{i_n}}}(M_{i_n})$-module M_{i_n}. Then $xP_{i_n} = 0$, so that $xr_{n-1} = 0$, i.e., $x \in (0 :_M r_{n-1})$. But $r_n \notin P_{i_n}$, hence r_n acts as an automorphism on the indecomposable P_{i_n}-homogeneous module M_{i_n}, and thus $xr_n \neq 0$, that is, $x \notin (0 :_M r_n)$. This proves that the chain (10.5) is strictly descending, which is a contradiction.

If $\mathcal{P}$ has a strictly descending chain

$$P_{i_0} \supset P_{i_1} \supset P_{i_2} \supset \dots,$$

then $P_{i_n} \supseteq \mathrm{ann}_R(M_{i_n})$, and $\mathrm{ann}_R(M_{i_n}) \supseteq P_{i_{n+1}}$ by Proposition 10.18(d). Therefore the set $\mathcal{N} = \{\, \mathrm{ann}_R(M_i) \mid i \in I \,\}$, partially ordered by set inclusion, is not artinian. This contradicts Lemma 10.17. □

We are ready to prove the main result of this section. It says that the Σ-pure-injective modules over a serial ring R are exactly the R-modules artinian over their endomorphism ring.

Theorem 10.20 *A right module over a serial ring is Σ-pure-injective if and only if it is artinian as a left module over its endomorphism ring.*

Proof. We already know that if $\mathrm{End}(M_R)M$ is artinian, then M_R is Σ-pure-injective (Example 1.41). Conversely, let M_R be a Σ-pure-injective module with R serial and let $S = \mathrm{End}(M_R)$. If $\{e_1, \dots, e_n\}$ is a complete set of orthogonal primitive idempotents in R, then $_S M = \oplus_{i=1}^n {}_S M e_i$, so that in order to prove that $_S M$ is artinian it is sufficient to prove that every $_S M e_i$ is artinian. Every endomorphism $f \in S = \mathrm{End}(M_R)$ induces by restriction an endomorphism of the right $e_i R e_i$-module $M e_i$. Hence if $S_i = \mathrm{End}((M e_i)_{e_i R e_i})$, every S_i-submodule of $(M e_i)_{e_i R e_i}$ is an S-submodule of $_S M e_i$. Thus in order to prove that each $_S M e_i$ is artinian it suffices to show that each $_{S_i} M e_i$ is artinian. The modules $(M e_i)_{e_i R e_i}$ are Σ-pure-injective (Corollary 1.44). Hence in order to prove the theorem we may suppose $R = e_i R e_i$, that is, we may suppose that R is a chain ring.

Let R be a chain ring and let M_R be a Σ-pure-injective module. By Theorem 10.19 there is a direct sum decomposition $M_R = M_0 \oplus M_1 \oplus \dots \oplus M_k$ with the property that for each $j = 0, 1, \dots, k$ the module M_j is a P_j-homogeneous Σ-pure-injective R-module for some completely prime ideal P_j of R. Since $\mathrm{End}(M_R) \supseteq \mathrm{End}(M_0) \times \mathrm{End}(M_1) \times \dots \times \mathrm{End}(M_k)$, in order to show that M is a left artinian module over $\mathrm{End}(M_R)$ it suffices to show that it is left artinian over $\mathrm{End}(M_0) \times \mathrm{End}(M_1) \times \dots \times \mathrm{End}(M_k)$. Hence we may assume that M_R is P-homogeneous for some completely prime ideal P of R. By Proposition 10.18 we may suppose that R is a noetherian serial ring. By Theorem 2.29 M_R is a direct sum of indecomposable modules M_i, $i \in I$, and by Lemma 10.17 the set $\mathcal{N} = \{\, \mathrm{ann}_R(M_i) \mid i \in I \,\}$ is well-ordered by set inclusion. As R is noetherian, the set $\mathcal{N}$ is finite. Hence there is a direct sum decomposition $M_R = N_0 \oplus N_1 \oplus \dots \oplus N_\ell$ with the property that each N_t is a direct sum of indecomposable faithful Σ-pure-injective $R/\mathrm{ann}(N_t)$-modules. Again $\mathrm{End}(M_R) \supseteq \mathrm{End}(N_0) \times \mathrm{End}(N_1) \times \dots \times \mathrm{End}(N_\ell)$, so that we may assume that M is a direct sum of indecomposable faithful Σ-pure-injective modules over the noetherian chain ring R. By Proposition 10.14 every indecomposable faithful Σ-pure-injective module is injective, so that M_R itself is injective because R is noetherian [Anderson and Fuller, Proposition 18.13].

Now the module M is uniserial as a left module over its endomorphism ring $S = \text{End}(M_R)$, because if $x, y \in M$, then either $(0 :_R x) \subseteq (0 :_R y)$ or $(0 :_R x) \supseteq (0 :_R y)$ (R is a chain ring), so that either there is a homomorphism $f : M_R \to M_R$ such that $f(x) = y$ or there is a homomorphism $f : M_R \to M_R$ such that $f(y) = x$ (because M_R is injective). And the uniserial module $_S M$ has the d.c.c. on cyclic submodules (Theorem 10.11). Hence $_S M$ is artinian. $\square$

10.6   Krull dimension and Σ-pure-injective modules

In this section we shall show that if $_S M$ is an artinian module over an arbitrary ring S and $R = \text{End}(_S M)$ is a chain ring, then R has finite Krull dimension.

Lemma 10.21 *Let $P \neq P'$ be two completely prime ideals of a chain ring R. Assume that there exists a P-homogeneous Σ-pure-injective R-module M_R and a P'-homogeneous Σ-pure-injective R-module M'_R such that $\text{Hom}_R(M, M') \neq 0$. Then $P \subset P'$ and there do not exist completely prime ideals properly between P and P'.*

Proof. Suppose that there exists an element $p \in P$, $p \notin P'$. Since $\text{Hom}_R(M, M') \neq 0$, there is a homomorphism $f : M \to M'$ and an element $x \in M$ such that $f(x) \neq 0$. By Proposition 10.18(c) there exists $n \geq 1$ such that $xp^n = 0$. Then $f(x)p^n = f(xp^n) = 0$. As $p \notin P'$ and M' is P'-homogeneous, the element p acts as a monomorphism on the abelian group M'. Thus $f(x) = 0$, a contradiction. This proves that $P \subseteq P'$. Since $P \neq P'$, it follows that $P \subset P'$.

Let Q be a completely prime ideal of R such that $P \subset Q \subset P'$. Then $MQ = M$ and $M'Q = 0$ (Proposition 10.18(d)). It follows that $\text{Hom}_R(M, M') = 0$, a contradiction. This concludes the proof of the lemma. $\square$

Recall that a right R-module M_R is *balanced* if the natural homomorphism $\rho : R \to \text{End}(_{\text{End}(M_R)}M)^{\text{op}}$ is surjective [Anderson and Fuller, §4]. Hence M_R is faithful and balanced if and only if ρ is an isomorphism.

Proposition 10.22 *Let R be a chain ring. Suppose that there exists a faithful, balanced, Σ-pure-injective module M_R. Then R has finite Krull dimension.*

Proof. By Theorem 10.19 there is a direct sum decomposition

$$M_R = M_0 \oplus M_1 \oplus \cdots \oplus M_m$$

of M_R such that for each $i = 0, 1, \ldots, m$ the module M_i is a non-zero P_i-homogeneous Σ-pure-injective R-module for some completely prime ideal P_i of R. Obviously we may assume $P_0 \subset P_1 \subset \cdots \subset P_m$.

We claim that every completely prime ideal Q of R is either equal to zero or equal to P_i for some index $i = 0, 1, \ldots, m$. In order to prove this claim, let Q be a completely prime ideal of R. If $Q \supset P_m$, then there exists an element $r \in Q \setminus P_m$. Since $r \notin P_i$ for every $i = 0, \ldots, m$, multiplication by r is an automorphism of the abelian group M_i for every i. Hence it is a group automorphism of M. Therefore multiplication by r is an automorphism of the module $\mathrm{End}_{(M_R)} M$. As the natural homomorphism $\rho \colon R \to \mathrm{End}(\mathrm{End}_{(M_R)} M)$ is an isomorphism, the element r is invertible in the ring R. This is a contradiction because $r \in Q$, whence $Q \subseteq P_m$.

If $Q \subset P_0$, then $M_i Q = 0$ for every $i = 0, \ldots, m$ by Proposition 10.18(d). Thus $MQ = 0$. But M is faithful, hence $Q = 0$. Therefore in order to prove our claim we may assume

$$P_0 \subseteq Q \subseteq P_m$$

and we must prove that $Q = P_i$ for some $i = 0, \ldots, m$. Suppose the contrary. Then there exists an index $i = 0, \ldots, m - 1$ such that $P_i \subset Q \subset P_{i+1}$. Set $M' = M_0 \oplus M_1 \oplus \cdots \oplus M_i$ and $M'' = M_{i+1} \oplus \cdots \oplus M_m$. Then M' and M'' are non-zero and $M_R = M' \oplus M''$.

If the abelian groups M', M'' are submodules of $\mathrm{End}_{(M_R)} M$, then

$$M = M' \oplus M''$$

is a direct sum decomposition of left $\mathrm{End}(M_R)$-modules, so that

$$\mathrm{End}(\mathrm{End}_{(M_R)} M) \cong R$$

has a non-trivial idempotent. This is a contradiction, because R is a chain ring. Hence either M' or M'' is not a submodule of $\mathrm{End}_{(M_R)} M$, that is, either $\mathrm{End}(_R M) M' \not\subseteq M'$ or $\mathrm{End}(_R M) M'' \not\subseteq M''$. In other words, either

$$\mathrm{Hom}_R(M', M'') \neq 0 \quad \text{or} \quad \mathrm{Hom}_R(M'', M') \neq 0.$$

This shows that there are two indices k and l such that $0 \leq k \leq i < l \leq m$ and either $\mathrm{Hom}_R(M_k, M_l) \neq 0$ or $\mathrm{Hom}_R(M_l, M_k) \neq 0$. From Lemma 10.21 it follows that $\mathrm{Hom}_R(M_l, M_k) = 0$ and there are no completely prime ideals properly between P_k and P_l. But $P_k \subseteq P_i \subset Q \subset P_{i+1} \subseteq P_l$, a contradiction. This concludes the proof of the claim. Hence R has either $m + 1$ completely prime ideals $P_0 \subset P_1 \subset \cdots \subset P_m$ or $m + 2$ completely prime ideals $0 \subseteq P_0 \subset P_1 \subset \cdots \subset P_m$.

We shall now prove that

$$\mathrm{ann}_R(M_i) = P_{i-1}$$

for every $i = 1, 2, \ldots, m$. From Proposition 10.18(d) we know that

$$P_{i-1} \subseteq \mathrm{ann}_R(M_i).$$

Assume that

$$P_{i-1} \subset \mathrm{ann}_R(M_i).$$

Then there exists an element $r \in \operatorname{ann}_R(M_i) \setminus P_{i-1}$. Set

$$M' = M_0 \oplus M_1 \oplus \cdots \oplus M_{i-1} \quad \text{and} \quad M'' = M_i \oplus \cdots \oplus M_m.$$

Then the abelian group M decomposes as a direct sum $M = M' \oplus M''$ of non-zero subgroups. Note that $M'r = M'$ and $M''r = 0$. There are three possible cases, according to whether M' and M'' are or are not submodules of $\operatorname{End}(M_R) M$. If both M' and M'' are submodules of $\operatorname{End}(M_R) M$, then $\operatorname{End}(M_R) M = M' \oplus M''$ has a non-trivial direct sum decomposition, so that its endomorphism ring $\operatorname{End}(\operatorname{End}(M_R) M) \cong R$ has a non-trivial idempotent, contradiction. If M' is not a submodule of $\operatorname{End}(M_R) M$, then $\operatorname{End}(_R M) M' \not\subseteq M'$, i.e., $\operatorname{Hom}_R(M', M'') \neq 0$. Hence there exist indices k, l such that $0 \leq k < i \leq l \leq m$ and $\operatorname{Hom}_R(M_k, M_l) \neq 0$. But $M_k r = M_k$ and $M_l r = 0$, so that for every $f \in \operatorname{Hom}_R(M_k, M_l)$ one has $f(M_k) = f(M_k r) = (f(M_k))r \subseteq M_l r = 0$. Thus $\operatorname{Hom}_R(M_k, M_l) = 0$, a contradiction. Similarly if M'' is not a submodule of $\operatorname{End}(M_R) M$. In this third case $\operatorname{End}(_R M) M'' \not\subseteq M''$, i.e., $\operatorname{Hom}_R(M'', M') \neq 0$. Thus there exist indices k, l such that $0 \leq k < i \leq l \leq m$ and $\operatorname{Hom}_R(M_l, M_k) \neq 0$. Lemma 10.21 implies that $P_l \subset P_k$, that is, $l < k$, a contradiction again. Therefore in every case we get a contradiction, and this shows that

$$\operatorname{ann}_R(M_i) = P_{i-1}$$

for every $i = 1, 2, \ldots, m$.

In particular, for each $i = 1, 2, \ldots, m$ the module M_i is a Σ-pure-injective faithful module over the ring R/P_{i-1}. It is easily seen that it is a P_i/P_{i-1}-homogeneous R/P_{i-1}-module, so that the ring $(R/P_{i-1})_{P_i/P_{i-1}}$ is noetherian by Proposition 10.18(b).

The ring R has the a.c.c. on right annihilators by Corollary 10.2. If the completely prime ideals of R are $P_0 \subset P_1 \subset \cdots \subset P_m$, then R has Krull dimension m by Theorem 7.35.

If the completely prime ideals of R are $0 \subset P_0 \subset P_1 \subset \cdots \subset P_m$, in order to conclude the proof applying Theorem 7.35, we must also prove that the ring R_{P_0} is noetherian. Now M_0 is a faithful R-module, because if r is an element of R such that $M_0 r = 0$, then $r \in \operatorname{ann}_R M_0 \subseteq P_0 \subset P_i$, so that from Proposition 10.18(d) we get that $M_i r = 0$ for every $i = 1, 2, \ldots, m$. Thus $Mr = 0$, i.e., $r = 0$ because M_R is faithful. Hence M_0 is a faithful, Σ-pure-injective, P_0-homogeneous R-module. From Proposition 10.18(b) it follows that the ring R_{P_0} is noetherian. This allows us to conclude that the chain ring R has finite Krull dimension $m + 1$. $\qquad \square$

10.7   Serial rings that are endomorphism rings of artinian modules

The aim of this section is to characterize the serial rings that belong to the class $\mathcal{A}$, that is, the serial rings that are isomorphic to endomorphism rings of artinian modules.

Theorem 10.23 *The following conditions are equivalent for a serial ring R with prime radical N:*

(a) $R \in \mathcal{A}$.

(b) *The ring R has finite Krull dimension and its classical quotient ring is artinian.*

(c) *The ring R has finite Krull dimension and $\mathcal{C}_R(0) = \mathcal{C}_R(N)$.*

(d) *The ring R is a ring of finite Krull dimension with the a.c.c. on left annihilators.*

Proof. Let $\{e_1, \ldots, e_n\}$ be a complete set of orthogonal primitive idempotents of R.

(a) $\Rightarrow$ (d). Suppose that $R \in \mathcal{A}$, so that $R \cong \mathrm{End}(M_S)$ for a ring S and an artinian S-module M_S and $_RM_S$ is a bimodule via the isomorphism $R \cong \mathrm{End}(M_S)$. Every chain ring $e_i R e_i$ is isomorphic to the endomorphism ring of the artinian S-module $e_i M_S$. Hence $e_i M$ is a faithful, balanced, Σ-pure-injective left $e_i R e_i$-module. By Proposition 10.22 the rings $e_i R e_i$ have finite Krull dimension, so that R has finite Krull dimension (Proposition 7.31(c)), and R has the a.c.c. on left annihilators by Corollary 10.5.

(b) $\Leftrightarrow$ (c). By Theorem 7.23.

(b) $\Leftrightarrow$ (d). By Theorem 6.54.

(d) $\Rightarrow$ (a). We claim that if R is a serial ring of finite Krull dimension m with the a.c.c. on left annihilators, then there exists a faithful, balanced, Σ-pure-injective, left R-module $_RM$ that contains a submodule isomorphic to $R/J(R)$. If we prove the claim, then M is a right module over its endomorphism ring $S = \mathrm{End}(_RM)$, M_S is artinian by the left-right dual of Theorem 10.20, and $R \cong \mathrm{End}(M_S)$, hence (a) holds.

The proof of the claim is by induction on m. If $m = 0$, the serial ring R is artinian. Then it is easily verified that the left R-module $_RM = R \oplus R/J(R)$ has the required properties.

Suppose $m > 0$ and that the claim holds for serial rings of Krull dimension $m - 1$ with the a.c.c. on left annihilators. Let R be a serial ring of Krull dimension m with the a.c.c. on left annihilators, let K be the derived ideal of R (page 190) and let $A = R/K$, so that A is a noetherian serial ring. As we saw on page 190, the maximal Ore set corresponding to the semiprime Goldie ideal K of R is $\mathcal{C}(K)$ and the elements of $\mathcal{C}(K)$ are regular in R, so that if T denotes the quotient ring of R with respect to $\mathcal{C}_R(K)$, we may assume $R \subseteq T$. By Theorem 7.37 T is a serial ring of finite Krull dimension $m - 1$ with the a.c.c. on left annihilators. Moreover, as we remarked on page 190, the ideal K is the Jacobson radical of T, T/K is a semisimple artinian ring, and T/K is the classical two-sided quotient ring of $A = R/K$. Set $Q = T/K$. As Q is a semisimple artinian ring, the module Q_Q is artinian. Hence the module $_R Q$ is Σ-pure-injective. By Proposition 6.48 the right A-module $Q/J(A)$ is artinian, so that the left A-module $Q/J(A)$ is Σ-pure-injective. In particular, $Q/J(A)$ is Σ-pure-injective as a left R-module.

Since we can apply the induction hypothesis to T, there exists a faithful, balanced, Σ-pure-injective, left T-module $_T N$ that contains a submodule isomorphic to T/K. In particular, N is a Σ-pure-injective left R-module. Thus if we set

$$_R M = N \oplus Q \oplus (Q/J(A)),$$

the left R-module M is Σ-pure-injective because it is the direct sum of three Σ-pure-injective left R-modules. As $_T N$ is faithful, it follows that $_R N$ is faithful, so $_R M$ is faithful. By Lemma 7.36(a) $K \subseteq J(R)$, so that

$$A/J(A) = (R/K)/J(R/K) = (R/K)/(J(R)/K) \cong R/J(R).$$

Hence $_R M \supseteq Q/J(A) \supseteq A/J(A) \cong R/J(R)$. Thus in order to conclude the proof of the claim it is sufficient to show that $_R M$ is a balanced R-module.

Let $S = \operatorname{End}(_R M)$ and let $\varphi \in \operatorname{End}(M_S)$. We must prove that φ is left multiplication by an element of R. If $f \in S = \operatorname{End}(_R M)$ is an idempotent endomorphism of $_R M$, then $\varphi(Mf) = (\varphi M)f \subseteq Mf$. Since all the direct summands of $_R M$ are of the form Mf for some idempotent $f \in S$, we have that every direct summand of $_R M$ is stable under the action of φ. Hence there exists an endomorphism α of the right $\operatorname{End}(_R N)$-module N, an endomorphism β of the right $\operatorname{End}(_R Q)$-module Q, and an endomorphism γ of the right $\operatorname{End}(_R(Q/J(A)))$-module $Q/J(A)$ such that $\varphi(x, y, z) = (\alpha(x), \beta(y), \gamma(z))$ for every $x \in N$, $y \in Q$, $z \in Q/J(A)$. As N is a left T-module and T is the quotient ring of R with respect to a set of regular elements, it follows that $\operatorname{End}(_R N) = \operatorname{End}(_T N)$. Hence α is an endomorphism of the right $\operatorname{End}(_T N)$-module N. As $_T N$ is balanced, there exists $t \in T$ such that $\alpha(x) = tx$ for every $x \in N$.

We shall now prove that $\beta(y) = ty$ for every $y \in Q = T/K$. Since $_T N$ contains a submodule isomorphic to $T/K = Q$, there exists a monomorphism of T-modules $f \colon {_T Q} \to {_T N}$. Let $f' \colon {_R M} \to {_R M}$ be the homomorphism defined by $f'(x, y, z) = (f(y), 0, 0)$ for every $(x, y, z) \in N \oplus Q \oplus (Q/J(A)) = {_R M}$. As $f' \in S$ and $\varphi \in \operatorname{End}(M_S)$, we have that $(\varphi(0, y, 0))f' = \varphi((0, y, 0)f')$ for every $y \in Q$, from which $f(\beta(y)) = \alpha(f(y))$. Thus $f(\beta(y)) = \alpha(f(y)) = t(f(y)) = f(ty)$. Since f is injective, it follows that $\beta(y) = ty$ for every $y \in Q$.

Now we prove that $t \in R$ and $\gamma(z) = tz$ for every $z \in {_R(Q/J(A))}$. Let $p \colon {_R Q} \to {_R(Q/J(A))}$ denote the canonical projection and let $p' \colon {_R M} \to {_R M}$ be the homomorphism defined by

$$p'(x, y, z) = (0, 0, p(y))$$

for every $(x, y, z) \in N \oplus Q \oplus (Q/J(A)) = {}_R M$. As $p' \in S$ and $\varphi \in \text{End}(M_S)$, we obtain that $(\varphi(0, y, 0))p' = \varphi((0, y, 0)p')$ for every $y \in Q$. Then

$$p(\beta(y)) = \gamma(p(y))$$

for every $y \in Q$. In particular, for $y = 1_Q$ we have that

$$\gamma(1_Q + J(A)) = \gamma(p(1_Q)) = p(\beta(1_Q)) = p(t \cdot 1_Q) = p(t + K).$$

Since γ is an endomorphism of the right $\text{End}({}_R(Q/J(A)))$-module $Q/J(A)$, γ is *a fortiori* an endomorphism of the right A-module $(Q/J(A))_A$. Hence the socle of $(Q/J(A))_A$ is invariant under the action of γ. By Proposition 6.48 the socle of $(Q/J(A))_A$ is $A/J(A)$, and therefore $\gamma(A/J(A)) \subseteq A/J(A)$, so that $p(t + K) = \gamma(1_Q + J(A)) \in A/J(A)$. Thus $t + K \in A = R/K$, that is, $t \in R$.

Finally, for every $z \in Q/J(A)$ there exists $y \in Q$ with $p(y) = z$. Then $\gamma(z) = \gamma(p(y)) = p(\beta(y)) = p(ty) = t(p(y)) = tz$. This shows that α, β and γ are left multiplication by the element $t \in R$. Hence φ is left multiplication by the element t of R, which concludes the proof of the claim. $\square$

In particular, every serial ring that is either left noetherian or right noetherian belongs to $\mathcal{A}$, because every such ring has Krull dimension at most one (Corollary 7.32) and has the a.c.c. on left annihilators (Theorem 6.54).

In the remaining part of this section we shall give an example that shows that the conditions which characterize the serial rings belonging to $\mathcal{A}$ (Theorem 10.23) do not hold for arbitrary rings belonging to $\mathcal{A}$.

Example 10.24 *There exists an indecomposable artinian right module M_T over a ring T such that $E = \text{End}(M_T)$ is a commutative ring of infinite Goldie dimension. In particular, E fails to have Krull dimension.*

Proof. We shall construct the ring E making use of Proposition 8.17. Let $K \subset F$ be fields such that F is a field extension of infinite degree over K. Let $F[x]$ be the ring of polynomials in one indeterminate x with coefficients in F and let $A = F[x]/(x^2)$, so that A can be viewed as the trivial extension $F \propto F$, that is, the abelian group $F \oplus F$ with multiplication defined by

$$(a, b)(c, d) = (ac, ad + bc)$$

for every $(a, b), (c, d) \in F \oplus F$. Then A is a commutative artinian ring that contains the trivial extension $E = K \propto F$ as a subring. It is easily seen that E is a local ring with maximal ideal $J = 0 \propto F$. Then $N = {}_E(E/J)_E$ is a bimodule, which is obviously simple both as a left and as a right E-module. The left E-module ${}_E(A/E)$ is semisimple, so it is cogenerated by ${}_E N$. By Proposition 8.17 there exists an artinian right module M_T over a ring T such that $E \cong \text{End}(M_T)$. Since E is local, M_T is indecomposable, and E has infinite Goldie dimension, because F is a vector space of infinite dimension over K. In particular, E fails to have Krull dimension by Proposition 7.13. $\square$

Observe that the endomorphism ring E of Example 10.24 is a commutative local ring with maximal ideal J for which the semisimple E-module J/J^2 is not finitely generated. Also note that E coincides with its classical quotient ring.

10.8 Localizable systems and Σ-pure-injective modules over serial rings

In Section 10.4 we described the structure of indecomposable Σ-pure-injective modules over chain rings making use of localization at completely prime ideals. In this section we shall extend those results to serial rings by describing the structure of Σ-pure-injective modules over serial rings. We apply the theory of localizable systems developed in Section 6.5.

In this section R will denote a serial ring and $\{e_1, \ldots, e_n\}$ will denote a complete set of orthogonal primitive idempotents of R.

Theorem 10.25 *Let M_R be an indecomposable Σ-pure-injective module over a serial ring R. For every index $i = 1, \ldots, n$ set $P_i = \{r \in e_i R e_i \mid (0 :_{Me_i} r) \neq 0\}$ and $\mathcal{P} = (P_1, \ldots, P_n)$. Then*

(a) $\mathcal{P} = (P_1, \ldots, P_n)$ *is a localizable system for R.*

(b) *For every $i = 1, \ldots, n$ and every $r \in e_i R e_i \setminus P_i$, right multiplication by r is an automorphism of the uniserial artinian left $\mathrm{End}(M_R)$-module Me_i.*

(c) *The R-module structure over M_R extends uniquely to an $R_{\mathcal{P}}$-module structure over M, and $M_{R_{\mathcal{P}}}$ is Σ-pure-injective and indecomposable.*

(d) *If M_R is faithful, then $M_{R_{\mathcal{P}}}$ is faithful.*

Proof. The left $\mathrm{End}(M_R)$-modules Me_j are uniserial artinian modules (Corollary 10.12). Hence (a) is an immediate application of Proposition 6.27. Part (b) holds because every injective endomorphism of an artinian module is an automorphism (Lemma 2.16(b)). In order to prove (c) we argue as in the proof of Theorem 10.16(b) applying the universal property of $R_{\mathcal{P}}$ (Proposition 6.22). For this it suffices to show that for every $x \in X_{\mathcal{P}}$ right multiplication by x is an automorphism of the abelian group M. This follows from (b) and the equality $x = \sum_{i=1}^{n} e_i x e_i$. Therefore M is an $R_{\mathcal{P}}$-module. The remaining part of (c) follows from Lemma 10.15. Finally, if M_R is faithful and q is an element of $R_{\mathcal{P}}$, then $q = rx^{-1}$ for suitable $r \in R$, $x \in X_{\mathcal{P}}$ (Section 6.5). Thus $Mq = 0$ implies $Mr = 0$, so $r = 0$. This proves (d). □

In Theorem 10.16(a) we saw that it is possible to associate a completely prime ideal P to every indecomposable Σ-pure-injective module over a chain ring. In Theorem 10.25(a) we have just seen that it is possible to associate a localizable system $\mathcal{P} = (P_1, \ldots, P_n)$ to every indecomposable Σ-pure-injective indecomposable module over a serial ring. In this notation $\mathcal{P} = (P_1, \ldots, P_n)$ is said to be *the localizable system of* M_R. We now show how to extend to modules over arbitrary serial rings the definition of homogeneous module we gave in Section 10.5 for modules over chain rings. Let M_R be a Σ-pure-injective module over a serial ring R. Then M_R is a direct sum of indecomposable modules M_i, $i \in I$, in an essentially unique way (Theorem 2.29). Let $\mathcal{P} = (P_1, \ldots, P_n)$ be a localizable system for R. The Σ-pure-injective module $M = \oplus_{i \in I} M_i$ over the serial ring R is said to be $\mathcal{P}$-*homogeneous* (and $\mathcal{P}$ is said to be the *localizable system of* M_R) if $\mathcal{P}$ is the localizable system of M_i for every $i \in I$.

Example 10.26 Let $\mathcal{P} = (P_1, \ldots, P_n)$ be a localizable system for a serial ring R, let $R_{\mathcal{P}}$ be the localization of R at $\mathcal{P}$ and let M_R be a $\mathcal{P}$-homogeneous Σ-pure-injective R-module. From Theorem 10.25(c) it is easily seen that M is an $R_{\mathcal{P}}$-module, which is Σ-pure-injective by Lemma 10.15. A complete set of primitive orthogonal idempotents for the ring $R_{\mathcal{P}}$ is given by the set

$$\{\, e_i \mid i = 1, 2, \ldots, n, \ P_i \neq \emptyset \,\}$$

(Proposition 6.24(c)). As an example we shall prove that the module M is homogeneous as a Σ-pure-injective $R_{\mathcal{P}}$-module and that the localizable system for the ring $R_{\mathcal{P}}$ associated with the Σ-pure-injective $R_{\mathcal{P}}$-module M is the localizable system given by all maximal ideals of the chain rings $e_i R_{\mathcal{P}} e_i$ (where i ranges over the set of all the indices $i = 1, 2, \ldots, n$ with $P_i \neq \emptyset$).

In order to prove this, we may suppose without loss of generality that M_R is indecomposable. Let i be an index such that $P_i \neq \emptyset$. We must prove that $Q_i = \{\, q \in e_i R_{\mathcal{P}} e_i \mid (0 :_{Me_i} q) \neq 0 \,\}$ coincides with the maximal ideal of the chain ring $e_i R_{\mathcal{P}} e_i$. Now $P_i \neq \emptyset$ implies that $Me_i \neq 0$, so $Q_i \neq \emptyset$. Thus Q_i is a completely prime ideal of $e_i R_{\mathcal{P}} e_i$. To show that Q_i is the maximal ideal of $e_i R_{\mathcal{P}} e_i$ it is sufficient to prove that every element q of $e_i R_{\mathcal{P}} e_i$ not in Q_i is invertible in $e_i R_{\mathcal{P}} e_i$. Now $q \in e_i R_{\mathcal{P}} e_i$ can be written in the form $q = e_i s e_i y^{-1}$ for suitable $s \in R$ and $y \in X_{\mathcal{P}}$ (Lemma 6.23(b)). If $q \notin Q_i$, then $(0 :_{Me_i} q) = 0$, so that $(0 :_{Me_i} e_i s e_i) = 0$. Hence $e_i s e_i \in e_i R e_i \setminus P_i$. Then $x = (1 - e_i) + e_i s e_i \in X_{\mathcal{P}}$. Hence x is invertible in $R_{\mathcal{P}}$. If x^{-1} is the inverse of x in $R_{\mathcal{P}}$ it is easily seen that $e_i y x^{-1} e_i$ is a right inverse for $q = e_i s e_i y^{-1}$ in $e_i R_{\mathcal{P}} e_i$. Thus q is right invertible in the chain ring $e_i R_{\mathcal{P}} e_i$. But in a local ring every right invertible element is invertible. $\qquad\square$

Now we shall prove that every $\mathcal{P}$-homogeneous Σ-pure-injective module over a serial ring R is a module over the ring $R_{\mathcal{P}}/\mathrm{ann}_{R_{\mathcal{P}}}(M)$ and that this ring is serial and noetherian. This generalizes Theorem 10.16, in which the case of a chain ring R was considered. We begin by showing that $R_{\mathcal{P}}/\mathrm{ann}_{R_{\mathcal{P}}}(M)$ is right noetherian.

Lemma 10.27 *Let* R *be a serial ring and let* $\mathcal{P} = (P_1, \ldots, P_n)$ *be a localizable system for* R. *If* M_R *is a* $\mathcal{P}$-*homogeneous* Σ-*pure-injective* R-*module, then* $R_{\mathcal{P}}/\mathrm{ann}_{R_{\mathcal{P}}}(M)$ *is a serial right noetherian ring.*

Proof. If M is a $\mathcal{P}$-homogeneous Σ-pure-injective R-module, then M is a homogeneous Σ-pure-injective $R_{\mathcal{P}}$-module and the localizable system for the ring $R_{\mathcal{P}}$ associated with the Σ-pure-injective $R_{\mathcal{P}}$-module M is the localizable system given by all the maximal ideals of the chain rings $e_i R_{\mathcal{P}} e_i$ (Example 10.26). Hence we may assume without loss of generality $R = R_{\mathcal{P}}/\mathrm{ann}_{R_{\mathcal{P}}}(M)$, that is, we may suppose that M_R is a $\mathcal{P}$-homogeneous Σ-pure-injective faithful module over the serial ring R and $\mathcal{P} = (P_1, \ldots, P_n)$ is the localizable system for R in which P_i is the maximal ideal of $e_i R e_i$ for each $i = 1, 2, \ldots, n$, and we have to prove that the ring R is right noetherian.

For this it suffices to show that for every fixed index i the right module $e_i R$ is noetherian. As $e_i R$ is uniserial, it is noetherian if and only if it has the a.c.c. on cyclic submodules. Let

$$e_i r_0 R \subseteq e_i r_1 R \subseteq e_i r_2 R \subseteq \cdots \tag{10.6}$$

be an ascending chain of cyclic submodules of $e_i R$, where $r_0, r_1, r_2, \ldots \in R$. Since $e_i R$ is uniserial and $e_i r_t R = \sum_{j=1}^n e_i r_t e_j R$, for every $t \geq 0$ there exists $j(t) = 1, 2, \ldots, n$ such that $e_i r_t R = e_i r_t e_{j(t)} R$. Consider the n subchains $\{ e_i r_t e_{j(t)} R \mid t \geq 0, \ j(t) = j \}$, $j = 1, 2, \ldots, n$, of the chain (10.6). If we prove the all these n subchains are stationary, then the chain (10.6) is necessarily stationary. Hence we must prove that for every index $j = 1, 2, \ldots, n$ every ascending chain of the type $e_i s_0 e_j R \subseteq e_i s_1 e_j R \subseteq \ldots$, $s_0, s_1, \ldots \in R$, is stationary. As M_R is a Σ-pure-injective R-module, the descending chain of finitely definable subgroups $(0 :_M e_i s_0 e_j) \supseteq (0 :_M e_i s_1 e_j) \supseteq \ldots$ of M_R is stationary. Hence in order to conclude the proof it is sufficient to show that $(0 :_M e_i s_k e_j) = (0 :_M e_i s_{k+1} e_j)$ implies $e_i s_k e_j R = e_i s_{k+1} e_j R$.

Suppose $(0 :_M e_i s_k e_j) = (0 :_M e_i s_{k+1} e_j)$. Since $e_i s_k e_j R \subseteq e_i s_{k+1} e_j R$, if $e_i s_{k+1} e_j = 0$, then $e_i s_k e_j R = e_i s_{k+1} e_j R$. Suppose $e_i s_{k+1} e_j \neq 0$. As M_R is faithful, there exists an indecomposable direct summand N of M_R such that $N e_i s_{k+1} e_j \neq 0$. In particular, $N e_j \neq 0$ and $N e_i \neq (0 :_{N e_i} e_i s_{k+1} e_j)$. From $e_i s_k e_j R \subseteq e_i s_{k+1} e_j R$ it follows that there exists $s' \in R$ such that

$$e_i s_k e_j = e_i s_{k+1} e_j \cdot e_j s' e_j.$$

Right multiplication by $e_i s_k e_j$ induces a monomorphism of left $\mathrm{End}(N_R)$-modules $N e_i/(0 :_{N e_i} e_i s_k e_j) \to N e_j$. Similarly, right multiplication by $e_i s_{k+1} e_j$ induces a monomorphism of left $\mathrm{End}(N_R)$-modules $N e_i/(0 :_{N e_i} e_i s_{k+1} e_j) \to N e_j$ and right multiplication by $e_j s' e_j$ induces an endomorphism of the left $\mathrm{End}(N_R)$-module $N e_j$. Note that $(0 :_M e_i s_k e_j) = (0 :_M e_i s_{k+1} e_j)$ forces $(0 :_{N e_i} e_i s_k e_j) = (0 :_{N e_i} e_i s_{k+1} e_j)$. Now $N e_i, N e_j$ are uniserial left $\mathrm{End}(N_R)$-modules (Corollary 10.12). From the equality $e_i s_k e_j = e_i s_{k+1} e_j \cdot e_j s' e_j$ and Lemma 6.26(a) we obtain that the endomorphism of $N e_j$ given by right multiplication by $e_j s' e_j$ is a monomorphism. Hence $e_j s e_j \notin P_j$. Thus $e_j s e_j$ is invertible in $e_j R e_j$, and $e_i s_k e_j R = e_i s_{k+1} e_j R$, as desired. □

We know that projective modules over a serial ring are serial (Theorem 3.10(b)). Similarly, finitely presented modules over a serial ring are serial (Corollary 3.30). We shall now show that Σ-pure-injective modules over a serial ring are serial as well.

Theorem 10.28 *Every indecomposable Σ-pure-injective module over a serial ring is uniserial. Every Σ-pure-injective module over a serial ring is serial.*

Proof. If M_R is an indecomposable Σ-pure-injective module over a serial ring R, then there exists a localizable system $\mathcal{P} = (P_1, \ldots, P_n)$ with the property that M is a faithful, indecomposable, Σ-pure-injective module over the serial right noetherian ring $R_{\mathcal{P}}/\mathrm{ann}_{R_{\mathcal{P}}}(M)$ (Theorem 10.25 and Lemma 10.27). By Proposition 10.14 the $R_{\mathcal{P}}/\mathrm{ann}_{R_{\mathcal{P}}}(M)$-module M is injective, and so uniserial (Lemma 5.17). By Corollary 6.25 the uniserial $R_{\mathcal{P}}$-module M is uniserial as an R-module. This proves the first part of the statement. The second part follows from the first one and Theorem 2.29. $\qquad\square$

From Theorem 10.28, the notion of homogeneous Σ-pure-injective module given for modules over chain rings in Section 10.5 is compatible with that for modules over serial rings given in this section. This is proved in the following corollary.

Corollary 10.29 *Let R be a serial ring and let $\mathcal{P} = (P_1, \ldots, P_n)$ be a localizable system for R. If M_R is an indecomposable $\mathcal{P}$-homogeneous Σ-pure-injective right R-module and $i = 1, 2, \ldots, n$, then either $Me_i = 0$ or Me_i is an indecomposable P_i-homogeneous Σ-pure-injective right $e_i R e_i$-module.*

Proof. Let M_R be a Σ-pure-injective right R-module. Suppose $Me_i \neq 0$. By Corollary 1.44 the right $e_i R e_i$-module Me_i is Σ-pure-injective. If M_R is indecomposable, then M_R is uniserial by Theorem 10.28. Then Me_i is a uniserial module over $e_i R e_i$ (if $x, y \in M$, there exists $r \in R$ such that either $xe_i = ye_i r$ or $ye_i = xe_i r$ because M_R is uniserial, so that either $xe_i = ye_i \cdot e_i r e_i$ or $ye_i = xe_i \cdot e_i r e_i$). In particular, the $e_i R e_i$-module Me_i is necessarily indecomposable.

Finally, if M_R is indecomposable and $\mathcal{P}$-homogeneous, then

$$P_i = \{\, r \in e_i R e_i \mid (0 :_{Me_i} r) \neq 0 \,\}.$$

Hence Me_i is P_i-homogeneous. $\qquad\square$

In Lemma 10.27 we saw that if R is a serial ring and M_R is a $\mathcal{P}$-homogeneous Σ-pure-injective R-module, then the serial ring $R_{\mathcal{P}}/\mathrm{ann}_{R_{\mathcal{P}}}(M)$ is right noetherian. In the next theorem we show that it is also left noetherian.

Theorem 10.30 *Let R be a serial ring and let $\mathcal{P} = (P_1, \ldots, P_n)$ be a localizable system for R. If M_R is a $\mathcal{P}$-homogeneous Σ-pure-injective R-module, then $R_{\mathcal{P}}/\mathrm{ann}_{R_{\mathcal{P}}}(M)$ is a serial noetherian ring.*

Proof. Let M be a $\mathcal{P}$-homogeneous Σ-pure-injective R-module. We have to show that $R_{\mathcal{P}}/\mathrm{ann}_{R_{\mathcal{P}}}(M)$ is left noetherian. From Example 10.26 we know that M is a homogeneous Σ-pure-injective $R_{\mathcal{P}}$-module and the localizable system for the ring $R_{\mathcal{P}}$ associated with the Σ-pure-injective $R_{\mathcal{P}}$-module M is the localizable system given by all maximal ideals of the chain rings $e_i R_{\mathcal{P}} e_i$. Hence we may assume $R = R_{\mathcal{P}}/\mathrm{ann}_{R_{\mathcal{P}}}(M)$ without loss of generality, that is, may assume that M_R is a $\mathcal{P}$-homogeneous Σ-pure-injective faithful module over the serial ring R and $\mathcal{P} = (P_1, \ldots, P_n)$ is the localizable system for R in which P_i is the maximal ideal of $e_i R e_i$ for each $i = 1, 2, \ldots, n$. We have to prove that the ring R is left noetherian. As in the proof of Lemma 10.27 it suffices to show that for every pair of indices $i, j = 1, 2, \ldots, n$, every ascending chain of the type $Re_j s_0 e_i \subseteq Re_j s_1 e_i \subseteq \ldots$, with $s_0, s_1, \ldots \in R$, is stationary. Suppose the contrary, that is, suppose that there exists a sequence $s_0, s_1, \ldots$ of elements of R such that $Re_j s_0 e_i \subset Re_j s_1 e_i \subset \ldots$. For every $m \geq 0$ there exists $t_m \in R$ such that $e_j s_m e_i = t_m e_j s_{m+1} e_i$. Then $e_j t_m e_j \in P_j$ for every $m \geq 0$, because otherwise $e_j Re_j t_m e_j = e_j Re_j$, so that $e_j s_m e_i = e_j t_m e_j s_{m+1} e_i$ forces $e_j Re_j s_m e_i = e_j Re_j s_{m+1} e_i$. Hence $e_j s_{m+1} e_i \in Re_j s_m e_i$, a contradiction. Thus $e_j t_m e_j \in P_j$ for every m, so that

$$e_j s_1 e_i = (e_j t_1 e_j)(e_j t_2 e_j) \cdots (e_j t_m e_j)(e_j s_{m+1} e_i) \in P_j^m (e_j s_{m+1} e_i)$$

for every $m \geq 0$. Now from $Re_j s_0 e_i \subset Re_j s_1 e_i$ it follows that $e_j s_1 e_i \neq 0$, so that $Me_j s_1 e_i \neq 0$ because M_R is faithful. Hence there exists $x \in M$ with $xe_j s_1 e_i \neq 0$. By Proposition 10.18(c) applied to the P_j-homogeneous Σ-pure-injective $e_j Re_j$-module Me_j it follows that there exists an integer $k \geq 0$ such that $xe_j P_j^k = 0$. Thus $xe_j s_1 e_i \in xe_j P_j^k (e_j s_{k+1} e_i) = 0$, contradiction. $\square$

10.9   Notes on Chapter 10

Proposition 10.1 was proved by [Fisher 72, Theorem 1.5] in the case in which R is the endomorphism ring of an artinian module and generalized by [Zimmermann, Satz 6.2] to the case of a faithful Σ-pure-injective module M_R (Corollary 10.2). The proof of Proposition 10.1 we have given here is essentially the original proof of [Fisher 72, Theorem 1.5]. In [Goldie and Small] it was proved that every nil subring of the endomorphism ring of a noetherian module is nilpotent too. [Procesi and Small] had previously proved that each nil subring of the endomorphism ring of a noetherian module over a ring that satisfies a polynomial identity is nilpotent.

The proof that (b) $\Rightarrow$ (c) in Example 10.3 is taken from [Chatters and Hajarnavis, Lemma 1.16].

If R and T are Morita equivalent rings, then R is an exchange ring if and only if T is an exchange ring [Nicholson].

It is easily seen that if R is a commutative ring and M_R is an artinian module, then $\mathrm{End}(M_R)$ is a left and right noetherian ring [Camps and Facchini,

Theorem 2.2]. Propositions 10.6 and 10.7 are due to [Quanshui and Golan], who proved them for the endomorphism ring of an S-module which is τ-torsionfree τ-artinian with respect to a torsion theory $\tau = (\mathcal{T}, \mathcal{F})$ on S-Mod.

Distributive modules may be viewed as a generalization of uniserial modules. A module M over a commutative ring R is distributive if and only if the localization M_P is a uniserial R_P-module for every maximal ideal P of R [Albu and Năstăsescu, § 1]. Theorem 10.8 and Corollary 10.9 are due to [Stephenson]. Theorem 10.11 and Corollaries 10.12 and 10.13 are essentially due to [Eklof and Herzog] and [Puninski 94]. These authors used techniques of model theory, and their results are more general than ours, because they proved them for pure-injective indecomposable modules over serial rings [Eklof and Herzog, Corollary 2.5], while our Corollary 10.12 concerns only Σ-pure-injective indecomposable modules over serial rings. (From this it follows that Proposition 6.27 and localization with respect to a localizable system can be applied to the study of arbitrary pure-injective indecomposable modules over serial rings, and not only to the study of Σ-pure-injective indecomposable modules over serial rings as we have done in this chapter.) Similarly, Puninski proved that every pure-injective module over a chain ring is distributive as a module over its endomorphism ring [Puninski 94, Corollary of Theorem 1], while from Theorem 10.11 it only follows that every Σ-pure-injective module over a chain ring is distributive as a module over its endomorphism ring. The technique we have used in proving Theorem 10.11 is also due to [Puninski 94, Proof of Theorem 1].

Most of the results from Proposition 10.14 to the end of the chapter come from [Facchini and Puninski 95] and [Camps, Facchini and Puninski]. Example 10.24 is due to [Camps and Facchini], and following the construction in the proof of Proposition 8.17 it would be easily seen that the ring T in Example 10.24 is a ring that satisfies the polynomial identity $(xy - yx)^2 = 0$. For further information about Σ-pure-injective modules over serial rings see [Facchini and Puninski 95].

The idea we used in Section 10.7 to characterize the serial rings that belong to the class $\mathcal{A}$, that is, the serial rings that are isomorphic to endomorphism rings of artinian modules, was the following. We wanted to determine whether a serial ring R belongs to $\mathcal{A}$. If R belongs to $\mathcal{A}$ and M_S is an artinian module over a suitable ring S such that $R \cong \mathrm{End}(M_S)$, then $_RM_S$ is a bimodule, and $_RM$ is necessarily a faithful, balanced, Σ-pure-injective R-module. Conversely, suppose we are able to describe all faithful, balanced, Σ-pure-injective left modules over a ring R. If one of these modules is artinian as a right module over its endomorphism ring, then $R \in \mathcal{A}$. Hence we were able to determine the serial rings in $\mathcal{A}$ because we had a sufficiently good description of Σ-pure-injective modules over serial rings.

The same idea can be used to determine other classes of rings contained in $\mathcal{A}$. This has been done, for instance, in [Camps and Facchini], where a complete characterization of the commutative Prüfer rings that belong to $\mathcal{A}$

is given. Recall that a commutative ring R is a *Prüfer ring* (or an *arithmetical ring*) if the localization R_P is a chain ring for every maximal ideal P of R, that is, if R is a distributive R-module. For a complete classification of Σ-pure-injective modules over a Prüfer ring see [Prest and Puninski].

Chapter 11

Open Problems

1. [Crawley and Jónsson, p. 854] Does a module with the finite exchange property have the exchange property? For some partial information on this problem see the notes on Chapter 2 on page 71.

2. Is every direct summand of a direct sum of modules with local endomorphism rings a direct sum of modules with local endomorphism rings? See Corollary 2.55 and the remark which follows it.

3. [Matlis, p. 517] Let R be a ring and $M = \oplus_{i \in I} M_i$ a right R-module which is a direct sum of indecomposable, injective submodules M_i. Is every direct summand of M also a direct sum of indecomposable, injective modules? By Theorem 2.12 this is equivalent to: "Let $M = \oplus_{i \in I} M_i$ be a direct sum of indecomposable, injective submodules M_i. Does every direct sum decomposition of M refine to a decomposition into indecomposable direct summands?" If R is right noetherian the answer is "yes", because in this case the module M is injective [Anderson and Fuller, Proposition 18.13], so that every direct summand of M is a direct sum of indecomposable, injective modules [Anderson and Fuller, Theorem 25.6].

4. In Section 4.2 we encountered some properties of modules with semilocal endomorphism rings. Determine further properties of such modules.

5. In [Herbera and Shamsuddin] some classes of modules with semilocal endomorphism ring were determined (see Section 4.3 in this book). Determine further classes of modules with semilocal endomorphism rings.

6. Extend Corollary 4.17 to serial modules of arbitrary Goldie dimension. More precisely, if E is the endomorphism ring of a serial module of finite Goldie dimension, then $E/J(E)$ is a semisimple artinian ring, i.e., the endomorphism ring of a semisimple module of finite length. Now let E be the endomorphism ring of a serial module of infinite Goldie dimension.

What can one say about $E/J(E)$? Which properties of endomorphism rings of semisimple modules (of infinite length) hold for $E/J(E)$? For instance, is $E/J(E)$ a right self-injective ring? Is it a Von Neumann regular ring? Is $E/J(E)$ of Type I in the sense of [Goodearl 91, Chapter 10]?

7. Does the Krull-Schmidt Theorem hold for artinian modules over a local ring R? Recall that it holds if R is either right noetherian or commutative (Section 2.12) and it does not hold if R is an arbitrary (non-local) ring (Section 8.2).

8. Let $\mathcal{A}$ be the class of rings that are isomorphic to endomorphism rings of artinian modules. Krull's problem was solved in Section 8.2 by showing that every ring in the class $\mathcal{F}$ of all module-finite algebras over commutative noetherian semilocal rings can be realized as the endomorphism ring of an artinian module over a suitable ring, i.e., by showing that $\mathcal{F} \subseteq \mathcal{A}$ (Proposition 8.18).

   Determine other significant classes of rings $\mathcal{F}$ with $\mathcal{F} \subseteq \mathcal{A}$.

9. Let M_R be a serial module of finite Goldie dimension. Is every direct summand of M_R a serial module? Equivalently, is every indecomposable direct summand of a serial module M_R of finite Goldie dimension a uniserial module? The answer is "yes" in the following cases:

   (a) when M_R is a finite direct sum of uniserial modules of type 1 (Theorem 2.12).

   (b) when the base ring R is serial and the module M_R is finitely presented (because if R is a serial ring and M_R is a finitely presented module, then every direct summand of M_R is finitely presented, and so serial by Corollary 3.30).

   (c) when the base ring R is either commutative or right noetherian (Corollary 9.25).

   (d) if $M \cong U^n$ for some uniserial module U [Dung and Facchini a, Theorem 2.7].

   For some other cases in which the answer is "yes" see [Dung and Facchini a].

10. Is every direct summand of a serial module a serial module?

    This generalizes Problem 9 to serial modules of possibly infinite Goldie dimension. The answer is "yes" if the base ring is either commutative or right noetherian (Corollary 9.25). Note that by Corollary 2.49 it suffices to consider direct summands of direct sums of *countable* families of uniserial modules.

11. Is every pure-projective module over a serial ring serial? Is every inde-composable pure-projective module over a serial ring uniserial? This is a particular case of Problem 10.

12. Let M be a serial module and let N be a non-zero direct summand of M. Does N contain an indecomposable direct summand?

13. (This is a particular case of Problem 12.) Let U be a uniserial module, let I be an index set and let N be a non-zero direct summand of the direct sum $U^{(I)}$ of copies of U. Does N contain an indecomposable direct summand?

14. Most questions about direct summands of serial modules can be reduced to problems about endomorphism rings of serial modules. We have solved some problems about direct sum decompositions of artinian modules realizing large enough classes of rings as endomorphism rings of artinian modules (Section 8.2). Hence describe the structure of endomorphism rings of serial modules (both in the case of serial modules of finite Goldie dimension and in the case of serial modules of infinite Goldie dimension), and determine sufficiently large classes of rings that can be realized as endomorphism rings of serial modules.

15. Do there exist uniserial modules that are not quasi-small? For some characterization of these modules see [Dung and Facchini 97, §4].

16. Consider the following two properties:

    (1) *n-th root property*: if A and B are two modules and A^n is isomorphic to B^n for some positive integer n, then A is isomorphic to B.

    (2) *$\aleph_0$-th root property*: if A and B are two modules and the direct sum $A^{(\aleph_0)}$ of countably many copies of A is isomorphic to the direct sum $B^{(\aleph_0)}$ of countably many copies of B, then A is isomorphic to B.

    It follows easily from the Krull-Schmidt-Remak-Azumaya Theorem, that the n-th root property and the $\aleph_0$-th root property hold for modules A and B with local endomorphism rings. We saw in Proposition 4.8 that the n-th root property holds for modules with a semilocal endomorphism ring. In particular, the n-th root property holds for serial modules of finite Goldie dimension. Professor Lawrence Levy (private communication) has found a nice example that shows that the $\aleph_0$-th root property does not hold for indecomposable modules A and B with semilocal endomorphism rings. Does the $\aleph_0$-th root property hold for uniserial modules A and B? The answer is "yes" if either A or B is a cyclic uniserial module [Dung and Facchini a, Proposition 2.11].

17. Let M_S be an artinian module over an arbitrary ring S and let $R = \operatorname{End}(M_S)$. Does R only have a finite number of minimal prime ideals? The answer is "yes" if R is a commutative semiprime ring (Example 10.3).

18. In Section 10.2 we determined some properties that hold for rings in the class $\mathcal{A} = \{\, R \mid R \cong \mathrm{End}(M_S)$ for some artinian module M_S over some ring $S \,\}$, i.e., properties of the endomorphism rings of arbitrary artinian modules over arbitrary rings. For instance, we have seen that every ring R in the class $\mathcal{A}$ has the following properties: (1) it is semilocal, (2) it satisfies the ascending chain condition on left annihilators, (3) each nil subring of R is nilpotent of bounded index of nilpotency. Determine further properties that hold for endomorphism rings of artinian modules.

19. Every nil subring of the endomorphism ring of an artinian module is nilpotent (Corollary 10.5). Nil subrings of a ring with right or left Krull dimension are nilpotent (Corollary 7.22). And every serial ring that is the endomorphism ring of an artinian module has finite Krull dimension (Theorem 10.23). Is there any relation between a ring being isomorphic to the endomorphism ring of an artinian module and having (finite) Krull dimension? Are there other properties that hold both for endomorphism rings of artinian modules and for rings with right or left Krull dimension? For an example of an endomorphism ring of an artinian module that fails to have Krull dimension see Example 10.24.

20. Give an example of an artinian module M_S over a ring S such that $\mathrm{End}(M_S)$ is not contained in any right artinian ring.

21. Let $\mathcal{P}_1$ be the class of all serial rings. In Theorem 10.23 we found a complete characterization of the rings in $\mathcal{P}_1$ that are endomorphism rings of artinian modules, that is, a characterization of the rings in the class $\mathcal{P}_1 \cap \mathcal{A}$. Similarly, let $\mathcal{P}_2$ be the class of all commutative Prüfer rings. A complete characterization of the rings in $\mathcal{P}_2$ that are endomorphism rings of artinian modules is given in [Camps and Facchini]. This is a characterization of the rings in $\mathcal{P}_2 \cap \mathcal{A}$.

Determine characterizations of the rings in $\mathcal{P} \cap \mathcal{A}$ for other significant classes $\mathcal{P}$ of rings.

Bibliography

[Albu and Năstăsescu] T. Albu and C. Năstăsescu, *Modules arithmétiques*, Acta Math. Acad. Sci. Hungar. **25** (1974), 299–311.

[Anderson and Fuller] F. W. Anderson and K. R. Fuller, "Rings and categories of modules", Second edition, Springer-Verlag, New York, 1992.

[Ara 96] P. Ara, *Strongly π-regular rings have stable range one*, Proc. Amer. Math. Soc. **124** (1996), 3293–3298.

[Ara 97] P. Ara, *Extensions of exchange rings*, J. Algebra **197** (1997), 409–423.

[Armendariz, Fisher and Snider] E. P. Armendariz, J. W. Fisher and R. L. Snider, *On injective and surjective endomorphisms of finitely generated modules*, Comm. Algebra **6** (1978), 659–672.

[Asano] K. Asano, *Über verallgemeinerte Abelsche Gruppen mit hyperkomplexen Operatorenring und ihre Anwendungen*, Jap. J. Math. **15** (1939), 231–253.

[Azumaya 50] G. Azumaya, *Corrections and supplementaries to my paper concerning Krull-Remak-Schmidt's theorem*, Nagoya Math. J. **1** (1950), 117–124.

[Azumaya 51] G. Azumaya, *On maximally central algebras*, Nagoya Math. J. **2** (1951), 119–150.

[Baba] Y. Baba, *Note on almost M-injectives*, Osaka J. Math. **26** (1989), 687–698.

[Baccella] G. Baccella, *Right semiartinian rings are exchange rings*, preprint, 1997.

[Bass 60] H. Bass, *Finitistic dimension and a homological generalization of semi-primary rings*, Trans. Amer. Math. Soc. **95** (1960), 466–488.

[Bass 64] H. Bass, *K-theory and stable algebra*, Publ. Math. I. H. E. S. **22** (1964), 5–60.

[Bass 68] H. Bass, "Algebraic K-theory", Benjamin, New York, 1968.

[Bessenrodt, Brungs and Törner] C. Bessenrodt, H. H. Brungs and G. Törner, "Right chain rings, Part 1", Schriftenreihe des Fachbereichs Math. **181**, Universität Duisburg, 1990.

[Bican] L. Bican, *Weak Krull-Schmidt theorem,* preprint, 1997.

[Brookfield] G. Brookfield, *Direct sum cancellation of noetherian modules,* preprint.

[Brungs] H. Brungs, *Generalized discrete valuation rings,* Canad. J. Math. **21** (1969), 1404–1408.

[Camps and Dicks] R. Camps and W. Dicks, *On semilocal rings,* Israel J. Math. **81** (1993), 203–211.

[Camps and Facchini] R. Camps and A. Facchini, *The Prüfer rings that are endomorphism rings of artinian modules,* Comm. Algebra **22**(8) (1994), 3133–3157.

[Camps, Facchini and Puninski] R. Camps, A. Facchini and G. Puninski, *Serial rings that are endomorphism rings of artinian modules,* in: "Rings and radicals", B. J. Gardner, Liu Shaoxue and R. Wiegandt eds., Pitman Research Notes in Math. Series **346**, Longman, Harlow, 1996, pp. 141–159.

[Camps and Menal] R. Camps and P. Menal, *Power cancellation for artinian modules,* Comm. Algebra **19**(7) (1991), 2081–2095.

[Cedó and Rowen] F. Cedó and L. H. Rowen, *Addendum to examples of semiperfect rings,* to appear.

[Chatters] A. W. Chatters, *Serial rings with Krull dimension,* Glasgow Math. J. **32** (1990), 71–78.

[Chatters and Hajarnavis] A. W. Chatters and C. R. Hajarnavis, "Rings with chain conditions", Pitman, London, 1980.

[Cohn] P. M. Cohn, *On the free product of associative rings,* Math. Z. **71** (1959), 380–398.

[Crawley and Jónsson] P. Crawley and B. Jónsson, *Refinements for infinite direct decompositions of algebraic systems,* Pacific J. Math. **14** (1964), 797–855.

[Crawley-Boevey] W. W. Crawley-Boevey, *Modules of finite length over their endomorphism ring,* in "Representations of algebras and related topics", H. Tachikawa and S. Brenner eds., London Math. Soc. Lecture Notes Ser. **168**, Cambridge Univ. Press, Cambridge, 1992, pp. 127–184.

[Dean and Stafford] C. Dean and J. T. Stafford, *A nonembeddable noetherian ring*, J. Algebra **115** (1988), 175–181.

[Dischinger] F. Dischinger, *Sur les anneaux fortement π-régulier*, C. R. Acad. Sci. Paris Sér. A-B **283** (1976), A571–A573.

[Dress] A. Dress, *On the decomposition of modules*, Bull. Amer. Math. Soc. **75** (1969), 984–986.

[Drozd] Yu. A. Drozd, *Generalized uniserial rings*, Math. Zametki **18** (1975), 705–710.

[Dung and Facchini 97] N. V. Dung and A. Facchini, *Weak Krull-Schmidt for infinite direct sums of uniserial modules*, J. Algebra **193** (1997), 102–121.

[Dung and Facchini a] N. V. Dung and A. Facchini, *Direct summands of serial modules*, to appear in J. Pure Appl. Algebra 1999.

[Eckmann and Schopf] B. Eckmann and A. Schopf, *Über injektive Moduln*, Arch. Math. **4** (1953), 75–78.

[Eilenberg] S. Eilenberg, *Homological dimension and syzygies*, Ann. of Math. **64** (1956), 328–336.

[Eilenberg and Nakayama] S. Eilenberg and T. Nakayama, *On the dimension of modules and algebras, V*, Nagoya Math. J. **11** (1957), 9–12.

[Eisenbud and Griffith] D. Eisenbud and P. Griffith, *Serial rings*, J. Algebra **17** (1971), 389–400.

[Eklof and Herzog] P. C. Eklof and I. Herzog, *Model theory of modules over a serial ring*, Ann. Pure Appl. Logic **72** (1995), 145–176.

[Elliger] S. Elliger, *Über direkte Summanden von Azumaya-Moduln*, Note Mat. **11** (1991), 127–134.

[Estes and Guralnick] D. R. Estes and R. M. Guralnick, *Module equivalences: local to global when primitive polynomials represent units*, J. Algebra **77** (1982), 138–157.

[Evans] E. G. Evans, Jr., *Krull-Schmidt and cancellation over local rings*, Pacific J. Math. **46** (1973), 115–121.

[Facchini 81] A. Facchini, *Loewy and artinian modules over commutative rings*, Ann. Mat. Pura Appl. **128** (1981), 359–374.

[Facchini 84] A. Facchini, *Lattice of submodules and isomorphism of subquotients*, in "Abelian groups and modules", R. Göbel, C. Metelli, A. Orsatti and L. Salce eds., CISM Courses and Lectures **287**, Springer-Verlag, Vienna, 1984, pp. 491–501.

[Facchini 96] A. Facchini, *Krull-Schmidt fails for serial modules*, Trans. Amer. Math. Soc. **348** (1996), 4561–4575.

[Facchini and Herbera] A. Facchini and D. Herbera, K_0 *of a semilocal ring*, preprint, 1998.

[Facchini, Herbera, Levy and Vámos] A. Facchini, D. Herbera, L. S. Levy and P. Vámos, *Krull-Schmidt fails for artinian modules*, Proc. Amer. Math. Soc. **123** (1995), 3587–3592.

[Facchini and Puninski 95] A. Facchini and G. Puninski, Σ-*pure-injective modules over serial rings,* in "Abelian groups and modules", A. Facchini and C. Menini eds., Kluwer Acad. Publ., Dordrecht, 1995, pp. 145–162.

[Facchini and Puninski 96] A. Facchini and G. Puninski, *Classical localizations in serial rings,* Comm. Algebra **24** (1996), 3537–3559.

[Facchini and Salce] A. Facchini and L. Salce, *Uniserial modules: sums and isomorphisms of subquotients,* Comm. Algebra **18**(2) (1990), 499–517.

[Faith and Herbera] C. Faith and D. Herbera, *Endomorphism rings and tensor products of linearly compact modules,* Comm. Algebra **25** (1997), 1215–1255.

[Faith and Utumi] C. Faith and Y. Utumi, *Quasi-injective modules and their endomorphism rings,* Arch. Math. **15** (1964), 166–174.

[Fisher 70] J. W. Fisher, *On the nilpotency of nil subrings,* Canad. J. Math. **22** (1970), 1211–1216.

[Fisher 72] J. W. Fisher, *Nil subrings of endomorphism rings of modules,* Proc. Amer. Math. Soc. **34** (1972), 75–78.

[Fitting] H. Fitting, *Die Theorie der Automorphismenringe Abelscher Gruppen und ihr Analogon bei nicht kommutativen Gruppen,* Math. Ann. **107** (1933), 514–542.

[Fleury] P. Fleury, *A note on dualizing Goldie dimension,* Canad. Math. Bull. **17** (1974), 511–517.

[Fuchs 69] L. Fuchs, *On quasi-injective modules,* Ann. Scuola Norm. Sup. Pisa **23** (1969), 541–546.

[Fuchs 70a] L. Fuchs, "Infinite abelian groups", vol. I, Academic Press, 1970.

[Fuchs 70b] L. Fuchs, *Torsion preradicals and the ascending Loewy series of modules,* J. reine angew. Math. **239** (1970), 169–179.

[Fuchs and Salce] L. Fuchs and L. Salce, *Uniserial modules over valuation rings*, J. Algebra **85** (1983), 14–31.

[Fuller] K. R. Fuller, *On indecomposable injectives over artinian rings*, Pacific J. Math. **29** (1969), 115–135.

[Fuller and Shutters] K. R. Fuller and W. A. Shutters, *Projective modules over non-commutative semilocal rings*, Tôhoku Math. J. **27** (1975), 303–311.

[Gabriel] P. Gabriel, *Des catégories abéliennes*, Bull. Soc. Math. France **90** (1962), 323–448.

[Goldie 58] A. W. Goldie, *The structure of prime rings under ascending chain conditions*, Proc. London Math. Soc. (3) **8** (1958), 589–608.

[Goldie 60] A. W. Goldie, *Semi-prime rings with maximum condition*, Proc. London Math. Soc. (3) **10** (1960), 201–220.

[Goldie 64] A. W. Goldie, *Torsion-free modules and rings*, J. Algebra 1 (1964), 268–287.

[Goldie and Small] A. W. Goldie and L. W. Small, *A note on rings of endomorphisms*, J. Algebra **24** (1973), 392–395.

[Gómez Pardo and Guil Asensio] J. L. Gómez Pardo and P. A. Guil Asensio, *Indecomposable decompositions of finitely presented pure-injective modules*, J. Algebra **192** (1997), 200–208.

[Goodearl 87] K. R. Goodearl, *Surjective endomorphisms of finitely generated modules*, Comm. Algebra **15** (1987), 589–609.

[Goodearl 91] K. R. Goodearl, "Von Neumann regular rings", Krieger Publishing Company, Malabar, 1991.

[Goodearl and Warfield 76] K. R. Goodearl and R. B. Warfield, Jr., *Algebras over zero-dimensional rings*, Math. Ann. **223** (1976), 157–168.

[Goodearl and Warfield 89] K. R. Goodearl and R. B. Warfield, "An introduction to noncommutative noetherian rings", Cambridge Univ. Press, Cambridge, 1989.

[Gordon and Robson 73a] R. Gordon and J. C. Robson, *Krull dimension*, Mem. Amer. Math. Soc. **133** (1973).

[Gordon and Robson 73b] R. Gordon and J. C. Robson, *Semiprime rings with Krull dimension are Goldie*, J. Algebra **25** (1973), 519–521.

[Grätzer] G. Grätzer, "General lattice theory", Academic Press, New York, 1978.

[Gregul' and Kirichenko] O. E. Gregul' and V. V. Kirichenko, *On semihereditary serial rings*, Ukranian J. Math. **39**(2) (1987), 156–161.

[Gruson and Jensen 73] L. Gruson and C. U. Jensen, *Modules algébriquement compacts et foncteurs* $\lim_{\leftarrow}^{(i)}$, C. R. Acad. Sci. Paris Sér. A-B **276** (1973), A1651–A1653.

[Gruson and Jensen 76] L. Gruson and C. U. Jensen, *Deux applications de la notion de L-dimension*, C. R. Acad. Sci. Paris Sér. A-B **282** (1976), A23–A24.

[Gruson and Jensen 81] L. Gruson and C. U. Jensen, *Dimensions cohomologiques reliées aux foncteurs* $\lim_{\leftarrow}^{(i)}$, in "Seminaire d'Algèbre Paul Dubreil and Marie Paule Malliavin", Lecture Notes in Math. **867**, Springer-Verlag, Berlin-Heidelberg-New York, 1981, pp. 234–294.

[Grzeszczuk and Puczyłowski] P. Grzeszczuk and E. R. Puczyłowski, *On Goldie and dual Goldie dimensions*, J. Pure Appl. Algebra **31** (1984), 47–54.

[Harada] M. Harada, *On almost relative injectivity on artinian modules*, Osaka J. Math. **27** (1990), 963–971.

[Hart] R. Hart, *Krull dimension and global dimension of simple Ore extensions*, Math. Z. **121** (1971), 341–345.

[Herbera and Shamsuddin] D. Herbera and A. Shamsuddin, *Modules with semi-local endomorphism ring*, Proc. Amer. Math. Soc. **123** (1995), 3593–3600.

[Herstein and Small] I. N. Herstein and L. Small, *Nil rings satisfying certain chain conditions*, Canad. J. Math. **16** (1964), 771–776.

[Herzog] I. Herzog, *A test for finite representation type*, J. Pure Appl. Algebra **95** (1994), 151–182.

[Jensen and Lenzing] C. U. Jensen and H. Lenzing, "Model theoretic algebra", Gordon and Breach Science Publishers, New York, 1989.

[Johnson 51] R. E. Johnson, *The extended centralizer of a ring over a module*, Proc. Amer. Math. Soc. **2** (1951), 891–895.

[Johnson 57] R. E. Johnson, *Structure theory of faithful rings II. Restricted rings*, Trans. Amer. Math. Soc. **84** (1957), 523–544.

[Kaplansky 49] I. Kaplansky, *Elementary divisors and modules*, Trans. Amer. Math. Soc. **66** (1949), 464–491.

[Kaplansky 50] I. Kaplansky, *Topological representation of algebras II*, Trans. Amer. Math. Soc. **68** (1950), 62–75.

[Kaplansky 54] I. Kaplansky, "Infinite abelian groups", Ann Arbor, 1954.

[Kaplansky 58] I. Kaplansky, *Projective modules*, Ann. of Math. **68** (1958), 372–377.

[Klatt] G. P. Klatt, *Projective modules over semiperfect rings*, unpublished.

[Krause] G. Krause, *On the Krull-dimension of left noetherian left Matlis-rings*, Math. Z. **118** (1970), 207–214.

[Krause and Lenagan] G. Krause and T. H. Lenagan, *Transfinite powers of the Jacobson radical*, Comm. Algebra **7** (1979), 1–8.

[Krull 25] W. Krull, *Über verallgemeinerte endliche Abelsche Gruppen*, Math. Z. **23** (1925), 161–196.

[Krull 28] W. Krull, *Zur Theorie der allgemeinen Zahlringe*, Math. Ann. **99** (1928), 51–70.

[Krull 32] W. Krull, *Matrizen, Moduln und verallgemeinerte Abelsche Gruppen im Bereich der ganzen algebraischen Zahlen*, Heidelberger Akademie der Wissenschaften **2** (1932), 13–38.

[Lanski] C. Lanski, *Nil subrings of Goldie rings are nilpotent*, Canad. J. Math. **21** (1969), 904–907.

[Lemonnier 72a] B. Lemonnier, *Déviation des ensembles et groupes abéliens totalement ordonnés*, Bull. Sci. Math. **96** (1972), 289–303.

[Lemonnier 72b] B. Lemonnier, *Sur une classe d'anneaux définie à partir de la déviation*, C. R. Acad. Sci. Paris Sér. A-B **274** (1972), A1688–A1690.

[Lenagan 73] T. H. Lenagan, *The nil radical of a ring with Krull dimension*, Bull. London Math. Soc. **5** (1973), 307–311.

[Lenagan 77] T. H. Lenagan, *Reduced rank in rings with Krull dimension*, in "Ring Theory", Proceedings of the conference held in Antwerp (1977), F. Van Oystaeyen ed., Lecture Notes in Pure and Appl. Math., Marcel Dekker, pp. 123–131.

[Lenagan 80] T. H. Lenagan, *Modules with Krull dimension*, Bull. London Math. Soc. **12** (1980), 39–40.

[Levitzki] J. Levitzki, *Prime ideals and the lower radical*, Amer. J. Math. **73** (1951), 25–29.

[Levy] L. S. Levy, *Krull-Schmidt uniqueness fails dramatically over subrings of* $Z \oplus Z \oplus \cdots \oplus Z$, Rocky Mountain J. Math. **13** (1983), 659–678.

[Łoś] J. Łoś, *Abelian groups that are direct summands of every Abelian group which contains them as pure subgroups*, Fund. Math. **44** (1957), 84–90.

[Malcev] A. I. Malcev, *On the immersion of an algebraic ring into a field*, Math. Ann. **113** (1936/37), 686–691.

[Matlis] E. Matlis, *Injective modules over noetherian rings*, Pacific J. Math. **8** (1958), 511–528.

[McConnell and Robson] J. C. McConnell and J. C. Robson, "Noncommutative Noetherian rings", John Wiley & Sons, Chichester, 1987.

[Mohamed and Müller] S. H. Mohamed and B. J. Müller, *On the exchange property for quasi-continuous modules*, in "Abelian groups and modules", A. Facchini and C. Menini eds., Kluwer Acad. Publ., Dordrecht, 1995, pp. 367–372.

[Monk] G. S. Monk, *A characterization of exchange rings*, Proc. Amer. Math. Soc. **35**(2) (1972), 349–353.

[Müller 70] B. J. Müller, *On semi-perfect rings*, Illinois J. Math. **14** (1970), 464–467.

[Müller 92] B. J. Müller, *The structure of serial rings*, in "Methods in Module Theory", G. Abrams, J. Haefner and K. M. Rangaswamy eds., Marcel Dekker, New York, 1992, pp. 249–270.

[Müller and Singh] B. J. Müller and S. Singh, *Uniform modules over serial rings*, J. Algebra **144** (1991), 94–109.

[Nakayama] T. Nakayama, *On Frobeniusian algebras II*, Ann. of Math. **42** (1941), 1–21.

[Năstăsescu and Van Oystaeyen] C. Năstăsescu and F. Van Oystaeyen, "Dimensions of ring theory", D. Reidel Publishing Company, Dordrecht, 1987.

[Nicholson] W. K. Nicholson, *Lifting idempotents and exchange rings*, Trans. Amer. Math. Soc. **229** (1977), 269–278.

[Noether] E. Noether, *Hyperkomplexe Grössen und Darstellungstheorie*, Math. Zeitschrift **30** (1929), 641–692.

[Oberst and Schneider] U. Oberst and H. J. Schneider, *Die Struktur von projektiven Moduln*, Invent. Math. **13** (1971), 295–304.

[Ore 31] O. Ore, *Linear equations in non-commutative fields,* Ann. of Math. **32** (1931), 463–477.

[Ore 33] O. Ore, *Theory of noncommutative polynomials,* Ann. of Math. **34** (1933), 480–508.

[Oshiro and Rizvi] K. Oshiro and S. T. Rizvi, *The exchange property of quasi-continuous modules with the finite exchange property,* Osaka J. Math. **33** (1996), 217–234.

[Osofsky 68] B. L. Osofsky, *Endomorphism rings of quasi-injective modules,* Canad. J. Math. **20** (1968), 895–903.

[Osofsky 70] B. L. Osofsky, *A remark on the Krull-Schmidt-Azumaya theorem,* Canad. Math. Bull. **13** (1970), 501–505.

[Prest] M. Prest, "Model theory and modules", Cambridge Univ. Press, Cambridge, 1988.

[Prest and Puninski] M. Prest and G. Puninski, Σ-*pure-injective modules over a commutative Prüfer ring,* to appear in Comm. Algebra.

[Procesi and Small] C. Procesi and L. Small, *Endomorphism rings of modules over PI-algebras,* Math. Z. **106** (1968), 178–180.

[Prüfer] H. Prüfer, *Untersuchungen über die Zerlegbarkeit der abzählbaren primären abelschen Gruppen,* Math. Z. **17** (1923), 35–61.

[Puninski 94] G. Puninski, *Indecomposable pure-injective modules over chain rings,* Trudy Moskov. Mat. Obshch. **56** (1994), 1–13.

[Puninski 95] G. Puninski, *Pure-injective modules over right noetherian serial rings,* Comm. Algebra **23** (1995), 1579–1591.

[Quanshui and Golan] Wu Quanshui and J. S. Golan, *On the endomorphism ring of a module with relative chain conditions,* Comm. Algebra **18** (1990), 2595–2609.

[Rentschler and Gabriel] R. Rentschler and P. Gabriel, *Sur la dimension des anneaux et ensembles ordonnés,* C. R. Acad. Sci. Paris Sér. A-B **265** (1967), A712–A715.

[Roux] B. Roux, *Sur les anneaux de Köthe,* An. Acad. Brasil. Cienc. **48** (1976), 13–28.

[Rowen 86] L. H. Rowen, *Finitely presented modules over semiperfect rings,* Proc. Amer. Math. Soc. **97** (1986), 1–7.

[Rowen 88] L. H. Rowen, "Ring theory", vol. I, Academic Press, San Diego, 1988.

[Schmidt] O. Schmidt, *Über unendliche Gruppen mit endlicher Kette*, Math. Z. **29** (1928), 34–41.

[Sharpe and Vámos] D. W. Sharpe and P. Vámos, "Injective modules", Cambridge Univ. Press, Cambridge, 1972.

[Shores] T. S. Shores, *Loewy series of modules*, J. reine angew. Math. **265** (1974), 183–200.

[Shores and Lewis] T. S. Shores and W. J. Lewis, *Serial modules and endomorphism rings*, Duke Math. J. **41** (1974), 889–909.

[Siddoway] M. F. Siddoway, *On endomorphism rings of modules over Henselian rings*, Comm. Algebra **18** (1990), 1323–1335.

[Singh] S. Singh, *Serial right noetherian rings*, Canad. J. Math. **36** (1984), 22–37.

[Small] L. Small, *Artinian quotient rings: addenda and corrigenda*, J. Algebra **4** (1966), 505–507.

[Stenström] B. Stenström, *Pure submodules*, Ark. Mat. **7** (1967), 159–171.

[Stephenson] W. Stephenson, *Modules whose lattice of submodules is distributive*, Proc. London Math. Soc. (3) **28** (1974), 291–310.

[Stock] J. Stock, *On rings whose projective modules have the exchange property*, J. Algebra **103** (1986), 437–453.

[Swan 60] R. G. Swan, *Induced representations and projective modules*, Ann. of Math. **71** (1960), 552–578.

[Swan 62] R. G. Swan, *Vector bundles and projective modules*, Trans. Amer. Math. Soc. **105** (1962), 264–277.

[Talintyre] T. D. Talintyre, *Quotient rings with minimum condition on right ideals*, J. London Math. Soc. **41** (1966), 141–144.

[Upham] M. H. Upham, *Serial rings with right Krull dimension one*, J. Algebra **109** (1987), 319–333.

[Utumi 56] Y. Utumi, *On quotient rings*, Osaka Math. J. **8** (1956), 1–18.

[Utumi 67] Y. Utumi, *Self-injective rings*, J. Algebra **6** (1967), 56–64.

[Vámos 90] P. Vámos, *Decomposition problems for modules over valuation domains*, J. London Math. Soc. (2) **41** (1990), 10–26.

[Vámos 95] P. Vámos, *The Holy Grail of algebra: seeking complete sets of invariants,* in "Abelian groups and modules", A. Facchini and C. Menini eds., Kluwer Acad. Publ., Dordrecht, 1995, pp. 475–483.

[Varadarajan] K. Varadarajan, *Dual Goldie dimension,* Comm. Algebra **7** (1979), 565–610.

[Vasershtein] L. N. Vasershtein, *Stable rank of rings and dimensionality of topological spaces,* Funct. Anal. Appl. **5** (1971), 102–110.

[Warfield 69a] R. B. Warfield, Jr., *A Krull-Schmidt theorem for infinite sums of modules,* Proc. Amer. Math. Soc. **22** (1969), 460–465.

[Warfield 69b] R. B. Warfield, Jr., *Purity and algebraic compactness for modules,* Pacific J. Math. **28** (1969), 699–719.

[Warfield 69c] R. B. Warfield, Jr., *Decompositions of injective modules,* Pacific J. Math. **31** (1969), 263–276.

[Warfield 72] R. B. Warfield, Jr., *Exchange rings and decompositions of modules,* Math. Ann. **199** (1972), 31–36.

[Warfield 75] R. B. Warfield, Jr., *Serial rings and finitely presented modules,* J. Algebra **37** (1975), 187–222.

[Warfield 79] R. B. Warfield, Jr., *Bezout rings and serial rings,* Comm. Algebra **7** (1979), 533–545.

[Wiegand] R. Wiegand, *Local rings of finite Cohen-Macaulay type,* preprint, 1997.

[Wong and Johnson] E. T. Wong and R. E. Johnson, *Self-injective rings,* Canad. Math. Bull. **2** (1959), 167–173.

[Wright 89] M. H. Wright, *Krull dimension in serial rings,* J. Algebra **124** (1989), 317–328.

[Wright 90] M. H. Wright, *Links between prime ideals of a serial ring with Krull dimension,* in "Non-commutative ring theory", S. K. Jain and S. R. López-Permouth eds., Lecture Notes in Math. **1448**, Springer-Verlag, 1990, pp. 33–40.

[Yakovlev] A. V. Yakovlev, *On the direct decomposition of artinian modules,* preprint, 1997.

[Zimmermann] W. Zimmermann, *Rein injektive direkte Summen von Moduln,* Comm. Algebra **5**(10) (1977), 1083–1117.

[Zimmermann and Zimmermann-Huisgen 78] W. Zimmermann and B. Zim-
    mermann-Huisgen, *Algebraically compact rings and modules*, Math. Z. **161**
    (1978), 81–93.

[Zimmermann-Huisgen and Zimmermann 84]   B. Zimmermann-Huisgen and
    W. Zimmermann, *Classes of modules with the exchange property*, J. Alge-
    bra **88** (1984), 416–434.

[Zimmermann-Huisgen and Zimmermann 90]   B. Zimmermann-Huisgen and
    W. Zimmermann, *On the sparsity of representations of rings of pure global
    dimension zero*, Trans. Amer. Math. Soc. **320** (1990), 695–711.

Index